FROM COMPASS TO COMPUTER

A History of Electrical and Electronics Engineering

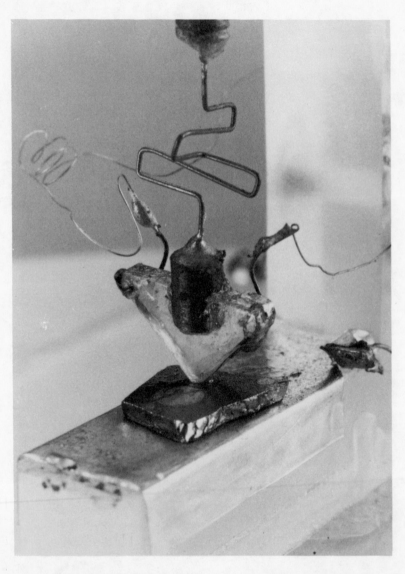

The start of the transistor era; the first point-contact transistor made on 23 December 1947 (Courtesy of Bell Laboratories)

From Compass to Computer
A History of Electrical and
Electronics Engineering

W. A. Atherton

San Francisco Press, Inc.
Box 6800, San Francisco, CA 94101-6800, USA

First published 1984 by
THE MACMILLAN PRESS LTD
London and Basingstoke
Companies and representatives
throughout the world
and in the United States of America and Canada by
San Francisco Press, Inc.
Box 6800, San Francisco, CA 94101-6800, USA

Printed in Hong Kong

ISBN 0 333 35266 1 (Macmillan Press hc)
ISBN 0 333 35268 8 (Macmillan Press pbk)
ISBN 0 911 30248 4 (San Francisco Press hc)
ISBN 0 911 30249 2 (San Francisco Press pbk)

Library of Congress Card Catalogue Number 83–61209

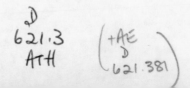

CONTENTS

FOREWORD

Our awareness of electricity, of all the forces of nature, developed entirely from man's innate curiosity about how certain materials behaved. From a slow beginning, electrical knowledge was a product of the 1800s, as nuclear energy is a product of the 1900s. Why did amber attract feathers and other light bodies? The answer, first attempted by Gilbert in 1600, is still in the process of formulation. Electrical science and engineering, one of the youngest disciplines, has become as extensive as it has grown universal, and all its flowering is crowded into only one century.

The Renaissance was followed by the Enlightenment, the Industrial Revolution, the Age of Power, the Electrical Age, and now the Age of Electronics. Each age has had a survival period of successively shorter duration, which reflects the erupting increases in channels of communication and knowledge retrieval. Our lives have about doubled in length in a century and our numbers have quadrupled. The survival factor, the sense of security, has kept pace, in spite of the present build-up of nuclear armaments and threats of violence in some locations. The slow but constant penetration of electronic intelligence does not make headlines, but its effect has been long lasting.

Fathered by scientific thought and mothered by experimental research, electrical progress is fittingly illustrated by the invention of the transistor. The last two centuries have provided mankind with a heritage of technological advances of greater beneficence that all the theocratic or political promises of the past ages. Through it all are woven the contributions of the electrical and electronics engineers with generators and controls so powerful and so delicate as to affect every product and process in which man is engaged.

Two dates—both easy to remember—mark the advent of the electrical age. In 1600 William Gilbert, physician to Queen Elizabeth I of England, published a treatise on the magnet in which electrical attraction is described. In 1800 Alessandro Volta, professor of physics at Pavia, published, in London, his invention of the electric battery, a source of continuous (direct) current. The magnetic component of this current became the source of today's electrical generation and the trigger to the great power supply that moves our world. Invisible and subtle, unknown two centuries ago, electric power has become the basic influence in the life of mankind.

The application of electromagnetic waves followed Hertz's demonstration of their reflection, refraction, and radiation. That occurred only a century ago, in the year when the term *electron* was first used. From it flowered a variety of communication systems that have permanently changed our habits and traditional institutions. The drama of reporting a *Titanic* disaster, an air raid on a metropolis, or an assassination attempt on a pope or a president brings home to a citizen a flavour unmatched by literature or oratory. Reporting from the moon or probing into outer space via electronics provides a credibility value exceeding the talents of a Jules Verne or Henry Stanley. We now view the universe's vastness and complexity undreamed of by Giordano Bruno or Edwin Hubble. We are still receiving reports from probes shot into space seven years ago from spacecraft travelling at speeds of over 25 000 miles per hour, which had long ago passed through the orbits of Jupiter and Saturn.

What has been accomplished by radio, radar, television, satellite telephony, and lasers in penetrating space and disseminating events in the macrocosm has been matched by the marvels of penetration into the structure of matter in the microcosmic domain. There emerge the scores of particle and energy components revealed by the electron microscope, x rays, and the particle accelerator. What Einstein, Bohr, and Fermi started has grown into an arcane and complex body of knowledge and conjecture for which billions are now budgeted by nations and universities so as to extend the frontiers of electromagnetic understanding. We have watched the development of the silicon chip with resulting computer and data bank so compact and comprehensive that nations compete in providing the maximum data in least space. Recent designs of computers and integrated circuits so shrank the electronic components that what can now be held in the palm of a hand functions better than earlier types that filled a space of a hundred cubic meters. This miniaturization in turn opened up dozens of new avenues of application, such as radar surveying, proximity fuzing in ballistic weaponry, radar exploration, and radioastronomy. What the thermionic valve (vacuum tube) had done for electronic radio, the silicon chip now does for integrated circuitry and data storage. The speed at which new inventions are absorbed into our industrialized economy is indicated by the growth of integrated circuitry in the USA, which rose from their introduction in 1964 to a value of over $5 billion in a dozen years.

Like the abacus and the mechanical calculator, the electronic computer has joined the myriad other devices for doing things more quickly, better, and more cheaply. Computer design and fabrication have become the world's third largest industry. The first industrial revolution improved physical tools, the second revolution evolved better mental tools; data processing and ready computing facility characterize the present fast-moving period of time. Data banks are to be made available for all who seek information in our coming society. Electrical science and engineering are now making their bid to satisfy man's hunger to learn. Memory capacities double and triple with each new model series. A wafer can now hold several hundred chips, and each chip contains some 250 000 transistors and similar elements. Each transistor is about 0.003 inch across. A television assembly is now crowded into a wristwatch space. We are truly crossing a threshold.

Norwalk, Connecticut, 1983   Bern Dibner

PREFACE

Most electrical and electronics engineers and technicians know very little about the history of their chosen profession. During their education and training they are offered minimal information on this subject—what they do learn often comes in the form of anecdotal footnotes in textbooks or as the reminiscences of an older generation. This book is offered to engineers, technicians, and students, not so much as a textbook but more as an account that may be read for enjoyment and relaxation as well as for enlightenment.

In common with the history of other branches of engineering, and with those of physics and science in general, the history of electrical and electronics engineering has become a profession in itself. Indeed much of the material on the subject seems to be written by historians for historians, which is a pity. I am not an historian and make no claim to be one. This book is not written for historians; they have their own sources, methods, and standards to which those of the engineering fraternity do not often refer. My aim has merely been to bring to a wider audience that which is already known to a select few. The result contains little original material but, I hope, much that engineers and interested laymen will find interesting and informative.

There is a growing awareness of the history of their discipline among electrical engineers. In America the Institute of Electrical and Electronics Engineers, Inc., welcomes papers on historical topics in its Transactions and has established a History Center in New York. In Britain the Institution of Electrical Engineers hold an annual meeting on history. Some Colleges and Universities offer short courses in the history of electrical engineering to engineering students. If this book serves to foster that awareness in some small way it will have been worthwhile.

Colyton, Devon, 1983

W. A. ATHERTON

ACKNOWLEDGMENTS

Many people and organizations have encouraged this work by providing help, information and advice; though too numerous to mention individually, to all I express my thanks. Especially I thank the University of Hong Kong, where much of this work was done, and the staff of the History and Electrical Engineering Departments; also Professor Charles Süsskind of the University of California, Berkeley, who edited the manuscript and made many useful suggestions.

Figure credits
Illustrations and photographs courtesy of, or by permission of:

Bell Laboratories: Frontispiece, Figures 1.1, 1.2, 5.9, 5.10, 9.1, 10.2(a).
Cork B. V., Netherlands: Cartoon chapter 9.
A. J. Croft, Clarendon Laboratory, University of Oxford: Figure 2.5.
Digital Press/Digital Equipment Corporation: Figures 11.8, 11.9, 11.10.
Edison National Historic Site, West Orange, N. J.: Figure 8.4.
Ferranti Company: Figure 7.5.
General Electric: Figures 6.2, 6.4, 6.5, 6.6, 6.7, 6.8(b), cartoon chapter 6.
Harvard University Press: Figures 2.4, 4.6 (redrawn).
Hewlett-Packard Company: Figure 12.1.
Illustrated London News: Figure 5.6 (redrawn).
Institution of Electrical Engineers: Figures 1.5, 2.1, 2.2, 2.3.
Intel Corporation: Figures 11.11, 11.12.
International Business Machines Corporation: Figures 11.1, 11.3, 11.7.
International Computers Ltd.: Figure 11.6.
Marconi Company Ltd.: Figures 1.3, 8.1, 8.5(c).
McGraw-Hill Inc., *Electronics*, Copyright © 1948: Figure 10.3.
Oxford University Press: Figure 7.2(a).
Pergamon Press, London (G. W. A. Dummer, *Electronic Inventions, 1745–1946*, 1977): Figure 10.9(a).
Punch: Figures 7.1, 8.8, 8.9.

Public Record Office, Crown Copyright: Figures 1.4, 11.5.
RCA Corporation: Figure 8.10(b).
RCA Laboratories: Figures 10.5, 8.12.
Royal Institution, London: Figure 4.2.
Science Museum, London, Crown Copyright: Figures 7.2(b), 8.6(a).
Photo. Science Museum, London: Figures 8.10(a), 10.1.
Siemens Museum, Munich: Figures 5.5, 7.3, 7.7.
Smithsonian Institution, Washington, D. C.: Figure 5.7.
J. J. Suran, GE, USA: Figure 10.8.
Texas Instruments Inc.: Figure 10.6.
Thorn EMI Lighting: Figure 6.8(a).

1 INTRODUCTION

Most students of electrical engineering, and that includes practising engineers, know precious little about its history. Many see no need to know about the past and may agree with Henry Ford's famous quip that history is bunk, despite the observation that since the quip itself has now passed into history it must presumably also be regarded as bunk. Maybe they view the history of their subject as old hat and irrelevant to our modern understanding. If that is the case, they are at odds with some of the men who made the most important contributions to their chosen subject of study.

James Clerk Maxwell, the originator of the electromagnetic theory of light, derived great benefit from reading Faraday's original publications and rightly advocated the study of original papers. "It is of great advantage to the student of any subject to read the original memoirs on that subject," he wrote, "for Science is always most completely assimilated when it is in the nascent state."[1] Although that is generally sound advice for the researcher it is not always practical for the average student. If studying modern flip-flop circuits, for example, one would gain little by searching out and reading the original publications by W. H. Eccles and F. W. Jordan. Nor would most students have time to do it. Nevertheless, a general introduction to the history of one's subject of study can be rewarding in terms of both interest and usefulness.

Sir Oliver Lodge, one of the pioneers of radio science and engineering, put it this way; "Early pioneering work is too often overlooked and forgotten in the rush of a brilliant new generation, and amid the interest of fresh and surprising developments. The early stages of any discovery have, however, an interest and fascination of their own; and teachers would do well to immerse themselves in the atmosphere of those earlier times, in order to realise more clearly the difficulties which had to be overcome, and by what steps the new knowledge had to be dovetailed in with the old. Moreover, for beginners, the nascent stages of a discovery are sometimes more easily assimilated than the finished product. Beginners need not, indeed, be led through all the controversies which naturally accompany the introduction of anything new; but some familiarity with those controversies and discussions on the part of

1

the teacher is desirable, if he is to apprehend the students' probable difficulties."[2]

It is in the spirit of Lodge's observations that the following chapters should be read; as an introduction to (or as a reminder of) some of the pioneering work in electrical and electronics engineering that has an interest and fascination of its own.

The present state of the art can excite and sometimes bewilder engineers, technicians, and students whether it be the latest microprocessor, speech synthesizer, communications satellite, or large generating station that is being studied. The same has been true for centuries and we should try to avoid the chronological snobbery that dictates that only the events of one's own lifetime are worthy of interest. One can imagine oneself in China around 2500 BC experiencing the thrill of watching an early magnetic compass seek out the north pole, or move oneself through time and space to California in 1976 when motion along the San Andreas fault was measured by a laser beam. Benjamin Franklin once wrote of his studies of electricity, "I have never before engaged in any study that so totally engaged my attention and my time as this has lately done." And Dr. Johnson in 1756 commented that, "Electricity is the great discovery of the present age, and the great object of philosophical curiosity." And both of these comments date from before the invention of the voltaic battery!

Excitement can sometimes be tinged with apprehension, as Michael Faraday, the 19th century patron saint of electrical engineers, found when he was on the verge of his discovery of electromagnetic induction. "It may be a weed instead of a fish that, after all my labour, I may at last pull up," he wrote. Despite his fear his discovery was no weed. Even Max Planck, the originator of quantum theory, viewed his brainchild with apprehension as he unleashed the events which turned the world of physics inside out. On the other hand, just a few years later, J. A. Fleming had what he called a sudden very happy thought, a thought that resulted in the thermionic valve and the beginning of vacuum-tube electronics. Many years later John Bardeen and Walter Brattain discovered transistor action when Brattain found that, "if I wiggled it just right," he had a solid-state audio-frequency amplifier with a gain of up to a hundred (Fig. 1.1).

Such exciting events continue to take place although their true worth may not always be recognized immediately. The negative-feedback amplifier, whose principle is fundamental to electronic control and whose application overcame a major obstacle in the progress of long-distance telephony, "had all the initial impact of a blow with a wet noodle," according to one report. Lee de Forest's audion, the first triode valve, was once described as worthless; and one news reporter dismissed the newly invented telephone with the words, "It can never be of any practical value."

Though advances continue to be made each one may be accompanied by a hundred or a thousand setbacks, and not all the great contributors receive their just reward. After news of Franklin's famous kite experiment with

Figure 1.1 Bardeen, Shockley, and Brattain, with equipment used in the invention of the transistor announced in 1948

lightning in 1752, the first electrical scientist to be killed by his experiments met his death in St. Petersburg while attempting to further the study of the lightning discharge. Ampère, whose name is now immortalized on every electrical plug and socket, once remarked that only two years of his life had brought him real happiness. In another sad case Edwin Armstrong, the man who gave so much to radio engineering, and who almost singlehandedly got FM radio to work, committed suicide in 1954.

Electrical engineering and electronics have brought a social revolution to all aspects of our lives, from business and commercial to educational and domestic. Writing in 1921 J. A. Fleming imagined a world in which the applications of electromagnetism had ceased to exist. He saw a world in which electric vehicles had stopped running; towns were plunged into darkness at night; telephones, telegraphs, electric bells, and railway signals were rendered inoperative. "In a month all large cities would be in a state of starvation," he wrote, "and the traffic and movement on which our commercial life depends would be destroyed."[3] Today the products of electrical and electronic engineering are even more important to our society. Our business world could not operate in the way it does without electrical communications, computers, photocopiers, and so on; architecture would be radically different without electric lighting and airconditioning; our home life would be almost unrecognizable without electric heating, cooking, television, refrigerators, and so on. "Politicians are apt to think that their labours are essential to the prosperity of the community," wrote Fleming. "They are, in truth," he decided, "not nearly so valuable as the work of the electrical engineer."[3]

The work that Fleming was then thinking of had begun just 50 years before; today, we have experienced just over a century of major electrical engineering.

The sciences of electricity and magnetism date back in primitive form to ancient times but the modern varieties, if that is the correct term, began in the 18th century. The discovery of electrical conduction in 1729 is a convenient starting point. Soon, two types of electricity had been identified and by the end of the century the chemical primary battery had been invented. It was truly an historic moment, for it liberated electrical science from the study of static electricity and electrical discharges and made possible the study of continuous currents. A new field of study, electrochemistry, was flourishing just ten years later and, after a further ten years, the most fundamental discovery in the history of the sciences of electricity and magnetism was made, the discovery of the single science of electromagnetism.

Thereafter came the solid foundation-laying work of men whose names are remembered in some of our basic laws of electricity: Ampère and electrodynamics; Ohm and the relationship between current, voltage, and resistance; Faraday and electromagnetic induction and electrolysis. From the discovery of electromagnetism it is possible to trace a continuous development of understanding spanning more than a century that incorporates the electromagnetic theory of light, the beginnings of relativity, and quantum theory and quantum mechanics. From the latter came our understanding of semiconductors and the path to the silicon chip.

Electrochemistry spawned the first electrical industry, electroplating, though it was pursued on only a small scale. The fruits of electromagnetism were greater. The first was the telegraph, which quickly spread to provide fast communications between towns and cities, and even between continents. The first telegrams between Europe and America were exchanged by cable in 1858. In the 1870s a great rush of applications of electricity began that led to the revolution experienced by society and that Fleming fondly saw as a contribution greater than that of the politicians. The invention of a good and reliable dynamo generator was pivotal. Its use with the newly invented electric incandescent lamp led to the building of central power stations and distribution systems that took electrical power into factories, offices, and homes. Electric motors could then be used in various locations for converting electrical power into mechanical power. The invention of the telephone at about the same time was to revolutionize at first local and then distant communications (Fig. 1.2).

CITY HALL, LAWRENCE, MASS.

Monday Evening, May 28

THE MIRACLE

WONDERFUL   DISCOVERY

OF THE AGE

Prof. A. Graham Bell, assisted by Mr. Frederic A.
Gower, will give an exhibition of his wonderful and
miraculous discovery **The Telephone,** before the people
of Lawrence as above, when Boston and Lawrence will
be connected via the Western Union Telegraph and vocal
and instrumental music and conversation will be transmitted a distance of 27 miles and received by the audience
in the City Hall.

Prof. Bell will give an explanatory lecture with this
marvellous exhibition.

Cards of Admission, 35 cents
Reserved Seats, 50 cents

Sale of seats at Stratton's will open at 9 o'clock.

Figure 1.2 Publicity for one of A. G. Bell's demonstrations of the telephone, 1877

The rapid upsurge in the availability and applicability of electricity led to
the rise of a new career, that of the electrical engineer. Education and training
began and professional institutions were founded. The first school in Britain
was the School of Telegraphy and Electrical Engineering in London.[4] The first
society was probably also in London, the Electrical Society founded in 1837. It
foundered after only six years with a debt of £85, mostly brought about by the
expense of printing papers and abstracts of foreign publications.[5] Its
successor, the Society of Telegraph Engineers, was founded in 1871. In 1888
this Society became the Institution of Electrical Engineers (IEE), a change of
title that reflected the broadening interests of its members. It remains today
one of the most important electrical institutes in the world. In the USA the
American Institute of Electrical Engineers, which later merged with the
Institute of Radio Engineers to form the Institute of Electrical and Electronics
Engineers (IEEE), was founded in 1884.

Some of the children of the time were to see the early pioneers at work and
be inspired to follow in their footsteps. One such person was C. W. Speirs, who
in his later years recollected seeing Swan's incandescent lamps when he was
aged about seven or eight. He recalled the gas lamps in the room being

dimmed, the electric ones being switched on, and "a general expression of surprise from the audience." He also remembered that the intensity of the lights "went up and down." He later learned the cause from Swan himself: the traction engine used to drive the dynamo had not enough steam to maintain it at a constant speed, which caused it to 'hunt.' Swan had deliberately switched off as much steam as possible to prevent the engine from speeding up as he feared a run-away engine would have burned out all the lamps that had taken him months to make.[6]

Speirs also recollected how as a boy he had used his knowledge of electricity to rid himself of the attentions of an old lady who would insist on kissing him. He sat on a stool with glass legs and connected himself to a bank of Leyden jar capacitors, which had been charged by a Wimshurst machine. When the old lady made contact she received the most shocking kiss of her life. Speirs subsequently received a shock of a different kind from his father. The story is reminiscent of that of Georg Boze of the University of Leipzig, who in 1743 amused himself and others by charging pretty girls with electricity so that they could dare men to kiss them and receive a shock that "broke their teeth."[7]

The young Speirs was in a fortunate position to be able to learn about electricity at a time when it was just beginning to reach the public, or at least a privileged few. To most people electricity was a mysterious new force that was not understood, as is illustrated by stories of its introduction to the public. In 1884 Crompton, one of the pioneers in Britain, told the story of a 'couple from the country' who bought an electric incandescent lamp and used a whole box of matches in trying to light it. After failing to get any light they declared the whole thing to be a swindle.[5] In a later tale a man was asked what he thought of the new electric lamp just installed in his home. "Marvellous," he replied. "I came home, hung my coat on a new peg they'd put on the wall, the light came on and it still hasn't gone out."

Such tales are amusing and serve to indicate how much we now take for granted concerning the use of electricity. And such stories are not always apocryphal. Wall plaques were used in 1880 to reassure the public that electric lamps were safe. "Do not attempt to light with a match," one such plaque advised. "Simply turn key on wall by the door." "The use of electricity for lighting is in no way harmful to health," it continued, "nor does it affect the soundness of sleep."[8] Despite such reassurances many families can still tell tales of an old granny who worried about the electricity dripping all over the floor if no light bulb was in the socket.

By the end of the 19th century the basic discoveries in electromagnetic science were being exploited to make radiotelegraphy a reality. Shortly afterwards Marconi succeeded in transmitting across the Atlantic (Fig. 1.3) and radio went on to find its first important commercial application in ship-to-shore telegraphy. In 1911 Marconi estimated that at least 3000 lives had been saved from shipwreck by wireless telegraphy. Yet radio's biggest contribution was to be in stimulating the development of the fledgling field of study we call electronics.

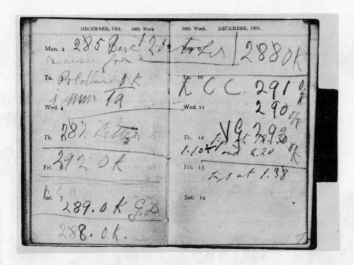

Figure 1.3 "Pips at 12.30, 1.10 and 2.20." A note in G. Marconi's diary for 12 December 1901 records the first transatlantic radio reception, the Morse letter 'S'

Vacuum-tube amplifiers and oscillators began to be applied to telephone and radio use just before World War I. That war greatly spurred the development of radio communications, which in turn meant considerable progress in such areas as frequency-division multiplexing and circuit design. Armies on both sides used telegraphs, telephones, and radio; and at sea the two great battle fleets were brought together for the Battle of Jutland with the help of radio monitoring. Postwar work built on the foundations that had been laid during the war. Network analysis spawned network synthesis. From radiotelephony came radio broadcasting and the start of consumer electronics on a large scale. Television experiments began and strengthened the call for broadband amplifiers, ramp generators, and better synchronization techniques. Stability criteria were worked out and, in general, linear electronics made continuing progress. The first electron microscope promised new information about the microscopic world and the first discovery by radio-astronomy was made. Radio beacons were the earliest radionavigation aids and the first primitive radar systems scanned the skies.

World War II accelerated the trend towards higher frequencies, the use of pulse techniques, and the dawn of digital electronics. Radar and radio-navigation techniques made rapid progress and new inventions like the cavity magnetron helped push frequencies and powers higher than ever. The war in the air over Britain and Northern Europe was fought as much with electronics as with guns, as radar and other radio techniques guided friendly aircraft to their targets and tracked the paths of enemy planes. Digital counters aided the nuclear program and special-purpose electronic computers were built in Britain to help crack the German codes (Fig. 1.4). Printed-circuit boards were

Figure 1.4 Colossus, the British special-purpose computer used from December 1943 for cracking German code in World War II. The pulleys at right guided punched paper tape

first used to reduce the amount of wiring in circuit assemblies; thick-film screen printing was employed to print passive components onto ceramic substrates.

By the end of that war the profession of the electrical engineer had acquired a new branch, digital electronics, that was to grow enormously and was also to achieve a new dimension. The invention of the transistor, and its offspring the integrated circuit, were to bring about an electronics revolution that few before the war could have conceived. Not only was electronics to be revolutionized but society itself was to feel changes which were on a par with those brought by the generation of electrical power at the end of the 19th century.

As computers shrank in size and grew in power, their price fell so much that machines offered by hobby shops merit comparison with the giants that struggled into operation just before 1950. By the end of the 1970s, after a century of electrical engineering on a large scale, it had become common to talk of mankind being on the verge of a second industrial revolution, a revolution that was expected to bring changes every bit as far reaching as those brought by the original industrial revolution and the agricultural revolution before it. Man's brain is to be aided as much as his muscle. "What we are doing in electronics will remake the world," said the incoming president of the

Table 1.1 Examples of Inventions and Innovations in Valve Electronics

	To about 1920	Between the Wars	World War II and After
Materials/ devices	Diode (1904) Triode (1906) CRT developed Tetrode (1916) High-power radio valves (1915)	Commercial tetrode (1926) Pentode (1926–1929) Iconoscope (1933)	Klystron (1939) Orthicon (1939), vidicon (early 1950s) Cavity magnetron (1940) TWT (1942) Nixie tube (1953) Miniature, subminiature valves
Techniques/ concepts	Amplification Positive feedback Radio circuitry Linear electronics Control of frequency bands	Network synthesis (1920s) Negative feedback (1927) Nyquist loop stability (1932) FM (1933) Standardization of octal base (1935) Increasing use of VHF Wave behaviour of electrons	Bode stabilization (1945) Printed circuit (1945?) Information theory (1948) Artificial satellites proposed (1946) Increasing use of microwaves Digital electronics Pulse techniques
Applications/ circuits	Amplifiers (c. 1911) Oscillators (c. 1911) Regenerative receivers (1912) Neutrodyne (1918) Superhet (1919) Multivibrators (1919) Flip-flop (1919) Filters (1920s) Carrier telephone circuits (1918)	Public broadcasting (1920) AGC (1926) Rectifying circuits, limiters, discriminators, AFC, saw-tooth deflection, synchronization, wideband amplifiers Demonstration of TV (1929) Radar (1928) Radio beacons (1930s) Radioastronomy (1932–1933) Electron microscope (1932)	Decade counter (1944) Analogue and digital computers Digital, pulse circuits Development of radar, radionavigation (Gee, Loran), colour television, feedback control systems, microwaves

Institution of Electronic and Radio Engineers in London in 1979. "The future and the past will be strangers to each other. What we have put our hand to will set the new patterns and styles for the world of the future."[9]

However, these words could have been equally well used on numerous occasions in the past, for electrical and electronics engineers have been remaking the world since 1837, when the first commercial electric telegraph went into operation. They helped remake the world when a cable crossed the Atlantic, when incandescent lamps glowed with a carbon filament, when central stations first sent power into peoples' homes, when electric trams rattled through the streets, when radio began broadcasting, when a transistor radio slipped into a pocket, when Telstar was launched, when the calculator displaced the slide rule, when computers were first blamed for human error—and humans for computer error. Electrical and electronics engineering have been revolutionizing the world for so long (see Table 1.1) that the real revolution would now occur only if they suddenly stopped doing so.

Electrical science, as well as engineering, has also contributed to the impact on various parts of society. Philosophy, for example, has been deeply affected by the theories of relativity and quanta, both of which are related to electrical theory. Electrical science is now so caught up with other physical phenomena that Ernst Mach called it "the theory of the general connexion of physical processes." Quantum electrodynamics, the theory of electrodynamic phenomena on the microscopic scale, has been described as "our greatest success so far in physics."[10]

Such enthusiasm for electrical theory and practice has been felt by men of vision ever since Robert Boyle wrote the first book devoted entirely to

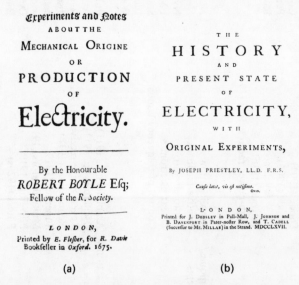

(a)   (b)

Figure 1.5 Title pages of early books on electricity; (a) by Robert Boyle (1675), (b) by Joseph Priestley (1767)

electricity in 1675 (Fig. 1.5). A century later Joseph Priestley described electrical experiments as "the cleanest and most elegant that the compass of philosophy exhibits."[5] One wonders how he would have felt if, after another century, and with the aid of a telephone, he could have listened with Willoughby Smith in London and "heard a ray of light fall on a bar of metal."[5] This event occurred soon after Smith had discovered the photo-electric properties of selenium and Hughes had invented the microphone. Photoelectricity is just one of several properties that distinguish semiconductors from other materials. Another century would have brought Priestley to modern times and, with the advent of semiconductor microelectronics, he would have seen how the intimate association of electrical science and electrical engineering have benefited society. He could watch as a doctor monitors a living fetus with ultrasonics, see the telephone that enables the heads of the two most powerful nations to speak directly to each other, study ships and aircraft as they navigate safely through the busiest sea and airlanes and, with a child's pocket money, buy an electronic machine that makes Pascal's and Leibniz's mechanical calculators—and Oughtred's slide rule too, all of which Priestley would have known—seem like something out of the Ark. He would also learn that mankind's basic instincts have not changed. Besides the benefits brought to society by electricians (as they were called in his day) he would also see the problems, including radio propaganda and computer-guided missiles.

In 1767 Priestley published what must have been the first book on the history of electricity.[11] One almost wonders what he found to put in it. Even 300 years ago, however, the practice of electricity held a fascination for enquiring minds. It has continued to do so to today.

References

1. J. C. Maxwell, *A Treatise on Electricity and Magnetism*, Clarendon Press, Oxford, 1904, 3rd edition, p. xi.
2. O. Lodge, *Nature* 111: 328, 1923.
3. J. A. Fleming, *Fifty Years of Electricity*, Wireless Press, London, 1921.
4. A. A. Bright Jr., *The Electric Lamp Industry*, Macmillan, New York, 1949.
5. R. Appleyard, *The History of the Institution of Electrical Engineers (1871–1931)*, IEE, London, 1939.
6. C. W. Speirs, *JIEE* 4: 384, 1958.
7. B. Dibner, *Elec. Eng.* 76: 592, 1957.
8. C. Phillips, *Electric Lamps: 100 Years On*, Thorn Lighting Ltd., London, 1979.
9. W. Gosling, Presidential address, *Radio and Electronic Engineer* 50: 1, 1980.

10. R. P. Feynmann, R. B. Leighton, and M. Sands, *The Feynmann Lectures on Physics*, Addison-Wesley, Reading, Mass., 1965, vol. 1, sec. 2–3.
11. J. Priestley, *The History and Present State of Electricity, With Original Experiments*, London, 1767.

2 ELECTRICITY AND MAGNETISM TO 1820

"At this point, I will set out to explain what law of nature causes iron to be attracted by that stone which the Greeks call from its place of origin magnet, because it occurs in the territory of the Magnesians."[1] So began the Roman poet Lucretius some 2000 years ago when he expounded his theory of magnetism, a theory to which some 17th century theories bore vague resemblance.

The unusual properties of lodestone, or magnetite, were known to the ancient Chinese and to the ancient Greeks. The Chinese are often credited with the invention of the magnetic compass, knowledge of which is thought to have reached Europe via the Arabs—an ancient example of the spread of technology. The Greeks not only gave us our word for magnetism but electricity as well, named after electron, the Greek word for amber. Ancient Greeks knew that amber when rubbed attracted to itself light bodies such as straw. This supposed magical property helped amber to become important in trade. In both magnetostatics and electrostatics the ancients knew about the powers of attraction, and Lucretius gave a clear description of magnetic repulsion: "It also happens at times that iron moves away from this stone; its tendency is to flee and pursue by turns."

Because some static phenomena are very easy to observe, electrostatics and magnetostatics are by far the oldest branches of our discipline. However, as long as investigations were restricted to static phenomena, the sciences of electricity and magnetism inevitably remained only curiosities in the back-waters of human interest. Dynamic phenomena are essential for any significant use of electricity and magnetism; it was only when a continuous flow of electricity could be easily generated that fascinating new areas of investigation were revealed. The first new area was electrochemistry. Like many new sciences it quickly yielded its easiest fruits. In this chapter we shall quickly review the development of electrostatics and magnetostatics and see how mankind first recognized, produced, and used electric currents. In the next chapter we shall see how the use of electric currents led to the realization that these two old sciences were in fact one united phenomenon.

Apart from ancient knowledge, the first important contribution came in 1269 from Petrus Peregrinus, a French military engineer. He used a compass to investigate the properties of a spherical lodestone and found that he could trace meridian lines that intersected at two points, which he named the North and South Poles, by analogy with the earth. He showed that like poles repel whereas unlike poles attract, and that cutting a magnet in two does not give two independent poles but leaves two complete magnets. Peregrinus used his newly found knowledge to construct a better compass that employed a graduated circular scale and a pivoted (rather than the customary floating) magnet.

In 1600 William Gilbert, physician to Queen Elizabeth, published his great book *De Magnete* and earned for himself acknowledgment as the 'father' of the experimental science of magnetism (Fig. 2.1). Written in Latin, the learned language of the day, it was the first detailed treatise on magnetostatics and electrostatics to be based on careful and accurate experiments. It was also the first great book on physical science to be published in England and preceded Newton's major work by eighty-seven years. Kepler and Galileo are said to have regarded it highly. Gilbert effectively brought electricity and magnetism out of the land of myth and legend, at least as far as experimental facts were concerned. He used asterisks to mark new discoveries and important experiments, large ones for the most important and small ones for the others. There was almost one for each page, twenty-one large and 178 small, so prolific was his work.

Gilbert was the first to record the idea of the earth acting as a giant magnet, since a compass needle was attracted to the earth's magnetic poles and not to a certain star as others believed. He investigated the angle of variation and the

Figure 2.1 Title page of Gilbert's 'De Magnete', 1600

angle of dip; the latter had been studied by Robert Norman in 1576. Magnetic and electrical attractions were often confounded and so Gilbert clearly restated the difference between them, though at times he was confused about magnetic and gravitational attractions himself. He coined the word 'electric' for materials that behave like amber (insulators) and 'nonelectric' for all others, and he greatly extended the list of known electrics. He checked the discoveries made by Peregrinus and examined the effects of temperature on magnetism. In his 'rays of magnetick virtue' some have seen the seed of the idea of the magnetic field, though this suggestion should be treated with scepticism. "Gilbert shall live till lodestones cease to draw," wrote Dryden.

On the theoretical side he debunked myths, such as the claim that garlic destroys the power of magnets, and he used experimental evidence to reject the theory of Giambattista della Porta, a contemporary Italian sage, that magnetic attraction was caused by eternal combat between magnet and iron within the lodestone. Pieces of iron, according to Porta, were attracted because they were called in as reinforcements for the iron in the stone since, he said, "all creatures defend their being." Porta himself had also attacked myths. By experiment he had proved the error of an old Greek theory that magnets fed on iron. Gilbert's own theory was not always better as it too rested in part on the fallacy of seeing animal attributes in inanimate objects. On the credit side he rejected magical concepts and instead used the concept of physical force. Lodestone, he considered, possessed life and hence a soul. His argument was that if the "dignity of life" can be given to "worms, ants, moths, plants and toadstools" it was inconceivable that the earth, and lodestone which is part of the earth, did not possess life; hence, the "magnetick virtue is animate."

Forty years after Gilbert the phenomenon of electrostatic repulsion was rediscovered and made known by Otto von Guericke, a German famous for his experiments on the production of vacuum. Thus by the mid-17th century the four basic features, electrostatic and magnetic attraction and repulsion, would be known by anyone interested in the subject, and as modern science continued to grow the pace of fact finding and theorizing quickened.

Experimental Discoveries

In 1670 Guericke should have made the pages of *New Scientist,* if it had existed, when he announced the first electrical machine, a friction generator (Fig. 2.2). It was simply a sulfur ball that could be spun on its axis so that it could be rubbed against the hand. It produced the first appreciable man-made sparks. Not long afterwards magnetic and electrical phenomena were shown to exist in vacuum and it was only a short step to speculation that lightning was an electrical phenomenon, a suggestion verified much later by Franklin. Vacuum experiments permitted the study of electrical glow discharges and it can be said that electric lighting had been produced and studied, first in France

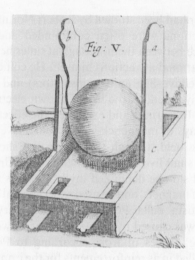

Figure 2.2 First electrical machine: Guericke's sulfur ball, about the size of a child's head (O. von Guericke, 'Experimenta nova Magdeburgica de vacuo spatio,' 1672, page 148)

by Jean Picard in 1675 but more thoroughly in England by Francis Hauksbee from 1705 to 1711. It was still some 200 years from being a practical and commercial proposition.

If any single event can be said to have been the real beginning of modern electrical science, it is probably the discovery of conduction by Stephen Gray in 1729. Gray found that charge, or 'electrick vertue' as he called it, could be transmitted from one body to another if they were connected by a nonelectric (conductor) but not if they were connected by an electric (insulator). In so doing he demonstrated electrical conduction and spotlighted the essential difference between conductors and insulators. Metals, and even the human body, were shown to be conductors. A second discovery made by Gray was that the charging of a body is associated with its surface, not its volume, which he did by charging two cubes of the same size, one hollow and the other solid, and demonstrating that their electrical properties were the same.

Soon after that, in 1733 in France, Charles Du Fay discovered that there were two types of electricity. One type was found in vitrified bodies such as glass or crystal, the other in resinous bodies like amber and sealing wax.[2] Accordingly he named them vitreous and resinous electricity, but Franklin later renamed them positive and negative. Du Fay repeated Gray's experiments, including one in which a human being was suspended by silk threads and electrically charged. The notable difference was that whereas Gray had charged someone else, Du Fay charged himself and was rewarded by seeing, in a darkened room, sparks leap from his body when another person approached.

Playing with static electricity generated by friction became an increasingly popular and amusing scientific pastime in Europe. In 1745 Ewald von Kleist, dean of a cathedral in Germany, charged a small bottle containing mercury and received a shock which he claimed "stuns my arms and shoulders." The following year Pieter van Musschenbroek, professor of physics at Leyden in Holland, independently performed a similar experiment, using water instead of mercury. His assistant received a severe shock. The experiment was repeated and Musschenbroek received his own shock. Later he is said to have remarked, "For the whole kingdom of France, I would not take a second shock." Another colleague claimed to have lost his breath for some minutes after the shock and feared permanent injury to his arm. Although the shocks may have been severe, one is left with a mental picture of these men standing in line to be shocked so as to be able to warn others against this terrible experience. Whatever the picture, the Leyden jar (or Kleist jar as it was first called), the first capacitor, had arrived and it provided the first means of storing electric charge. It was quickly shown that the charge stored was proportional to the thickness of the glass and the surface area of the conductors. For a while the glass was thought to be essential but that was proved to be wrong in 1762 when the first parallel-plate capacitor was made from two large boards covered with metal foil. It was the Leyden jar, used singly or grouped into banks of capacitors, that became the standard laboratory equipment.

That lightning is nothing more than a big electric spark was demonstrated in 1752 by Benjamin Franklin in Philadelphia in his famous kite experiment. Franklin, scientist, diplomat, and folk hero, had proposed using an iron rod 30 or 40 feet long mounted on top of a tall building, with the idea that the pointed end would draw charge from any thunderclouds and the charge could then be detected. This experiment was performed successfully in France but, unconvinced, Franklin decided to use a kite to carry the wire up into the clouds themselves. In doing so he managed to charge a Leyden jar and receive a shock from it. Further experiments showed that the clouds were usually negatively charged but occasionally positive.

The two experiments were repeated by others, not all of whom were as lucky as Franklin. "It is not given to every electrician to die in so glorious a manner as the justly envied Richmann," wrote Priestley in 1767 in one of the first books on the history of electricity.[3] The "justly envied Richmann" was Prof. G. W. Richmann, a Swede who had been working in St. Petersburg. Following Franklin's ideas he had experimented with what was becoming known as atmospheric electricity, a name chosen to distinguish it from frictional electricity. When Richmann hurried to his equipment during a thunderstorm there was a big flash and he was killed. Subsequently the organs of his body became specimens for scientific research. Despite Priestley's epitaph few would have sought to experience death by electricity, though Franklin used it to kill a turkey and kindle a fire on which to cook it. In 1890 electricity was used for the first time in a judicial execution when William Kemmler became the first victim of the electric chair.

The study of electricity moved from purely qualitative to at least partly quantitative with the work of Charles Coulomb, who succeeded in clarifying many of the ideas about electrostatics that had been expressed rather vaguely until then. With the aid of a sensitive torsion balance of his own design, he carried out experiments in 1785 that led him to announce that the inverse-square law applied to the force between two electrically charged bodies and to that between two magnetized bodies. Coulomb's laws were the first quantitative laws in the study of electricity and magnetism. Others before him had made attempts at measurements. In 1746 L. G. Le Monnier found the speed of electricity to be at least 30 times faster than that of sound. In 1767 Priestley made some rough measurements of conductivity and also discovered that in a hollow vessel all the electric charge resided on the outer surface. From this result he correctly inferred (but did not prove) the inverse-square law for electrostatic attraction; the force between two charges diminished in proportion to the square of the distance between them. In 1750 John Michell stated the inverse-square law for magnetism but could not give conclusive proof. Henry Cavendish proved the law for electrostatics (1772–1773) and performed detailed measurements on capacitance and conductivity. However, as with much of his work in which he anticipated many of the great men of electrical science, he did not publish it and it only came to light when Maxwell published it in 1879. By then it was history.

We are now on the brink of the discovery of the electric current by Luigi Galvani and the invention of the primary battery by Alessandro Volta, but before studying these two dynamic (in all senses of the word) events let us briefly review the development of electrical and magnetic theories.

Evolution of Theory

To a modern student of electricity, theory may sometimes seem to outshine experiment. Ranging from theories of antenna arrays to the theory of Zener diodes, it can appear to be complete and irrefutable. In unguarded moments students may be tempted to disregard experimental evidence as 'error' when it does not fit the simplified theory they have learned. However, that would be the greatest error because it leads to delusion and frustration. In truth, theories are made to fit observations, not the other way around, and even a brief study of the history of science shows that theories have to be remodelled or abandoned as evidence is accumulated and refined. As Maxwell once wrote, "Every student of science should be an antiquary in his subject," and that is just as applicable to engineering as it is to science.

It was in the 17th century that the first modern moves were made to achieve rational theories of electricity and magnetism; René Descartes, the famous Frenchman who gave us the Cartesian co-ordinate system, was one of the first contributors. He rejected the property of magnetic attraction, which he saw as belonging to the occult, and instead sought a mechanical explanation for what looked like attraction. He believed that grooved particles of matter or 'effluvia' were ejected from the magnet. They followed certain paths—our lines of force of which he probably produced the first drawing (Fig. 2.3)—and re-entered the magnet at the opposite pole. In the process they propelled iron towards the magnet. Though vastly more developed, this theory was closely akin to that of Lucretius.

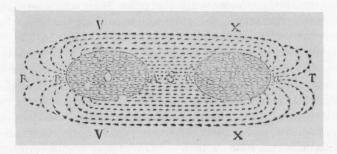

Figure 2.3 Paths of Descartes's grooved particles, our lines of force (R. Descartes, 'Principia Philosophiæ, Balviana,' Amsterdam, 1685, page 198)

Several other mechanical theories followed; in general, they all relied on one or two types of effluvia flowing around the magnet, attracting or repelling iron in various ways. Some modern dictionaries still give this supposed flow of minute particles as one meaning of the word effluvium. The idea of a stream of minute particles causing the observed effects was also applied to electricity and persisted well into the 18th century in the form of an electrical atmosphere surrounding a charged body. This concept resulted from an understandable confusion between what are now called the electric charge (matter) and the electric field (atmosphere). Similar confusion occurred in magnetism.

In the early 18th century the existence of an electrical matter was accepted by all. In America, Franklin made important contributions, one of which, that pointed objects are particularly good at "throwing off the electrical fire," led to the invention of the lightning conductor. More important was his one-fluid theory of electricity. In this basically correct theory electricity was thought to consist of a flow of only one type of particle. The particles repelled each other but were attracted to ordinary matter. Even if not explicitly stated, this theory involved the concept of electrical charge. William Watson in England independently made the same suggestion and it was advanced by F.U.T. Aepinus, who came to the then rather staggering conclusion that the particles of ordinary matter repelled each other. This result was to explain why two bodies repelled each other when drained of their charge.

Meanwhile, Robert Symmer developed a rival theory based on the assumption of two fluids, or types, of electricity. In this new theory a charged body contained unequal amounts of the two fluids, whereas a neutral body contained equal amounts. Both the one-fluid and two-fluid theories could describe the observed electrostatic phenomena; to some observers, there seemed little to chose between them.

Somewhat similar events took place in the theory of magnetism: both one-fluid and two-fluid theories were developed. In the one-fluid case the two poles were explained as being formed respectively by an accumulation and depletion of the fluid, whereas in the two-fluid case the fluids migrated to opposite ends of a magnetized body to form the different poles. There were several adherents to each theory, but others, the Swiss mathematician Leonhard Euler for example, stayed with theories of effluvia. Coulomb advanced the two-fluid theory and placed on it the all-important restriction, and in a way a very modern-sounding concept, that the fluids could migrate only to the opposite ends of their respective molecules, not to the opposite ends of the magnet. This molecular polarization was what made it impossible to obtain magnetic monopoles when a magnet was cut in half, and it also explained why there was no magnetic discharge, or flow of magnetism, similar to the electric discharge.

The fluid theories of both electricity and magnetism gradually overcame the effluvia ideas that matter could leave the body and travel through the space around it. With the loss of effluvia the concepts of electrical and magnetic atmospheres went too, which left a serious problem. In developing theories to explain electrostatic induction, magnetic poles, and the like, the scientists had lost their explanation of how two bodies at a distance from each other could be pulled together or pushed apart by magnetism or electricity. Mechanistic theories had been introduced to provide physical contact of some sort but fluid theories did not provide that. As a result the philosophy of action at a distance was adopted and the force of attraction, discarded by Descartes, came back into vogue. Coulomb was one of those responsible. Action at a distance was to remain in the sciences of electricity and magnetism until replaced in the mid-19th century by the field concept derived from the work of Faraday and Maxwell.

The 17th century witnessed a scientific revolution one aspect of which was the shift from qualitative to quantitative science in many branches of physics. Johannes Kepler's and Isaac Newton's work are probably the most famous examples of mathematical laws from this period. In electricity and magnetism this shift began a little later than in other branches of physics. Coulomb's experimental proof of the inverse-square laws may be taken as a convenient starting point, as it provided electrical engineering's oldest mathematical laws (1785). Coulomb's work was outstanding. Also by experiment, he established that magnetic phenomena cannot be caused by effluvia but are caused by forces of attraction and repulsion. The mathematics developed to exploit the inverse-square law of gravitational attraction could now be used to advance electrostatics and magnetostatics, and the path had been laid for S. D. Poisson and others to follow. In electricity Coulomb saw little to choose between the one and two-fluid theories; in magnetism, he began the modern era of molecular theories.

Following Coulomb, the mathematical theories of first electricity and then magnetism were advanced, early in the 19th century, by the French mathematician Poisson. J. L. Lagrange had already simplified the theory of attraction by using a mathematical function $V(x,y,z)$ that depended on all the attracting particles. This function was shown to satisfy Laplace's equation, a partial differential equation applicable to space free from matter. Poisson's role was to extend this solution to the case for space containing electrical charge. The result was the important Poisson's equation (1813). Poisson also pointed out that the function V is a constant over all points on the surface of a conductor. He next turned to magnetism and in 1824 presented a complete theory based on Coulomb's work. From it came the idea of magnetic moment per unit volume, magnetic intensity, the equivalent surface, and equivalent volume distributions of magnetization, and again the function V. Four years later the English mathematician George Green gave this function a name: the potential; hence the electric and magnetic potentials.

In this summary of the 17th and 18th century work on electrical and magnetic science only the more important points have been mentioned. "If I have seen further than other men," said Newton, "it is because I have stood on the shoulders of giants." Only the giants have been mentioned here and not even all of their work has been included.

Electric Current

The flow of an electric current had been studied by Gray and indeed by anyone who had discharged a Leyden jar capacitor. Still, such studies barely deserve being classed as electrodynamics since electrodynamics is mostly concerned with what happens when there is a continuous flow of current. Even so some progress had been made, on resistivity for example, and one man had all but stated Ohm's law (chap 3). The first source of a continuously flowing electric current was the primary cell invented by Volta in 1799, and that arose out of study that followed up observations made by an Italian anatomist, Luigi Galvani.

For many years Galvani had been interested in the motive power of muscles and had used frog's legs in his studies. As early as 1773 he had reported on purely mechanical investigations and, as muscular contraction caused by electric shock was well known, it may be possible that he was thinking of turning to electrical investigations also. However, there seems little doubt that the first effect was observed by accident in 1780. On the laboratory bench were a charged electrical machine and a dissected frog. When his assistant touched a nerve in the frog's leg with a metal scalpel the muscle moved violently. Another assistant thought a spark had been drawn from the machine at the same moment. Apparently the discharge had passed, via the scalpel, through the frog's muscle and caused it to convulse. Galvani set about a long and dedicated study of the phenomenon in an effort to trace its cause. Some of his work was published in 1786 and a fuller account in 1791. With hindsight it is obvious that at different times he was working with a host of phenomena, some of which were then unknown: electrochemically produced currents, the discharge of frictional electricity, electromagnetic induction, and even electromagnetic waves received from lightning flashes. It is little wonder that Galvani did not understand the processes involved. His erroneous conclusion was that he had been studying animal electricity, which was somehow caused by the animal's nerves. Galvani's work was only the second report of electricity being produced by electrochemical action; the first had been made by a Swiss, J. G. Sulzer, in 1762. Sulzer had been pursuing a theory of sensations and when placing two pieces of different metals on his tongue, lead and silver for example, he had experienced an unusual taste. The taste was caused by the flow of current from a primitive primary cell.

Both Sulzer and Galvani gave wrong explanations for the phenomena they had discovered. Yet Galvani's work was important because it led another Italian, Alessandro Volta, to investigate. Volta did not accept Galvani's animal electricity. He believed that the current flow was caused by the contact of two different metals. In that he also was wrong. It was another Italian, G.V. Fabroni, who got the right theory by pointing to a chemical action between the liquid, which always seemed to be present in Galvani's and Volta's work, and the two different metals; i.e., between the electrolyte and electrodes (words introduced by Michael Faraday in the 1830s). So two schools of thought arose to explain what became known as galvanism, the metal contact school and the chemical school. Galvani died in 1798 before the issue was settled and before Volta announced the invention of the electric battery.

Volta repeated Sulzer's experiment, this time with a silver and a gold coin, and set out to verify his own ideas. In one experiment he brought zinc and copper discs into contact, holding each with an insulated handle. With an electroscope he showed that on separation each disc was charged. Choosing by experiment silver and zinc as the metals best suited for his purpose, he arranged a series of them in a pile. Between each pair of metal discs he placed a piece of cardboard soaked in water or salt water, which he believed provided a conducting path from the top of one pair of metals to the bottom of the next but in such a way that the two pairs did not touch. If that happened Volta knew that the pile produced no more effect than a single pair of discs. In fact the cardboard contained the electrolyte. The finished pile of discs multiplied the effects of a single pair of discs many times and he was able to receive a shock from his pile similar to that obtained from a charged Leyden jar. The important difference was that Volta's pile, the first primary battery, did not need to be recharged.

The news was announced in a letter to the Royal Society in London. "The apparatus of which I speak," wrote Volta, "will doubtless astonish you."[5] Today the present generation of its offspring still do; the vast array of batteries available can be a little bewildering even to some engineers.

Volta's pile, "as high as can hold itself without falling," consisted of 30, 40,

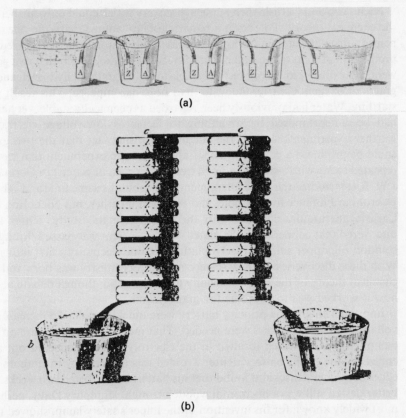

Figure 2.4 Volta's battery, 1800: (a) Crown of cups, (b) pile (Phil. Trans. Roy. Soc. London, 90: 430, 1800)

or 60 cells (Fig. 2.4). From such a primitive beginning grew today's huge international industry. As an alternative to the pile Volta also used pairs of metals soldered together with each end dipping into water or brine contained in goblets. This arrangement Volta called the crown of cups. Again 30 or more cells could be arranged in series to produce a battery. In this experiment Volta observed at least two phenomena that were at odds with his own contact theory. One was that a given liquid worked better with some metals than others; the second was that the metal–liquid contact should have a large surface area whereas "the rest of the arc [presumably including the metal–metal contact] may be as much narrower as we please, and may even be a simple metallic wire." With as much hindsight as we have it is temptingly easy, if hardly fair, to pick out such anomalies. However, only six weeks after the letter was written Volta's theory was under attack in England. William Nicholson and Anthony Carlisle set up a pile and Nicholson expressed surprise that "the chemical phenomena of galvanism, which had been so much insisted on by Fabroni"[5] played no part in Volta's observations.

Nicholson and Carlisle then made a discovery which could only further aid the chemical theory. To obtain a better electrical contact with the pile they placed a drop of water on the top plate and to their astonishment observed the production of a gas in the water. With further work they found that two gases were being produced. In one test they measured 142 grains of hydrogen and 72 grains of oxygen. Clearly, the water had been decomposed. This result was startling. Water had previously been regarded as chemically stable, yet here it had been decomposed by something as feeble as low-voltage electricity. Further experiments showed the equally surprising fact that the two gases could be produced a couple of inches apart. (It is an experiment that can be repeated today with a simple radio battery and a glass of water.) In Germany, J.W. Ritter studied the same phenomenon at about the same time and others, Fabroni in Florence for example, had observed it earlier, but Nicholson and Carlisle are usually credited with the first systematic study. Ritter also discovered that copper could be plated out if a current was passed through a solution of copper sulfate. Electroplating became electricity's first industry. With these discoveries the new science of electrochemistry was born and the chemical theory of the battery steadily gained ground, though debate about how it worked was to continue for many years.

Improvements to the primary battery were quickly made. To increase the voltage more cells or pairs were needed. That meant greater pressure on the damp cardboard, which resulted in the electrolyte being squeezed out and ended the life of the battery. Ritter avoided this problem by turning up the edges of his metal discs and found that his batteries then lasted two weeks.[6] A better design still was a horizontal wooden trough. Humphry Davy, perhaps most widely known for his invention of the miner's safety lamp, showed that pure water did not work as an electrolyte and the importance of the electrolyte was increasingly recognized. It was only a short step to fixing metal plates, say zinc, to a support by means of which they could be lowered into the electrolyte in the trough and positioned between the vertical plates of the other metal. In this way some control could be obtained over the battery's output current, though there was some confusion between voltage and current. Also, the zinc plates could easily be removed for cleaning or replacement. These trough arrangements have been suggested as the origin of our circuit symbol for a battery, a word which, incidentally, had been in common use for a bank of Leyden jars.

It became usual to describe a battery by the number of pairs of metal, a measure of the tension or voltage, and by the active surface area, a measure of the current available. A battery of 600 pairs was presented to the École Polytechnique in Paris by Napoleon, and there is a story that when he went to inspect it he seized the terminal wires and applied them to his tongue. "His Imperial Majesty was rendered nearly senseless by the shock," which is hardly surprising as he had probably applied over 600 V to his tongue.[6] When he recovered, it is said, "he walked out of the laboratory with as much composure as he could assume, not requiring further experiments to test the battery." In

1808 the Royal Institution in London obtained a monster of 2000 pairs with an active surface area of 128 000 inch^2. Both the London and Paris batteries were put to good use by the leading scientists of the day. At the opposite end of the scale W. H. Wollaston, an English chemist, made a tiny cell using a flattened thimble, sulfuric acid, and a small strip of zinc. He is even said to have fused fine platinum wire with it.

In 1809 Davy demonstrated a new and exciting use for the battery. With the Royal Institution's 2000-cell unit he produced and maintained a brilliant electric arc between two charcoal electrodes. The electrodes themselves were heated to incandescence. Sparks had of course been produced before but there is a world of difference between sparks and arcs. Davy's demonstration was the precursor of many investigations that took place later in the 19th century to harness the arc light and turn it into a practical, commercial lighting system. Davy's arc also gave an intense heat capable of melting platinum, quartz, and sapphire. Diamond and graphite evaporated.

Chemists decomposed various salts, ammonia, nitric acid, and other substances, and in 1807 Davy attacked the fixed alkalies, potash and soda, whose constituents were then unknown. Using electrolysis he isolated two new elements, which he called potassium and sodium. It was not long before he extended his success by isolating calcium, strontium, and magnesium. Electrochemistry was off to a flying start. At last electricity, now no longer a plaything, was proving itself useful.

Galvanism had demonstrated its powers in chemistry, lighting, and heating; but as yet only in chemistry did it have practical applications. Successful applications in heating and lighting, particularly commercial ones, had to await an even better source of electric power, the dynamo.

Though some startling applications had been found for them, batteries still presented some problems. Polarization led to a gradual deterioration in performance, and frequent dismantling, cleaning, and reconstruction was necessary for several reasons. Electrolysis had yet to be satisfactorily explained. It was tackled by T. von Grothuss and by Davy, and in due course the theory was advanced in Switzerland by Auguste de la Rive, in Sweden by J. J. Berzelius, and by the German physicist Rudolf Clausius.

New versions of the voltaic pile appeared. In 1812 another Italian, Giuseppe Zamboni, professor of natural philosophy at Verona, produced a dry pile made from discs of paper, tinned on one side and coated with manganese dioxide on the other. The electrolyte was apparently supplied by absorption of moisture from the atmosphere. Thousands of discs could be stacked so as to obtain a high voltage, though the current was tiny. Its main claim to fame was its durability. One Zamboni pile appears in the *Guinness Book of Records*: the one at Oxford University, which consists of about 2000 discs. It was set up in 1840 and arranged so as to operate a pendulum that oscillates about twice a second between two gongs. It is still working (Fig. 2.5). Its long life is thought to be a result of the depolarizing action of the manganese dioxide. In 1948 articles were published describing the construc-

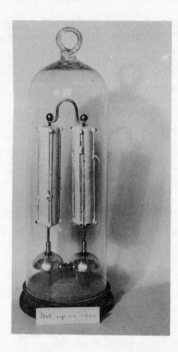

Figure 2.5 Zamboni pile, operating since 1840 at the University of Oxford

tion of a wartime Zamboni pile used as a power source in an infrared telescope. It was reported to give 2000–4000 V at about 10^{-9} A.[7]

The first big improvement in primary battery design came from A. C. Becquerel in Paris in 1829, the grandfather of A. H. Becquerel who discovered radioactivity. His two-fluid cell solved two problems. Polarization was one; the other was the gradual neutralization of the acid by the zinc. Amalgamation of the zinc was also found to give a significant improvement. This was discovered by William Sturgeon and was probably first done in the regular construction of batteries by M. Kemp in Edinburgh around 1828.

To follow the continuing improvement in battery design would now lead us through a long list of names and dates. A few outstanding designs do deserve mention, for example the Daniell and Leclanché cells.

The Daniell cell, invented by the professor of chemistry at Kings College in London, was first described in 1836. J. F. Daniell proudly noted that "the current was now perfectly steady for six hours together."[6] Like the Becquerel cell it used two fluids, dilute sulfuric acid, and copper sulfate solution, which were separated by a porous membrane—the windpipe of an ox. The electrodes were made from copper and amalgamated zinc. As with Volta's pile, the Daniell cell underwent much improvement. Following the professor of chemistry came a professor of physics, William Grove. The Grove cell of 1839, "a small voltaic battery of great energy," used sulfuric and hydrochloric acids, and the usual amalgamated zinc but now with expensive platinum for the other electrode. Grove also produced a gas battery, the first fuel cell. Fuel cells have long been expected to have a big impact in electrical engineering, but to date their only significant use has been in space vehicles. Grove's fuel cell used sulfuric acid, platinum, oxygen, and hydrogen.

R. W. Bunsen, of burner fame, replaced Grove's platinum with carbon and his cheaper battery found itself a market. Alfred Smee's cell of 1840, an improved single-fluid cell, also found a market and was widely used on the railways where it was left unattended for a year or so to get on with the job. However, the primary cell which has in modified form remained important to the present day was that invented by Georges Leclanché, a French chemist, in 1865.

Leclanché used a glass jar filled with a saturated solution of ammonium chloride and containing the usual amalgamated zinc rod. A porous pot was used, as had become common, but this time it held a carbon rod surrounded by a mixture of carbon and manganese dioxide powders—a significant change. The cell produced about 1.5 V, considerably more than Daniell's which gave about 1.08 V. Over the years the cell was improved. A lid made it more easily portable, and in 1876 the porous pot holding the powder was abandoned in favour of the mixture formed into a solid block by use of a binder. By about 1890 such dry cells were available commercially. One report in 1903 stated that production in Germany had reached large proportions.[8] The smallest cell, 30 mm high and 15 mm in diameter, gave about 0.25 to 0.3 A at 3.5 to 4.0 V; a suggested application was in small lamps for use by dentists, doctors, and military officers.

The modern dry cell is a direct derivative of the Leclanché cell. The liquid electrolyte has been replaced by a paste or jelly, the porous pot by a muslin bag, and the glass jar by the zinc rod, which has been converted into a can into which everything else is placed. The can is encased in a steel jacket to prevent leaks and is finished off with a paper or plastic wrapper on which is given the voltage, a multiple of the 1.5 V found by Leclanché in 1865. By 1868 some 20 000 Leclanché cells had been installed on railways and telegraphs, so quickly did it find favour.[9] Annual production of dry cells in 1909 was about 34 million units and had risen to a couple of thousand million by 1944. In the late 1970s that figure was approached by a single country. Japan, for example, produced 1885 million dry cells in 1977[10]; American shipments were over $600 million per annum in 1976. The sales of Volta's apparatus, which "will doubtless astonish you" as he wrote in 1800, would astonish anybody, so important has the simple dry cell become. Modern variations are prolific. For just one type, the silver oxide battery widely used in watches and cameras, the market for 1980 has been estimated at 500 million units.[12]

So much for primary cells. What about secondary, or storage, cells?

The first storage battery was probably that constructed by Ritter in 1803. However, a storage battery can only become important if there is a convenient method of charging it, and generators only appeared after the discovery of electromagnetic induction in 1831. The most important storage battery today is the lead–acid battery, the prototype of which was invented in 1859 by Gaston Planté in France and consisted of two sheet lead plates rolled up and dipped into dilute sulfuric acid. It was the outcome of a careful study of electrolytic polarization. However, Planté's battery had a serious defect from the standpoint of large-scale commercial production. It had to be 'formed,' a tedious process that involved repeated charging and discharging of the battery. This problem was overcome in 1881 by another Frenchman, Camille Fauré, who coated the plates with red lead. About the same time it was realized that grids could be used instead of plates, with the grid holes filled by a paste containing the active material. Various designs were tried and patented.

Storage batteries began to acquire a market in diverse areas: electric vehicles, submarines, railway signalling, power stations, and in general lighting. In 1902 an Italian electric railway was using them at its substations. Each one had a capacity of 1500 to 2500 A-hour for a 1-hour discharge rate and cost £2000 ($8000).[13] New types of storage battery appeared. For example, Thomas Edison introduced an alkaline battery that used nickel oxide and iron in an electrolyte of potassium hydroxide solution. The now popular nickel–cadmium battery first saw light of day in Sweden about 1900 in a form known as the Jungner cell. A variation, the sintered or Durac type, was developed in Germany during World War II.

The most common present-day use of storage batteries is in motor vehicles. After the coronation of King George VI in 1937 *The Electrician* reported that the "King unwittingly was sitting throughout the procession upon the storage battery which was used for the illumination of the state coach." The growth of the storage battery market can be judged from some American statistics. The value of units produced in 1909 was about $4.25 million.[14] By 1950 the figure has risen to $319 million, and was around $1504 (£750) million in the mid-1970s. In Japan, where nearly 200 000 tons of lead were used for storage batteries in 1973, about 90 per cent of the lead–acid battery production goes to the motor industry.

A less obvious use for batteries is as a practical standard for voltage. In the second half of the 19th century a growing need was felt for the standarization of units—all units, not just electrical ones. In 1861, at the suggestion of William Thomson (who became Lord Kelvin in 1892), the British Association appointed a committee on standards of electrical resistance. At the time no coherent system of electrical units had been accepted, although Wilhelm Weber's absolute system existed on paper. In 1851 Weber had shown that resistance could be expressed in terms of velocity but in practice arbitrary units were used. Charles Wheatstone, for instance, had suggested 1 foot of copper wire weighing 100 grains; another suggestion was 1 mile of copper wire 1/16 in. in diameter. The committee's first report in 1862 recommended a unit of resistance equal to 10^7 MGS electromagnetic units.[15] The name Ohmad was suggested but was shortened to Ohm. If voltage or current could be defined the other would automatically follow. The voltage of the Daniell cell was suggested as a standard but rightly rejected in favour of an arbitrary unit: 10^8 CGS units became the definition of the volt. The Daniell cell continued to be used as a practical standard cell and was stated to have an emf of 1.079 V. Any practical unit must give way if a better one is offered and the Daniell cell eventually gave way in 1891 to the Clark cell invented by Latimer Clark in 1872. In 1908 it in turn was replaced by the Weston Normal or cadmium cell invented in 1892 by Edward Weston. The Weston cell, still used as a standard, was provisionally stated to produce a voltage of 1.0184 V at 20°C.

Research to produce better batteries still continues and has become increasingly important with the dramatic rise in costs of primary energy sources. With the increasing price of oil, electrically propelled cars became more desirable. Storage batteries charged by current from coal or nuclear power stations are the critical element and efforts have been made to produce lower power/weight and power/volume ratios. A wide variety of electrodes and electrolytes have been investigated but, though many look promising, none has yet produced a market rival to the basic lead–acid storage battery. That is a late 20th century problem. Back at the beginning of the 19th century, the primary battery was still new. It might seem that with a more or less constant electric current available, the laws of electromagnetism and electrodynamics were unlikely to remain undiscovered for very much longer. However, twenty years were to pass before the flood of discoveries began.

References

1. Lucretius, *On the Nature of the Universe*, Penguin, London, 1951.
2. R. Taton, Ed., *A General History of the Sciences: The Beginnings of Modern Science, from 1450 to 1800*, Thames and Hudson, London, 1964.
3. J. Priestley, Chapter 1, Ref. 11.
4. E. T. Whittaker, *A History of the Theories of Aether and Electricity*, Nelson, London, 1951.
5. W. F. Magie, *A Source Book in Physics*, McGraw-Hill, New York, 1935.
6. J. J. Fahie, *A History of Electric Telegraphy to the Year 1837*, E. and F. N. Spon, London, 1884. Reprint edition: Arno Press, New York, 1974.
7. A. Elliott, *Elec. Eng.* 20: 317–319, 1948.
8. *Electrician* 50: 906–907, 1903.
9. G. W. Vinal, *Primary Batteries*, Wiley, New York, 1950.
10. Statistics Bureau, *Japan Statistical Yearbook, 1979*, Tokyo, 1979.
11. U.S. Bureau of the Census, *Statistical Abstract of the United States: 1978*, Washington, D.C., 1978.
12. W. Bond, *New Scientist* 72: 323–325, 1976.
13. *Electrician* 50: 256, 1902.
14. G. W. Vinal, *Storage Batteries*, Wiley, New York, 1955.
15. British Association, *Reports of the Committee on Electrical Standards*, Cambridge University Press, London, 1913.

3 THE 1820s: DAWN OF A NEW AGE

In his book published in 1600 William Gilbert dismissed Porta's theory of magnetism as "the ravings of a babbling old woman." Porta replied that Gilbert was "an Englishman with barbarous manners." Most scientists today when in disagreement with each other are a little more polite than that, at least in public. Despite their mutual condemnation both Gilbert and Porta are remembered for their contributions to experimental magnetism and for their (to modern ears) quaint theories. No theory of magnetism, however, whatever its weakness or strength, could begin to approach the truth as we know it while magnetism and electricity were regarded as separate, even if similar, phenomena. When the discovery of the united phenomenon of electromagnetism was announced by Oersted in 1820 a whole new world was opened for scientific exploration and the ground rules of our electrical science and engineering were made.

This chapter is mostly concerned with the 1820s, a golden period that was the dawn of a new era of electrical science and that saw the first glimpse of electrical engineering. The old electricity of electrostatics paled into insignificance when compared with the new era of 19th century electrodynamics. The new era began with the discovery of electromagnetism, the interplay between electricity and magnetism. This discovery led to the work of the famous men remembered in our systems of electrical units: Oersted, Ampère, Ohm, Henry, Faraday, and others. Coulomb and Volta had gone before. It culminated with the work of famous electrical engineers and the companies they founded: Bell, Edison, Marconi, Siemens, Ferranti. The 19th century was an era of what can almost be regarded as classical electrical engineering when compared with the electronics of the 20th century.

Even before that era began some evidence had already been accumulated that hinted at a close link between electricity and magnetism.[1] For example, both had two polarities and both exhibited forces of attraction and repulsion. It had been observed that some ferrous bodies became magnetized after a lightning strike, and Franklin had shown that lightning was an electrical phenomenon. Also, delicately suspended magnets were affected by the aurora borealis (northern lights), which itself seemed to resemble lightning. Steel sewing needles had been magnetized by the discharge from Leyden jars and it was known that the polarity depended on the direction of the discharge. In 1734 Emanuel Swedenborg, a Swedish scientist and theologian, argued that the two were closely related because they were both polar forces, and 14 years later Lorenzo Béraud of Lyons believed them to be different effects of the same force. G. B. Beccaria reviewed the situation in 1758 and suggested that natural currents in the earth were the indirect cause of the earth's magnetism. In 1774 the Royal Bavarian Academy posed a prize essay question to try to determine if indeed there was a "real and physical analogy" between the two. There was no agreed answer.

After the invention of the primary battery a continuous, rather than a momentary, discharge became available. As a discharge had been shown to be the key, why did it take nearly twenty years to discover the magnetic effects of an electric current? The answer seems to lie with Coulomb. After having shown that the inverse-square law is applicable to magnetic and electrical forces Coulomb continued his work on magnetism. Perhaps prompted by the inconclusive answers to the Bavarian Academy's question, he approached the problem using the inverse-square relations and concluded that electricity and magnetism had to have quite different causes. Therefore there was probably no close link between them. "This hypothesis," Ampère remarked in 1820, "was believed as though it were a fact." According to Ampère, the reason it took twenty years to discover electromagnetism was that everyone had believed Coulomb. No one had bothered to look.[2] Certainly no eminent scientist gave the matter any serious experimental investigation. Apart from a minor investigation in 1805 of a hypothetical link between voltage and magnetism, and a mistaken claim by G. D. de Romagnosi in 1802 (in which electrostatic phenomena were believed to be electromagnetic), it would appear that in most minds the question was dead and buried by the time the battery was invented. New problems, such as explaining the operation of the battery, or the investigation of electrochemistry, probably seemed more interesting than digging up a tired old question.

Oersted's Discovery

Not everyone, however, accepted the separation of these two sciences. To some the philosophy that natural phenomena are interrelated was more powerful than Coulomb's 'proof.' Oersted was one such person. During a visit to Germany in 1812 he published his work on the "identity of chemical and electric forces," a translation of which appeared in France the next year. Earlier, in 1806, his philosophy that "all phenomena are produced by the same power" had led him to suggest that this power "appears in different forms as, for example, Light, Heat, Electricity, Magnetism, etc."[2] Clearly Oersted had long held the view that electricity and magnetism are related. The experiment that finally revealed the intimate link between them occurred to him, as one biographer put it, "as a means of testing the soundness of the theory which he had long been meditating."[3] His discovery therefore cannot be classed as accidental.

Hans Christian Oersted was a 42-year-old professor at the University of Copenhagen when he discovered the magnetic field associated with an electric current while lecturing on electricity, galvanism, and magnetism to a group of students. The effect was described by him as being "very feeble," and it made no strong impression on the students who witnessed history being made. "The effect was certainly unmistakable," wrote Oersted, "but still it seemed to me so confused that I postponed further investigation to a time when I hoped to have more leisure."[4] Not until three months later did he return to, in Williams's words, "the discovery which was to give him immortality in the history of science."[2] When that time came, armed with a more powerful battery, he established the fundamental fact that the compass needle detector was affected only when a current was flowing and not by the mere presence of a voltage or charge. Open-circuit techniques had been tried before in vain. In his own words, "the galvanic circle must be complete, and not open."[5] Having made the discovery, Oersted examined it in some detail by positioning his compass needle in various places around the wire and by using a variety of materials. In modern terms, he concluded that the magnetic field was circular and was dispersed in the space around the wire, and also that the metal used for the wire was not critical but perhaps affected the magnitude, an effect explained later by Ohm's law. He found that the magnetic field passed through various media, and that a needle made from 'nonmagnetic' material was not affected. He was also quite certain that the phenomenon was not electrostatic. Although his experiments laid bare the basic effect, Oersted's explanation of it never gained ground. He talked of the "conflict of electricities," by which he meant the effects caused in and around the conductor by action between the supposed two types of electricity. When a current flowed, these two types of electricity did not give a uniform stream but "a kind of dynamic oscillation."[2] In his earlier book he claimed to have demonstrated that heat and light were associated with this electric conflict. Magnetism he saw as another effect of this conflict in and around the conductor, an effect capable of turning a compass needle on its pivot. His full explanation has a strong flavour of effluvium theories.

Oersted's discovery was of fundamental importance and was immediately recognized as such throughout Europe. Besides revealing the intimate link between magnetism and electricity it had also revealed what was thought to be the first known nonlinear force. Previously all forces were known to act in straight lines. This one appeared to go around in circles, and that seemed to contravene Newton's laws of mechanics. For a while it posed some severe problems.

The news of the discovery of electromagnetism was published in Latin in a pamphlet dated 21 July 1820, probably the last important scientific news to be issued in that language. An English translation was made the same year. A frenzy of activity followed, especially in Paris. By the end of the year, quantitative laws had been established and practical applications inaugurated.

Ampère and Others

D. F. J. Arago brought the news to the French Academy of Sciences on 11 September 1820, and after only a week André Marie Ampère had studied and extended it. Ampère has been called the father of electrodynamics, the Newton of electricity. He coined the word electrodynamics and in a mere six years laid its foundations. The old electricity he named electrostatics so as to highlight the essential differences between the two fields of study. Although Ampère's life as a scientist was an outstanding success, his personal life was marred by tragedy. Born in Lyon on 20 June 1775, just a few years before Galvani first saw a frog's leg twitch, he spent his early life in self-education aided by his father, a well-to-do retired merchant. Postponing the study of Latin he turned to mathematics and by his late teens was well up with the masters. France was now plunged into the bloodbath of revolution. The guillotine took its toll and Ampère's father became a victim. Struck by tragedy the young man became a recluse and abandoned study altogether; indeed it is said that he became a near idiot. A year passed before he began to rejoin the world, through a study of biology and by writing poetry. In 1799 he married and his wife bore him a son who became well known in his own right. But tragedy struck again. After less than four years of marriage his wife died. A second marriage broke up after only a few years. For much of his life Ampère seems to have been unhappy, even confessing late in life that only two years had brought him real happiness. He died in 1836 aged 61, and it might almost appear that he felt he had little left to live for after the great discoveries of the 1820s and early 1830s.

In the weeks that followed Arago's announcement of Oersted's news Ampère read many papers to the Academy of Sciences in Paris and became the outstanding contributor to the new knowledge of electrodynamics. Whereas Oersted discovered the subject, it was Ampère who tore it open and mathematically dissected it. He carefully defined what he meant by electric current and electric tension (voltage), though as yet he had no clear idea of the role played by resistance. As a result he saw the 'electromotive action,' of a battery for example, as producing a voltage but no current in an open circuit, and a current but no voltage in a closed circuit. In the second case he stated, "there is no longer any electric tension," or "tensions would disappear or at least would become very small."[5] A few years later G. S. Ohm recognized the true role of resistance but found it difficult to convince others; one reason, it has been claimed, was Ohm's "conceptual innovation" that the current and voltage were not independent but were related by the resistance of the circuit.[6]

Ampère went on to show, as Oersted had done before him, that it was the current, not the voltage, that caused the magnetic effects. This idea he extended to the decomposition of water: "The tensions are not the cause of the decomposition of water, or of the changes of direction of the magnetic needle." The direction of the deflection of the needle he expressed as a law: if an observer had a current flowing from his feet to his head then a needle placed in front of him would have its north-seeking pole deflected to his left. Later this law was expressed as the right-hand rule (Fig. 3.1).

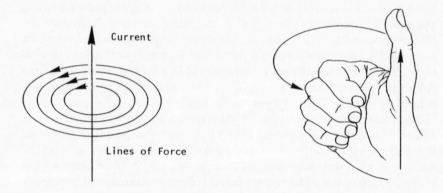

Figure 3.1 Right-hand rule to express Ampère's law governing direction of magnetic field caused by an electric current

With electricity and magnetism now known to be related the question arose as to which of the two was the fundamental phenomenon. Some argued that magnetism was at the core of electromagnetism. Ampère took the opposite view. To some extent he still sided with Coulomb, in the sense that he believed two essentially different phenomena would not interact; but as electricity and magnetism did interact it could be concluded that they were not essentially different. To Ampère electrical fluids seemed more likely than magnetic fluids, hence electricity was probably the cause of magnetism. When an electric current influenced a magnetic needle it must therefore be the result of electricity acting upon electricity. If that was true, he reasoned, then two electric currents should interact, and he found that they did.[2] Two wires carrying current attract one another when the currents flow in the same direction but repel when the currents are in opposite directions (Fig. 3.2). This discovery was made by clever experimentation and was suggested to Ampère by P. S. Laplace.[5] Ampère satisfied himself that these attractions and repulsions were not electrostatic phenomena and he described them as voltaic so as to distinguish them clearly from the electrostatic attraction and repulsion. Volta's name, like Galvani's, was passing into electrical terminology. To detect a current Ampère used an instrument he described as "similar to a compass, which, in fact, differs from it only in the use that is made

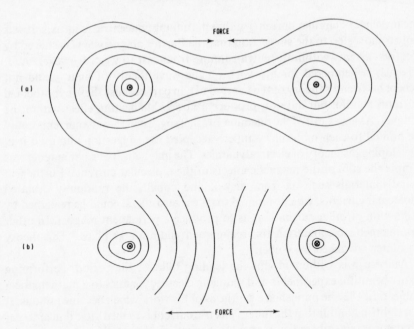

Figure 3.2 Attractive and repulsive forces between two current-carrying wires

of it." He called the instrument a galvanometer, after galvanism and Galvani. It was left for someone else to name its derivative after Ampère himself. The galvanometer was clearly distinguished from the electrometer, which was used to measure tension. Another use for the galvanometer as a current detector was suggested, "by employing as many conducting wires and magnetized needles as there are letters . . . we may form a sort of telegraph." Several years passed though before the electromagnetic telegraph became a reality. Codes were used so that one wire per letter was not needed, and improved electromagnets gave better detectors.

At the University of Halle, J. S. C. Schweigger reasoned that since a needle could be deflected in the same direction by a wire above it as by one below it, provided the currents were in the opposite sense, then twice the deflection should occur if the wire passed over the needle and then doubled back under it. He found that he did get almost twice the deflection. The result was Schweigger's multiplier, a more sensitive galvanometer which was simply a squared coil of wire, about 100 turns, with a compass needle inside. It was claimed to be "as sensitive to the action of the pile as the nerve of a frog," and this sensitivity helped it become important in the first electromagnetic telegraph systems.

Some of Ampère's galvanometers also used coils of wire, or solenoids, and he and others showed that such coils could "imitate all the effects of the magnet."[7] A bar magnet itself could now be explained by the assumption of the presence of circular currents within it, running concentric to its axis. It was only a short step to the suggestion that the Earth's magnetism was caused by electric currents within the earth running from east to west. However, A. J. Fresnel, remembered for his work on the wave theory of light, could not accept his friend's theory of the bar magnet, in part because these hypothetical currents should, but did not, produce a noticeable heating effect within the magnet. As a way out of the dilemma Fresnel suggested that the currents could be limited to each molecule. Ampère accepted the suggestion and used it in developing his theory of electrodynamics. The magnetism of a bar magnet was simply the sum of the magnetic effects of the molecular currents. Further, in some materials such as iron, nickel, and cobalt, the randomly oriented molecular currents, which summed to give a zero effect, could be realigned by the action of other currents so as to produce a permanent magnet. In other nonmagnetizable materials this realignment did not take place.[2] This theory was later advanced by Weber.

Ampère was without doubt the leading light of the period, performing many beautiful experiments and reducing electrodynamics to a mathematical subject. In 1827 he published a synthesis of his work which became famous. It is a rightful honour that the unit of electric current is named after him and that his name is remembered on every plug, socket, and fuse. Others, however, also helped to advance electrodynamics and electromagnetism.

A year older than Ampère was J. B. Biot, a man remembered for his work with meteorites and polarized light, as well as for his work with magnetism. His sense of adventure is also remembered. In 1795 he took part in a street riot which the young Napoleon Bonaparte put down with a "whiff of grapeshot," an event which is sometimes said to mark the end of the French Revolution. It also marked the temporary end of Biot's freedom as he was sent to prison. Then in 1804 he made a balloon flight to 13 000 feet or more with J. L. Gay-Lussac to test a suggestion that the earth's magnetism might be dependent on height. His contribution to electromagnetism came in 1820 when he announced (with Félix Savart) what has become known as Biot and Savart's law, though the mathematical expression was given by Laplace. Biot and Savart measured the force exerted on a magnet by a current carrying wire and found it to be inversely proportional to the distance from the wire. Their experimental technique involved timing the oscillation of magnets suspended by unspun silk; the effect of the earth's magnetism was compensated for by an artificial magnet.

More famous names were also involved in the months after the announcement of Oersted's discovery. Oersted himself suggested that if electricity affected a magnet, then a magnet might affect electricity. This effect was duly verified when Davy showed that an electric arc could be deflected by a magnet. D. F. J. Arago, T. J. Seebeck, G. G. Pohl, and Davy independently discovered that a current passing through a coil could magnetize iron or steel needles placed inside the coil. Fresnel claimed to have decomposed water with a current produced by a magnet. Perhaps he had witnessed the effects of a current produced by electromagnetic induction, but he soon found anomalies that led him to doubt his result and retract his claim. Michael Faraday, who eventually did track down electromagnetic induction, made his first important contribution in 1821 when he obtained "the revolution of the wire round the pole of the magnet"—the first, and very primitive, electric motor (Fig. 3.3). With his first big discovery Faraday also found himself facing the very unpleasant charge of plagiarism. W. H. Wollaston had surmised that it should be possible to make a current-carrying wire rotate about its own axis when a magnet was brought near it. Wollaston and Davy tried the experiment and met with no success. Faraday arrived as they were discussing the problem but apparently gave little attention to it. Later he was asked to write an historical account of this new branch of science and prepared himself by studying the experiments that had gone before. As he learned he came to realize the real possibilities of obtaining electromagnetic rotation and in elegant experiments he made a wire rotate around a magnet and a magnet rotate around a wire. The date was September 1821, five months after Wollaston and Davy's futile efforts. On Christmas Day he managed to dispense with the magnet and made a current-carrying wire rotate in the earth's magnetic field. Electricity had at last been made to perform mechanical work. Though the conversion had been demonstrated only on a trivial scale, the principle on which it was based was one which would later be developed to produce the vast range of electric motors that existed by the end of the 19th century. Wollaston had expected the

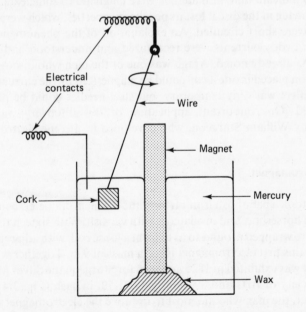

Figure 3.3 Faraday's conversion of electricity into mechanical motion, rotation of wire around magnetic pole

wire to rotate about its own axis, but Faraday plainly showed that this did not happen; the wire was forced away at right angles to the magnetic pole so that it moved in a circle round it. Faraday's discovery was not the one sought by Wollaston, but when his paper was published it attracted wide interest in England where Wollaston's ideas of rotation were known. Faraday was unjustly accused of stealing the idea without acknowledgment. Even Davy, Faraday's mentor, joined the accusers. Some say that Davy was growing jealous of the man who had once been his assistant; others point to Faraday as Davy's greatest discovery.

A particularly strange discovery was announced in Paris in 1824 by Arago, another French adventurer who has been described as a fiery republican. In 1806 Arago went to Spain to help in an experiment to determine the length of a degree of meridian. Two years later guerilla warfare erupted against Napoleon's France and Arago found himself in prison. He escaped to Algiers and 12 months after his imprisonment reached Marseilles, bringing with him the results of his work. He was immediately elected to the French Academy. In later years Arago was involved in the French Revolutions of 1830 and 1848, and he refused to take the oath of allegiance when Louis Napoleon made himself emperor in 1852. For a while, before his death in Paris in 1853, he served as minister of war and marine. The strange discovery he contributed to electromagnetism was that when a copper disc was spun about its axis it influenced a compass needle placed above it, so much so that the needle could be made to rotate. Arago's disc did not have to be made from copper and seemed to indicate that all materials were magnetic to some extent. If radial cuts were made in the disc it lost its peculiar properties, which were regained if the slots were short circuited. An explanation of the phenomenon was not found until eddy currents were recognized and understood and Lenz's law applied. As already noted, Arago was one of the men who discovered that a piece of iron placed inside a coil could be magnetized by the current. With soft iron the effect was only temporary, but steel needles could be permanently magnetized. One important application of this discovery was made in England by William Sturgeon, who developed it into the electromagnet.

The Electromagnet

Sturgeon took a bar of soft iron 1 ft long and 0.5 in. in diameter, bent it into the shape of a horseshoe, and insulated it with varnish. With sixteen turns of bare copper wire wrapped around so as to form a loose coil, with adjacent turns not touching, this first electromagnet lifted a mass of 9 lb. Together with straight magnets it was exhibited in 1825 and won for Sturgeon the Silver Medal of the Royal Society of Arts and 30 guineas (£31.50) in cash (Fig. 3.4).

However, the man who did most to improve the electromagnet was Joseph Henry in America. Henry, a man of modest financial means and living in the then small town of Albany, was a teacher whose time for research was largely

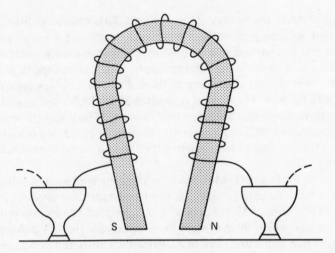

Figure 3.4 William Sturgeon's horseshoe electromagnet. The cups contained mercury for making electrical contacts

restricted to vacations. Late in 1827 he demonstrated an improvement of Sturgeon's magnet and declared his interest in investigating large-scale equipment using small currents, "the least expense of galvanism" as he put it, possibly an economy measure brought about by his shortage of money.[8] The next summer he was able to demonstrate two significant steps. Sturgeon's magnet had used bare wire coiled loosely, and therefore obliquely, over insulated iron. Henry's first improvement was to insulate the wire rather than the iron. That was not the first time insulated wire had been used, but it did enable him to take the second step of packing together tighter turns and so ensuring that the wire lay almost at right angles to the axis of the iron bar—an important feature. In Sturgeon's magnet each turn produced a magnetic field oriented at an angle to the axis, in Henry's version the field from each turn was correctly oriented along the axis, which yielded a stronger field for the same number of turns. In addition more turns could be packed together because the wire was insulated. How to insulate the wire was an example of one type of problem met by some early workers in electrical engineering, that of obtaining materials. Getting copper wire in quantity could itself be difficult, and then it had to be insulated. Henry is said to have sacrificed his wife's white silk petticoat to obtain the silk ribbons he needed. Linen thread was used later because it was cheaper, and Henry must have spent many boring hours insulating the wires for his experiments.

With 30 feet of wire wound into 400 turns Henry's magnet could lift 14 lb, about 25 times its own weight. For a time this was the popular way of measuring the strength of a magnet; there were as yet no magnetic units. Electrical units were also needed. The current supplied to an electromagnet is of obvious importance but the only measurement given for the current is the size of the plates in the battery, $2\frac{1}{2}$ in.[2] of zinc. To obtain more lifting power Henry tried wrapping more and more wire, but more wire meant more resistance and less current and Ohm's law was still unknown to Henry. One possibility was to increase the voltage supply, with a more expensive battery. Henry also explored the alternative method of using a second identical coil in parallel with the first. The magnet now lifted 28 lb, double the original figure and over three times that achieved by Sturgeon. Henry was the first to use multiple coils and is said to have startled his friends by leaping from his chair when the idea occurred to him, banging his hand on the table and exclaiming, "I have it."

In Utrecht, Professor G. Moll took the opposite approach of using bigger batteries, a brute-force technique that produced impressive results; 154 lb was lifted in 1830. News of Moll's achievements prodded Henry into publishing his own results, which by then were more advanced. Henry's publication of 1831 placed him in the first rank of American scientists and he was rewarded by an appointment as a professor at Princeton University.

In addition to reporting his small electromagnet, which lifted 28 lb, Henry also published details of experiments on a larger scale. A U-shaped bar of soft iron 2 in. square and 20 in. long, had been wound with nine coils each containing 60 ft of insulated copper wire. Across the pole faces a 7-lb armature had been added to which weights could be hung (Fig. 3.5). With the entire contraption supported by a wooden frame Henry was able to perform thorough experiments on the lifting powers achieved by parallel combinations of coils. One coil alone would lift 7 lb; all coils in parallel lifted 650 lb. By an increase in the size of his battery the maximum weight lifted was pushed to 750 lb.

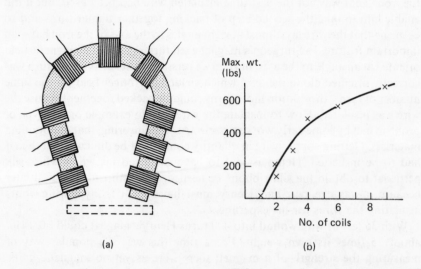

Figure 3.5 (a) Henry's nine-coil electromagnet; magnet 21 lb; armature 7 lb (after Ref. 8.). (b) Graph of Henry's results showing saturation effect

Besides making the world's most powerful magnet Henry also noticed the effects of the magnetic saturation of iron: with nearly all the coils in use an extra coil did not give as dramatic an increase as when few coils were in use. There was a limit to what a given magnet could lift. Henry was very much a qualitative scientist; he did not tabulate his results or plot them as graphs. If he had done so the magnetic saturation would have been more evident and he might well have been credited with its discovery, which he is not. Henry also came very close to recognizing the importance of the magnetic circuit, but this discovery also eluded him. Like many scientists of his day he did not use mathematics to help him. Yet to some of his generation, and increasingly so to later generations, mathematics was an indispensable tool.

Another important feature of Henry's paper of 1831 was his first moves towards understanding what he called 'quantity' and 'intensity' magnets and batteries, the first notion of the effects of parallel and series connections and of matching impedances. A quantity magnet was one with several short coils connected in parallel (low resistance) and was found to work best with a quantity battery, i.e., one with a single cell, or several cells connected in parallel. The cells of the day had a high internal resistance, which could be the limiting resistance in a circuit. The quantity battery paralleled these resistances and so reduced the total internal resistance, which allowed the maximum current to be obtained. Intensity magnets and batteries also worked well together. An intensity magnet had its coils connected in series, which presented a high resistance. A high-voltage or intensity battery was therefore needed, its cells connected in series. At that time Henry had no knowledge of Ohm's law and his results were empirical. His terminology survived for about a generation before the words series and parallel became accepted. An understanding of the practical effects of series and parallel connections was of prime importance in the design of the early telegraph systems. Both W. F. Cooke and Wheatstone in England, and S. F. B. Morse in America, the originators of the first successful commercial telegraphs, had to have this point explained to them by Henry.

Henry demonstrated the principle of the electromagnetic telegraph in 1831 to his students at Albany. About a mile of wire was strung round the classroom with a battery at one end and a small electromagnet at the other. When the magnet was energized it repelled a small permanent magnet which struck a bell and thus gave an audible signal. Whether or not this was a telegraph depends on how the word is defined, but Henry did not develop it as a communications medium. It has variously been described as a telegraph and as the first electric bell. What Henry had really shown was that mechanical motion could be produced at great distance by means of electricity. This principle could now be developed to produce telegraphs, electric bells, and relays. It is the basic idea behind much that we now take for granted, including loudspeakers, telephones, and electric motors.

Henry benefited little in the financial sense from his work. He was appointed to a position at Princeton but he did not patent any of his work, leaving it, he hoped, for the benefit of humanity. This action also left open the loophole that others might apply for the patents instead. Coulson reports that later in life Henry commented, "In this, I was perhaps too fastidious."

Other electromagnets were made under Henry's guidance. News of the big magnet attracted interest and the Penfield Iron Works became the owner of the first commercial electromagnets, which were used to extract iron from iron ore. Shortly afterwards the small town where the Iron Works was located changed its name to Port Henry in honour of the designer. The magnets were widely regarded as "a new wonder of Nature and Providence." In 1831 a monster was produced for Yale University that could lift 2300 lb, just over one ton. The horseshoe was a foot high and weighed $59\frac{1}{2}$ lb.

Others also built electromagnets, for example J. P. Joule in England. He is more commonly remembered for his work on the mechanical equivalent of heat which, incidentally, was refused publication by scientific journals and was eventually printed by a Manchester newspaper.[9] He is also credited with the discovery of magnetostriction in 1846, the slight change in the length of an iron bar when it is magnetized. In one of Joule's designs of the late 1830s two iron discs were used with teeth protruding at right angles. Wire was wound zig-zag fashion between the teeth. When energized, alternate teeth became north and south poles and the two discs were clamped firmly together. A magnet and armature weighing $11\frac{1}{2}$ lb could lift 2700 lb.

Typical present-day laboratory electromagnets have evolved from the theories (by J. Stefan and J. Ewing for example) and practices (especially of H. du Bois) developed late in the last century, but electromagnets are also made in a vast range of other designs aimed at their variety of uses in laboratories, industry, commerce, and the home. A few minutes work by the reader should produce a long list of applications, from tape recorder heads and TV picture tubes to mass spectrometers and electron microscopes.

Laws of Conduction

Ohm's law was published in 1826–1827 and is today the electrical law most widely known outside the electrical profession. Yet it was received with near total lack of enthusiasm by its discoverer's contemporaries. Indeed one, G. G. Pohl, described it as a "web of naked fancies" that had "no support even in the most superficial observation of facts." Ohm commented, "Pohl is well known to be arrogant and his blindness in despising my work is due only to his own attempt to restrain me. He is misguided by his own animosity and not led by the truth."[10] The exchange is reminiscent of Gilbert and Porta, yet Pohl was not alone in dismissing Ohm's work. In fact he was with the majority. Several attempts have been made to explain the unenthusiastic reception of Ohm's law; we shall look at them briefly later. The underlying reason appears to be that it was widely accepted in the 1820s that some conductors passed electricity better than others, yet the applied voltage and the current produced were viewed almost as two unconnected phenomena. The fact of electrical resistance was recognized; its major role in life was not.

We have already seen that Henry and Oersted had noted certain effects that were determined by resistance. They were not alone. Davy had made some measurements and notable work had been performed in the 18th century. Stephen Gray had made the basic discovery that some materials conducted electricity and others did not, and this work was extended by Du Fay and by J. T. Desaguliers, a Frenchman who fled to England to escape the religious persecution of Louis XIV. It was he who introduced the terms insulator, after the Latin for island, and conductor, to replace the old words 'electric' and 'nonelectric.' In 1753 G. B. Beccaria of Turin showed that when a discharge was made to pass along a path which included a tube filled with water, the shock received by the observer was more powerful if the cross section of the tube was increased. Possibly this was the first reference to the dependence of resistance on cross-sectional area. Henry Cavendish went considerably further, bringing forward the concept (though not the name) of resistivity. Like Beccaria, he performed his work long before the galvanometer was invented and so he too used his own body as an ammeter. He compared resistances by passing the discharge of Leyden jars through his body in series with first one, and then a second resistance, and judging the shock. Even Maxwell, who brought Cavendish's work to public notice so much later, was impressed by the accuracy of the results. One of those results was that "iron wire conducts about 400 million times better than rain or distilled water."[11] Cavendish not only understood the concept of resistivity but also demonstrated a clear grasp of the effects of length and cross-sectional area.

Later in the same century Joseph Priestley, who discovered oxygen and is mentioned in the last chapter for his book on the history of electricity, also performed "experiments on the conducting powers of various substances." Besides using the human ammeter technique he also devised a rather more reliable method of obtaining experimental results: measuring the length of an air gap across which he could just get a spark to jump.

Not so well known now as Cavendish or Priestley is Sir John Leslie, who went considerably further than either of them and all but stated Ohm's law in a paper written in 1791 but not published until 1824, which was still two years before Ohm's earliest correct statement of the law. Leslie's work has been compared with Ohm's and the reviewer commented that, "Both Ohm's and Leslie's statements can be represented by an equation of the form $I = V/R$."[12] However, Leslie did not actually state the equation.

On general philosophical grounds Leslie objected to the arbitrary classification of materials as conductors or nonconductors, and set out to determine the parameters that control the velocity of the transmission of electricity. He succeeded both in producing a theoretical treatment and in obtaining experimental verification. Ohm's later work, though, was far more rigorous. Leslie's theoretical treatment was based on an analogy drawn between the conduction of heat and the conduction of electricity, and on the assumption that there is an electrical law analogous to Newton's law of cooling. Ohm's treatment used the same analogy between heat and electricity but employed Fourier's analysis of heat flow (1822) as the analytical tool. Leslie's problem was complicated by the fact that, even under perfect experimental conditions, he had no constant-voltage source. His source was a statically charged body whose exponential discharge gave him his current.

One of Leslie's conclusions was, "The rate of communication of electricity is proportional to the intensity;" and that, in essence, is Ohm's law. Leslie discovered even more. Put into modern terminology, he argued that the current was inversely proportional to the conductor's length, directly proportional to its cross-sectional area, and proportional to its conductivity. He also found that a series combination of resistances of different conductivities or cross-sectional areas, or both, could be replaced by an equivalent uniform resistor. It then remained to verify some of his claims by experiment. He did so by comparing the time taken to discharge a Leyden jar through a slip of paper with the time required when the paper's length was halved, when its width was halved, and when it was coated with coal dust to change its resistivity.

Unfortunately Leslie's paper was not published at the time he performed the work, though it was read at two meetings of the Royal Society of Edinburgh in 1792. Leslie became indignant over a two-year delay in publishing and recalled the paper from the Society.[13,14] By the time it was published in 1824, apparently at the request of a friend who needed papers to support his journal, the march of science had moved the field of interest away from electrostatics. Leslie's own interests no longer lay with electricity, and the question of identifying static electricity with electricity from a battery was still undecided. It is to be regretted that his accurate work did not make a contribution to the growth of knowledge of the laws of electricity.

Three years before the publication of Leslie's paper, Sir Humphry Davy made known his own investigations into electrical conductivity. His technique was simple and elegant. The wire under test was placed in parallel with a conducting path of water so that the current from a primary battery flowed through two parallel resistors, one fixed (the water) and one variable (the wire). The resistance of the wire was such that current flowed through the water, dissociating it into hydrogen and oxygen. The length of wire was then reduced until the dissociation stopped, which happened when the wire's resistance reached a fixed (but unknown) limiting value. In this way Davy demonstrated that the resistance of a wire was directly proportional to its length, indirectly proportional to its cross-sectional area, and independent of the shape of the cross section, and that it increased with temperature and yielded different resistivities for different metals. That was a considerable achievement, since he used only a battery, bits of wire, and some water. With Davy's work the basic knowledge gained for electrostatics had been extended to electrodynamics.

Ampère, who had clarified some of the confusion between the old and new electricities, saw an 'electromotive action' as being the prime mover behind both. He believed that in a battery this action produced an accumulation of positive and negative charge at the terminals. This accumulation grew until a state of equilibrium was reached between the electromotive action and the mutual attraction of the positive and negative electricity. This state of equilibrium was a state of tension, electric tension. When the circuit was closed positive and negative currents flowed and the tension virtually disappeared. The electromotive action, in vainly trying to re-establish the tension, kept the currents flowing. The electromotive action was now an electromotive force and Ampère was led to his closest approach to Ohm's law. "The currents . . . are accelerated until the inertia of the electric fluids and the resistance which they encounter . . . make equilibrium with the electromotive force, after which they continue indefinitely with constant velocity so long as this force has the same intensity."[5]

Although Ampère believed that the battery voltage became virtually zero he clearly saw the current being dependent on the circuit resistance and an electromotive force supplied by the battery. His phrasing does not quite give us Ohm's law, but it is close. It is also interesting to see how some of our present-day terminology was used by Ampère. In later years, voltage, tension, emf, and potential difference were joined by the popular and descriptive term 'electric pressure.'

Ohm's law was finally stated by Georg Ohm in 1826 and again in 1827, after an initial attempt in 1825. The first two publications gave experimental derivations of the law; the 1827 version was a book that gave a mathematical derivation based on J. B. Fourier's analysis of the flow of heat along a wire. The first paper related the 'fractional loss of force' v to the length of wire x:

$$v = m \log \left[1 + (x/a) \right]$$

where m and a are constants.[6] The second corrected the first and gave Ohm's law in an easily recognizable form,

$$X = a/(b + x)$$

where X is the 'strength of the magnetic action' (current) in a conductor of length x (resistance), and a and b are constants dependent on the 'exciting force' (voltage) and the 'resistance of the rest of the circuit.'

Ohm must have been satisfied with the first equation as a summary of his experimental results which, though taken under difficult conditions, were approximately correct. Indeed it has been shown that the equation itself is approximately correct if the voltage source has a large internal resistance, as was probably the case with Ohm's battery. Even in this early form the law proved useful and Ohm was able to explain why Schweigger's multiplier did not increase its sensitivity in the manner expected when more and more turns, and hence more and more resistance, were added to it.

In his first experiment Ohm used several test resistors and a standard resistor whose resistance was considerably lower than the test samples. Using a Coulomb-type torsion meter, which measured force, he measured the difference between the current flowing when the standard, and then the test resistor, were in circuit. Good experimental technique enabled him to overcome two problems: current surges when switching the current on or off, and the steadily decreasing voltage of his battery. The latter was a common problem in those days and may have contributed to the slow acceptance of Ohm's law as others failed to verify the law, hindered perhaps by the same problem.

The second experiment was particularly elegant in its simplicity and the care with which it was performed. In 1822 T. J. Seebeck of Berlin discovered the thermoelectric effect now named after him and it was suggested that Ohm should use it to provide a constant voltage. He did so, using a copper–bismuth thermocouple as a 'thermoelectric battery.' It was from the results of this experiment that Ohm's law was derived and published. Ohm even took care to avoid tangling with the baffling phenomenon revealed by Arago's disc.

Boiling water and melting ice were used to obtain a constant temperature differential across the thermocouple and eight lengths of copper wire were the test resistors. The current surges and supply voltage fluctuations of the first experiment were eliminated; so too was the standard resistor and its associated measurements of current difference. Instead there was a constant voltage supply, a simple measurement of the resistor's length, and the force exerted on the torsion balance by the current's magnetic field. Davy's observation of the effects of temperature on conductivity was also confirmed and a caution issued that this dependence could lead to anomalous results.

In the book published in 1827 the mathematical derivation gave the law as

$$S = \gamma E$$

where S is the current, γ the conductivity, and E the 'difference of the electroscopic forces at the terminals," a term derived from electroscope measurements of the potential. In the analysis, electricity was assumed to move from one particle of the material directly to the one next to it. The magnitude of the current was assumed to be proportional to the difference between the "electric forces" of the two particles, "just as, in the theory of heat, the flow of caloric between two particles is regarded as proportional to the difference of their temperatures."[11] By this approach Ohm was able to show that the current depended only on the resistance of the conductor, and on a second variable whose relationship to electricity was the same as that of temperature to heat. The question that Ampère had tried to answer, concerning the state of the tension of a battery when the circuit was closed, was now answered by Ohm. The voltage was distributed around the circuit in a manner determined by the resistances of the parts of the circuit. He verified this result by electroscope measurements.

As already mentioned, Ohm's work was not easily accepted. Some have argued that his mathematics was too difficult to understand, that it was difficult to verify his work because of battery voltage fluctuations, and that people influenced by G. W. F. Hegel's philosophy were not interested in new experiments. Perhaps all three did contribute. Ohm was one of the first to use

Fourier's mathematics, and in his own mathematical derivation he never referred to his experimental work. That probably explains why many, for a long time, believed his law to have been obtained only by mathematics. It also withheld from some the details of how the experimental difficulties could be overcome. Hegel's philosophy has been described as dealing, "with reality, not solely with man's instruments for knowing or discussing it," whatever that means. Even if some were unduly influenced by it, it should not be forgotten that Germany was also the home of great experimenters like Schweigger, Seebeck, J. K. F. Gauss, and Weber. Also Ohm's book was sent to Paris where, it was hoped, Ampère and P. L. Dulong would examine it. Schagrin[6] has given a carefully prepared claim that it was Ohm's 'conceptual innovation' that lay at the root of the reaction to his work. "Ohm appeared to be confounding the well-recognized distinction between tension electricity and current electricity"—a distinction which, as we have seen, Ampère himself endorsed.

When Ohm performed his work he was a little-known schoolteacher who hoped his efforts would earn him a university appointment. Instead there was so much criticism that he resigned his schoolteaching position and for six years lived as a poor and badly disappointed man. Slowly his work became known and appreciated, first in private letters and then in print. Independent experimental verification was provided in 1831 and again several times afterwards. Ohm was recalled from obscurity in 1833 and appointed to a position at the Polytechnic at Nuremberg, the same year that Henry went on record as asking "where the theory of Ohm might be found." An English translation of his work appeared in 1841 and he was awarded the Copley Medal of the Royal Society of London in the same year. At last, in 1849, only five years before his death, he received a university appointment, the chair of physics and mathematics at Munich. After his death a statue was erected in Munich and a street named after him, and of course, his name has been immortalized as the name of the unit of resistance.

Despite the slow acceptance of this basic law further progress was made in understanding the flow of electric current. In 1833 S. H. Christie emphasized that the conductance of a wire was proportional to d^2/l and not, as some had come to believe, $d/\sqrt{l}$. Also in 1833 Christie derived the bridge principle for comparing resistances, a technique that Wheatstone made his own ten years later. In 1848 Gustav Kirchhoff, also remembered for his work with Bunsen on spectra, extended the use of Ohm's law to more complicated circuits; he too used the analogy with heat that had served Ohm and Leslie so well. It was also Kirchhoff who identified Ohm's electroscopic force with the electrostatic potential of Poisson and Green. The other basic law concerning current flow was supplied by J. P. Joule in 1841, when he found that the heat produced per unit of time is proportional to I^2R.

Fifty years after the first statement of Ohm's law, any remaining controversy was quenched by a report from the British Association for the Advancement of Science which stated that Ohm's law "must now be allowed to rank with the law of gravitation and the elementary laws of statical electricity as a law of nature in the strictest sense." By that time Ohm's law was in common use. Edison, for example, used both Ohm's and Joule's laws in deciding to use a high-resistance filament in his light bulb rather than a low-resistance filament. At least one university graduate, however, needed Edison's instruction on the use of Ohm's law, after completing his university education in electricity.

With the advent of telegraphy, particularly with the high-capacitance submarine cables, new problems were forced onto engineers. William Thomson (Lord Kelvin) takes much of the credit for solving the mysteries of the effects of capacitance with a telegraph equation which was famous in the last century, yet another advance in electrical theory made with the help of the analogy with Fourier's treatment of the diffusion of heat. Inductance was given its rightful place in the 1880s by Oliver Heaviside when he incorporated it into the equation of telegraphy and coined the word 'impedence'. The $R + j\,X$ notation for impedance was accepted internationally in 1911. With modifications to account for capacitance and inductance the fundamental importance of Ohm's law, and the electrodynamics of the 1820s, became even more apparent.

References

1. J. J. Fahie, Chapter 2, Ref. 6.
2. L. P. Williams, *Michael Faraday*, Chapman and Hall, London, 1965.
3. P. F. Mottelay, *Bibliographical History of Electricity and Magnetism*, Griffin, London, 1922.
4. B. Dibner, *Elec. Eng.* 80: 426–432, 1961.
5. W. F. Magie, Chapter 2, Ref. 5.
6. M. L. Schagrin, *Am. J. Phys.* 31: 536–547, 1963.
7. R. Taton, Ed., *Science in the Nineteenth Century*, Thames and Hudson, London, 1965.
8. T. Coulson, *Joseph Henry, His Life and Work*, Princeton University Press, Princeton, N. J., 1950.
9. I. Azimov, *Azimov's Biographical Encyclopaedia of Science and Technology*, Pan, London, 1972.
10. H. J. J. Winter, *Phil. Mag.* (Series 7) 35: 371–386, 1944.
11. E. T. Whittaker, Chapter 2, Ref. 4.
12. R. G. Olson, *Am. J. Phys.* 37: 190–194, 1969.
13. S. Lee, Ed., *Dictionary of National Biography*, Smith, Elder & Co., London, vol. 11, 1909.
14. J. Leslie, *Edinburgh Phil. J.* 11: 1–39, 1824.

4 ELECTROMAGNETISM

By 1980 the world's annual generation of electrical energy was almost 8×10^{12} kilowatt-hours. It was produced by exploitation of electromagnetic induction, a scientific effect discovered in 1831. Until then anyone who wanted to use electricity had to use batteries or friction generators, or even electric eels. Even long after 1831 dynamos exploiting induction were only used on a tiny scale, mostly for experiments.

Electromagnetic induction was the second major discovery in electromagnetism, a major science of the 19th century. Electromagnetism is the scientific base of traditional electrical engineering; power generation and its use, transmission lines and cables, radio and early electronics. It is a pillar of physics and the precursor of relativity and quantum theories. Quantum theories in turn led to an understanding of semiconductors and to their widespread use in modern electronics. Electromagnetism has thus made an enormous impact on electronics as well as on electrical engineering and physics. It began, as we have seen, with Oersted's discovery of the generation of a magnetic field by an electric current. Almost immediately researchers began to look for the opposite effect, an electric current induced in a wire by a magnet. The search was taken up by many people and continued for eleven years before it was successfully concluded by Michael Faraday and Joseph Henry.

In this chapter we shall examine the development of the science of electromagnetism over eighty-five years. It was a period that took science from Oersted to Einstein, and society from the first railways to the first aircraft, from the aftermath of the Napoleonic era to the rumblings that led to World War I. Engineering progressed from sailing ships to the *Dreadnought* and from the simplest electromagnet to wireless telegraphy. At the beginning of the period, bars of chocolate had just begun to be mass produced and Charles Mackintosh was developing his raincoat. By the end the thermionic diode had been patented, double-sided audio discs had been made, and Gillette razor blades were on sale.

One of the first to begin the search for electromagnetic induction was A. J.

Fresnel in Paris. It was known that one could magnetize a piece of iron by placing it inside a coil of wire and passing a current through the coil. Fresnel suggested reversing the experiment. Would a magnet inside a coil cause a current to flow in the coil? "Not that such a result is a necessary consequence of the original observation," he wrote.[1] Nevertheless he tried the experiment and in November 1820 reported to the Academy of Sciences that he had succeeded in decomposing water, a useful test for the presence of an electric current. The current, he claimed, had been induced in a coil wrapped around a magnet. Ampère corroborated the discovery; he too had noticed something in the way of feeble currents produced by a magnet. The search for electro-magnetic induction seemed to be over almost before it had begun. However, both men quickly retracted their claims. Fresnel could not repeat his work and Ampère had been uncertain of his results in the first place. He announced them only when Fresnel's report gave him the confidence to do so. The quick success proved to be an illusion and the scientists had to begin afresh.

Some of the philosophy of static phenomena still ran deep in their minds, as might be expected. Electromagnetism had been discovered when charge was on the move, not when it was stationary. A current was needed—a dynamic system, not a static one. Yet in searching for induction the scientists had made a false start by expecting a stationary magnet to produce a current. A system was expected to give out energy without any being put into it. (The discovery of the law of conservation of energy was still more than two decades away.) When electromagnetic induction was eventually discovered its dynamic nature came as a great surprise.

Yet to a certain extent logical thought should have hinted at this result.[2] Steel placed inside a coil was turned into a magnet when a current flowed through the coil. Why should not a magnet placed inside a coil produce a current in it? Repeated failures to observe this 'opposite' effect could be correctly taken as evidence that it did not happen. The true opposite experiment should have been the demagnetization of a magnet inside a coil. There would have been experimental difficulties, of course, but if the experiment had been tried an induced current would have been observed while demagnetization took place. A century and a half of hindsight colours our view and makes it difficult for us to know what was really in the minds of men in the 1820s. Yet as far as this author is aware, no one was led to make this deduction.

A lot of space would be needed to detail all the subsequent attempts to discover electromagnetic induction. Here are what were probably the two most interesting near misses.[1]

The first of the two experiments was devised by the great man himself, Ampère On the possibility of electromagnetic induction, he appears to have held opposing views at various times. In 1821 he devised an experiment to see whether an electric current could be induced, not by a magnet, but by another electric current. That is the electrodynamic equivalent of classical electrostatic induction, a principle widely exploited later, for example in transformers. The magnetic field of Ampère's induced secondary current, if there was one, would enable it to be detected. Ampère's equipment consisted of a light-weight ring made from a thin strip of copper and suspended so that it lay inside and almost touching a flat coil wound parallel to the ring (Fig. 4.1). A current through the primary coil was expected to induce a secondary current in the ring. A permanent magnet was used to help detect the induced current. A copper ring would not be affected by the magnet; any movement would have to be caused by the magnetic field of the secondary current reacting to the presence of the permanent magnet. Ampère saw no movement during his first attempts in Paris but later, during a demonstration with a more powerful magnet given to the young Auguste de la Rive in Switzerland, they both witnessed some slight movement. What was probably a true effect of electromagnetic induction had been discovered and Ampère recognized it as such. Why then do we give Faraday and Henry the honour of being its discoverers?

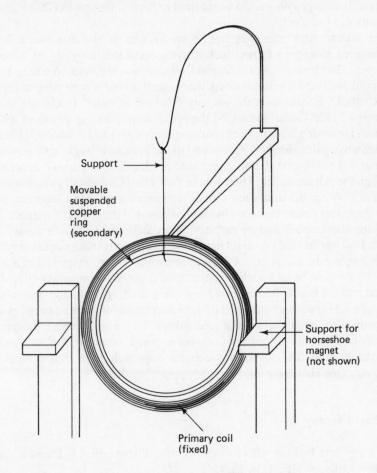

Support

Movable
suspended
copper
ring
(secondary)

Support for
horseshoe
magnet
(not shown)

Primary coil
(fixed)

Figure 4.1 Ampère–De la Rive experiment (1822) (after Ref. 1)

Ampère reported these results to the Academy of Sciences, but De la Rive had meanwhile published an account of the experiment and Ampère did not go into print. His version only appeared long after his death and long after Faraday's and Henry's publications of their own work. It is possible that the twenty-year-old De la Rive's account was inadequate and actually misled later researchers by attributing the effect to a temporary magnetization of a nonmagnetic body.[1] He did not suggest that this temporary magnetization was caused by an induced current and consequently the report lost its impact. Ampère did say that "the objective of the experiment was to learn whether an electric current could be produced by the influence of another current," but his report (when it was published at last) was brief and did not fully describe the work. He appeared to believe, wrongly, that the induced current was continuous. Despite the opinions of others Ampère seems to have regarded the experiment as relatively unimportant, and indirect evidence suggests that he actually changed his mind and reverted to his former opinion that induced currents did not exist.[1]

The second near miss was not so spectacular as the first but it is an outstanding example of cruel luck denying someone the glory of a major discovery. The Swiss engineer Daniel Colladon was the man on whom Lady Luck did not shine. He is not altogether forgotten, but is remembered for his work with C. F. Sturm on the velocity of sound in water. In Geneva in the summer of 1825 Colladon tackled the problem of inducing a current with a magnet. He made a large helix of insulated copper wire and connected the ends to a sensitive galvanometer. To one end of the helix he brought up a powerful magnet. Like everyone else he expected to produce a continuous current flow through the galvanometer. There is little doubt that Colladon's galvanometer, like Faraday's in his somewhat similar later experiment, must have recorded the momentary current induced by the movement of the powerful magnet. But by being too careful, and by not having an assistant, Colladon missed the effect. Fearing that the powerful magnet would directly influence his sensitive meter he placed it safely out of the way in another room, at the end of a 50 m lead. By the time he had walked to the other room the momentary effect had passed and the needle had returned to its zero reading. Colladon had missed his chance. He became convinced of the nonexistence of induced currents and his conviction apparently influenced others. Not even Ampère disagreed. Colladon, it would appear, would be a prime nominee for any award commemorating the closest unsuccessful approach to a major scientific discovery (see also Note 27).

Michael Faraday

The long searched-for effects were at last discovered by Faraday, and independently by Henry, in 1831.

Faraday's outstanding contributions earned him such epithets as 'the greatest experimental philosopher the world has ever seen' and 'the patron saint of electrical engineers.' Yet an even greater contribution than his discovery of electromagnetic induction lies in the methods he used to formulate his theories. He lacked mathematical training and was led to picture in his mind the physical phenomena he investigated. Iron filings lying on a sheet of paper above a magnet arrange themselves into a pattern suggestive of lines of force. It was from this simple concept that Faraday initiated that long train of thought which took the main line leading us, with innumerable branch lines, to field theory and Maxwell's equations, and from which sprang a feeder network to theories of relativity, quanta, and beyond.

Faraday was born of Yorkshire parents at Newington near London. His father was a blacksmith, one of the technicians of yesteryear. "My education," wrote Faraday, "was of the most ordinary description, consisting of little more than the rudiments of reading, writing and arithmetic at a common day school."[3] No university place would come his way. In his early teens he became apprenticed to a bookbinder and bookseller. It was this chance that inadvertently launched his scientific career by placing appropriate stimulating books within his reach. He later recalled that the *Encyclopædia Britannica* and a book of *Conversations on Chemistry* had given him his introductions to electricity and chemistry, respectively—the two fields to which he was to contribute most. His limited formal education may have been a blessing in disguise, as it may have forced him into original thinking; but it was also a handicap. His limited mathematics barred him from a most important approach to science. At the age of sixty-six he wrote to the mathematical physicist James Clerk Maxwell, his junior by forty years, to ask plaintively, "When a mathematician engaged in investigating physical actions and results has arrived at his conclusions, may they not be expressed in common language as fully, clearly, and definitely as in mathematical formulae? If so, would it not be a great boon to such as I to express them so?— translating them out of their hieroglyphics, that we also might work upon them by experiment."[15] Evidently he believed mathematicians could do so and praised Maxwell for always presenting to him a perfectly clear idea of his conclusions. There are engineers today who would echo Faraday's plea, while others hide behind the hieroglyphics and avoid plain language translations.

Fired by his ambition to do scientific work Faraday, then a nobody, wrote to Sir Joseph Banks, who was very much a somebody: president of the Royal Society. (It was through Banks that Volta had broken the news of the invention of the chemical battery.) Apparently Sir Joseph did not reply. Undaunted, Faraday wrote to Sir Humphry Davy and sent along carefully bound notes of Davy's lectures which he had attended. Davy was impressed, maybe flattered, and hired young Faraday as a laboratory assistant at the Royal Institution. There he stayed throughout his working life, later as director and finally as professor.

By 1831 Faraday had been searching for electromagnetic induction on and off for many years. Possibly he had got the idea from Davy early on in the long speculative search.[5] One of his early experiments, in 1824, resembled one of the later successful ones. A wire helix was connected to a battery through a galvanometer so that the current flow was indicated. A magnet was then placed in various positions to determine whether it would cause a change in the current. Apparently it did not. Presumably the currents induced by the magnet's movement were weak and were masked by the battery current. The following year Faraday attempted to induce a secondary current from a primary one, using two straight wires and a combination of a helix and a straight wire.

The historic discovery of electromagnetic induction was finally made, according to Faraday's laboratory notebook, on 29 August 1831. It was a famous victory, one that has been described repeatedly. Two coils of wire were

Figure 4.2 Faraday's electromagnetic induction ring

wound on opposite sides of a soft iron ring 6 in. in diameter (Fig. 4.2). One coil was connected to a battery; the other, to a wire that passed over a magnetic needle made to serve as a galvanometer. Nothing happened while the battery was in the circuit or out of it; but while the battery connection was being made or broken, Faraday observed a deflection of the needle. The primary current had induced a secondary current. Faraday, who was an excellent experimentalist, had made the observation that had eluded poor Colladon. Faraday noted both the transitory nature of the induced current, which occurred only when the primary current started or stopped, and that its direction was different in the two cases. He named the primary A and the secondary B and concluded that there "is no permanent or peculiar state of wire from B but effect due to a wave of electricity caused at moments of breaking and completing contacts at A side."[6] His use of the word 'wave' has led to some speculation that he may have even at this early stage anticipated the wave nature of electricity. Also, his thoughts were focussed on the state of the wire rather than on the current. We shall come back to both these points later. On September 24 he used two bar magnets to induce a current in a coil wrapped around an iron cylinder, "here distinct conversion of Magnetism into Electricity." Many more experiments followed. The breakthroughs had been made. Induced currents had been produced, but only when there was relative motion between the magnets and the coils.

Faraday read an account of his work to the Royal Society on 24 November 1831 and sent a preliminary report to the Academy of Sciences in Paris. His full account was published early in 1832. With Faraday's work the two methods of inducing a current, for so long the subject of speculation and experiment, had been found to be valid, but unexpectedly they did not produce a steady current. Induction by the use of moving magnets was soon applied to the generation of currents on a small scale from hand-driven generators. Gauss and Weber used a small magneto generator on their telegraph system from 1835. Later, big machine-driven magneto generators were used to drive arc lamps in lighthouses in some of the first large-scale applications of electrical engineering.

Meanwhile in America Joseph Henry performed an experiment, in essence the same as one of Faraday's, to produce an induced current in a secondary coil by making or breaking the primary circuit (Fig. 4.3). Though his work was independent of Faraday's, some news of Faraday's achievements reached him before he published his own account. (Transatlantic communications took months in the 1830s.) Nevertheless he is credited with having discovered electromagnetic induction independently, possibly even before Faraday. He described the discovery succinctly: "It appears that a current of electricity is produced, for an instant, in a helix of copper wire surrounding a piece of soft iron whenever magnetism is induced in the iron; and a current in an opposite direction when the magnetic action ceases; also that an instantaneous current in one or the other direction accompanies every change in the magnetic intensity of the iron."[7]

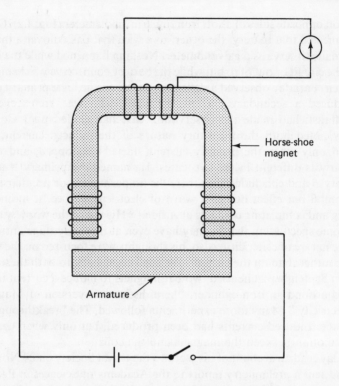

Figure 4.3 Henry's experimental discovery of induction (after Chapter 5, Ref. 2)

In the same paper Henry announced his discovery of self-induction (the current induced in part of a coil by the change in current in another part), resulting from the observation of a vivid spark produced when a long wire was disconnected from a weak battery. The announcement appears almost as an afterthought at the end of the paper: "I can account for this phenomenon only by supposing the long wire to become charged with electricity, which by its reaction on itself projects a spark when its connection is broken." Faraday also discovered self-induction a couple of years later.

The discovery of the induction of a secondary current by a primary current was the initial discovery which then led on to the induction of a current by a magnet. Both Faraday and Henry used what were essentially transformers with soft-iron cores. Faraday's apparatus used a ring; Henry's, an electromagnet with an armature. The equipment was nearly ideal and there has been much speculation about why each chose the type of equipment he did. Henry's choice came as a continuation of his work on electromagnets, but for Faraday the soft iron ring was a departure from his previous style.

One biographer, L. P. Williams,[8] has made a fascinating reconstruction of Faraday's 'mental evolution' that led to the ring experiment. Though he admits that his account contains more conjecture than is desirable, and that it may not be correct, it is worth recounting briefly for its insight into what may have happened. Faraday, it is known, was not happy with Ampère's theory of magnetism, which postulated electric currents existing around the axis of a magnet, or around each constituent particle of the magnet. Some experimental evidence seemed to tell against it. Faraday's own ideas assumed that electromagnetism was the result of some peculiar state into which the particles of the conductor were thrown (hence his comment noted above), but he was unable to detect this state. Further, his own discovery of electromagnetic rotation ruled out a simple rearrangement of particles since this rearrangement would not produce dynamic rotation, yet he was unwilling to accept a fluidic flow of particles. What he wanted was some means of transmitting force without transmitting matter—a transmission medium.

If Williams's reconstruction is valid Faraday found his answer in two analogies: one between electricity and the wave theory of light, and the other between electricity and the wave theory of sound. Analogies had already served electrical science well, in the mathematical derivation of Ohm's law for example, and they would certainly do so again.

Thomas Young's undulatory theory of light dated from 1801 but it was Fresnel's later theory that reached Faraday from 1827 to 1829. In 1830 Sir John Herschel had pointed to the analogy between sound and light: both depended on the vibratory motion of an elastic medium. Also, from 1828 to 1830, Faraday read a series of papers on acoustics on behalf of Charles Wheatstone. Wheatstone was interested in Chladni figures, figures produced in sand strewn on a vibrating plate, and showed that such figures could be produced on a second surface vibrating in resonance with the first, i.e., by acoustical induction. On 2 February 1831 Faraday recorded the beginning of his own 'extended and vigorous' research into acoustical figures that continued to the middle of July, just six weeks before he discovered electromagnetic induction.

The obvious conclusion, that Faraday was led to electromagnetic induction by his work on acoustical induction, is tempting. Even at the beginning of his work on acoustics he examined the effects of the medium on the propagation of sound, and the soft iron ring used in the electromagnetic experiment was an excellent medium and shape for transmitting the magnetic effects of the primary current to the secondary coil.

Williams's reconstruction is clever and carries conviction, but it is based on circumstantial evidence. Faraday left no indication that his thoughts had followed that line. His 'wave of electricity' may indeed have come by analogy with sound and light, or it may have been only a graphic description of the transient effects he observed. Six months later the situation left no room for doubt.

In March 1832 Faraday lodged a letter with the Royal Society to establish his priority claim to certain views.[8] This unusual action was perhaps impelled by ill-founded counterclaims to the discovery of electromagnetic induction made by badly informed people on behalf of others whose work was actually performed to verify prepublication news of Faraday's discovery. Perhaps Faraday still rankled a little over the ill-founded Wollaston affair some ten years earlier.

In the letter Faraday stated his position. "I am inclined to compare the diffusion of magnetic forces from a magnetic pole," he wrote, "to the vibrations upon the surface of disturbed water, or those of air in the phenomena of sound; i.e. I am inclined to think the vibratory theory will apply to these phenomena, as it does to sound and most probably to light." In that statement we begin to see how Faraday's thoughts on induction were to influence the future of electromagnetic theory.

Faraday published 158 papers and co-authored another four, a prodigious output.[9] (A modern-day university professor may publish thirty or forty during a busy and productive career.) Approximately half related to electrical science and just under a third, to chemistry. The rest were on a variety of topics including one 'On holding the breath for a lengthened period,' and another published in 1823 with the curious title, 'Change of musket balls in shrapnel shells: Action of gunpowder on lead,' which sounds almost like a defence contract. Among so much work his contribution on induction has been called the Mont Blanc of his achievements. What then of the rest of the Alps?

Sandwiched between his work on induction (1831–1832) and self-induction (1834) he advanced the knowledge of electrolysis and gave us a terminology still in use: anode, cathode, ion, anion, cation, and the word electrolysis itself. Avoiding theoretical preconceptions he formulated the laws of electrolysis and during the course of the work satisfied himself about a basic problem that had been rearing its head ever since different types of electricity (current, electrostatic, electrochemical, and so on) had been encountered. He concluded, "Electricity, whatever may be its source, is identical in its nature."

The famous concept of physical lines of force associated with a physical strain, which Faraday believed to exist in the propagating medium for electricity and magnetism, was particularly useful to him and has helped generations of students since. With it he was able to reject the old axiom of action at a distance, which assumed that all forces originated from point charges or poles, and replace it with a philosophy of force transmitted along lines of strain in adjoining particles, the starting point from which grew Maxwell's electromagnetic theory and modern field theory. As we shall see, he also raised the question of the nature of the propagating medium.

Faraday's concepts were not welcomed by everyone at the time. Sir George Airy, the Astronomer Royal, complained: "The effect of a magnet upon another magnet may be represented *perfectly* by supposing that certain parts act as if they are pulled by a string, and that certain other parts pushed as if by a stick. And the representation is not vague, but is a matter of strict numerical calculation. . . . I can hardly imagine anyone who practically and numerically knows this agreement, to hesitate an instant in the choice between this simple and precise action, on the one hand, and anything so vague and varying as lines of force, on the other hand."[10]

If only life were so simple.

The lines of magnetic force were seen by Faraday as the lines he could depict by iron filings, or by taking tangents to a group of tiny compass needles. Each line formed a closed curve which at some point passed through the parent magnet, a familiar picture that still appears in modern textbooks (Fig. 4.4). Groups of these lines of force could be thought of as a tube of force, or a Faraday tube as they are often called now. These lines could represent the magnitude of the magnetic intensity as well as its direction. It was only a small step to the idea of a unit tube whose product of magnitude and cross section was a constant. For simplicity a unit tube could be rendered as a unit line of force, and the concentration of such lines indicated the strength of the magnetic field in the region. Faraday eventually found that induction produced, not a current, but an electromotive force that depended on the relative motion of the conductor and the lines of force. Expressed mathematically in modern texts this rule is often called Faraday's law, though other laws also bear his name.

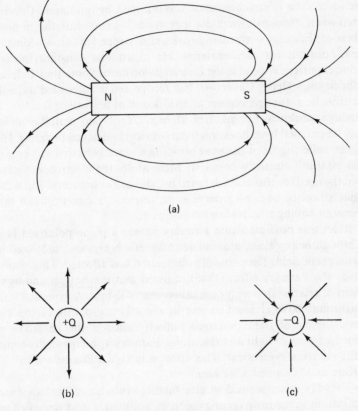

Figure 4.4 *Lines of force: (a) magnet; (b) positive charge; (c) negative charge*

With Faraday's conviction of the reality of the lines of force came the idea that they represented the lines of a strain within the propagating medium, whatever that medium was. In the induction-ring experiment the medium was the soft iron of the ring. The idea of a strain in the ring could be used to explain the momentary nature of the induced secondary current, and that was a notable victory. When the primary current was started the particles of the medium were thrown into a state of tension which Faraday called the electrotonic state, a hypothesis that served him well before it had to be abandoned. Stopping the primary current enabled the electrotonic state to relax back to normal. The secondary current, or wave of electricity, was produced by changes in the electrotonic state.

The rejection of the action at a distance philosophy was another success for this hypothetical state. Electric and magnetic lines of force were found to be curved, as demanded by the electrotonic state, whereas action at a distance required them to be straight. Faraday also used it to explain, to his own satisfaction, the previously separate phenomena of metallic induction, electrolytic conduction, and electrostatic induction. Each depended on the degree of strain the medium's particles could stand when subjected to an electric force. The electrotonic state was strained by induction. Conduction occurred when it broke down under that strain.[11] In an insulator, or dielectric (another of Faraday's words), breakdown depended on a constant—a different constant for each material. He called this constant the specific inductive capacity. We call it the dielectric constant. Some time before 1781 the publicity-shy Henry Cavendish had recognized it and called it the degree of electrification, though no one in 1837 knew of his work.

Faraday's work was not yet over, though a four-year period of ill health, possibly heightened by exhaustion from overwork, delayed progress. He may have been suffering from mercury poisoning, since mercury was commonly used to establish electrical contacts. Meanwhile, the electrotonic state had satisfied the need for the strain denoted by lines of electric force. But was there a similar strain of adjoining particles to transmit magnetic force through intervening nonmagnetic bodies?

To detect this possible strain Faraday passed a plane-polarized beam of light through heavy glass, a dense nonmagnetic body, and subjected it to a strong magnetic field. The plane of polarization was affected. This result, now known as the Faraday effect (1845), showed that magnetism and light had some sort of relationship with each other, one of the many experimental facts that historians have claimed as one of the starting points of the electromagnetic theory of light. Maxwell himself called it the keystone of the combined sciences of light and electricity. Faraday also repeatedly sought, in vain, the electrical equivalent. This effect was at last discovered in 1875 by John Kerr and is named after him.

The Faraday effect seemed to give further evidence of the existence of a physical strain in the propagating medium. Soon after its discovery Faraday realized that the bar of heavy glass he had used belonged to a new class of magnetic materials that, unlike other magnetic materials, aligned themselves across the line joining magnetic poles when freely suspended between them. He named this new class of materials diamagnetics (*dia* for across). The more usual type, which aligned themselves parallel with the line joining the poles, he called paramagnetics, and he reserved the term ferromagnetic for materials such as iron and steel. He now studied diamagnetism in detail and slowly discovered that the facts no longer fitted his hypothesis. The previously fruitful concept of the electrotonic state had to be abandoned.

The loss of the concept of the electrotonic state also meant the loss of Faraday's adjoining particles that transmitted the strain, which he still believed to be delineated by the curved lines of force. What exactly was strained, then? Faraday himself provided a tentative answer in a speculative paper published in 1852.[8] If the lines of magnetic force existed, he wrote, "it is not by a succession of particles . . . but by the condition of space free from such material particles." Six years earlier in his paper on 'Thoughts on ray vibrations,' he had suggested that radiation was a "high species of vibration in the lines of force which are known to connect particles and also masses of matter together."[8] If we take these statements together, Faraday can be seen heading towards some kind of electromagnetic theory of radiation, in which radiation was a vibration of lines of force marking some strain in space. Maxwell's electromagnetic theory began in Faraday's head.

Faraday retired from his several duties between 1861 and 1865 and spent his remaining years under royal patronage, living in a house at Hampton Court near London. Since his illness of the late 1830s his mind had been impaired and he suffered increasingly from intermittent loss of memory. His last few years were spent in a state of some mental confusion. In 1828, when his mental powers were strong, he had reasoned that the property we now know as magnetostriction might exist and he had sought it unsuccessfully. Five years before his death in 1867 this deep perception of nature still showed through on occasions and his last experiment was to look for some effect of a magnetic field on the spectral lines in emitted light. He found nothing. In 1897 Pieter Zeeman repeated the experiment with more sensitive equipment and discovered the effect Faraday had sought. Even near the end, with an impaired brain, Michael Faraday was still trying to break new ground.

Electromagnetic Theory

Faraday's ideas and theories about dielectric media and lines of force were given mathematical treatment in 1855–1856 in a paper entitled 'On Faraday's lines of force.' It was written by the young Scottish mathematical physicist James Clerk Maxwell, who was born just eleven weeks after Faraday discovered electromagnetic induction.

It was a lengthy paper of over 20 000 words, or roughly one-and-a-half times the length of this chapter. And it was published only about a year after

Maxwell's graduation from the university. In it Maxwell stated his intentions clearly. "I am not attempting to establish any physical theory," he wrote.[12] Instead he was attempting to take Faraday's processes of reasoning, which were usually regarded as rather vague when compared with those of the mathematical fraternity, and show that his ideas and methods could be expressed with mathematical rigour. That he succeeded may be evidenced by a remark from Faraday. "I was at first almost frightened when I saw such mathematical force made to bear on the subject, and then wondered to see that the subject stood it so well."[13]

Even before that another young Scot, seventeen-year-old William Thomson (later Lord Kelvin), had taken the first step in linking Faraday's conceptions to the laws of mathematics. In 1842 he showed that the equations that governed the action at a distance type of electricity were equivalent to the equations that described the interaction between adjoining particles in the theory of heat flow through a solid. Five years later Kelvin examined the analogy between electrical phenomena and elasticity and showed that the elastic displacement in an incompressible elastic body gave effects similar to the electric force in electrostatics. It was partly from Kelvin's, but mostly from Faraday's work, that Maxwell evolved the famous concept of displacement current and derived the set of equations and the electromagnetic theory named after him.

The labours of Maxwell and those who were inspired by him are the subject of the rest of this chapter, but Maxwell was not the only one to advance the cause of electromagnetic science. Before examining his work we must briefly describe the work of others.

On the Continent, after the enunciation of Heinrich Lenz's law (1834), Franz Neumann took this law and Ampère's model as starting points for his own attempt to discover the laws of induced currents. (Lenz's law states that induced current flows in a direction such that its effect opposes the change that produced it.) In 1845 he published the well-known Neumann formulas for mutual inductance in which he also introduced the concept of the vector potential. (Kelvin's vector for his elastic displacement could be identified with Neumann's vector potential, though he was not aware of that at the time.) Meanwhile Wilhelm Weber had been working on the law of force between moving electric charges (1846) and had used a constant of proportionality, which we would denote as c, whose dimensions were those of velocity. Weber's theory has been dubbed the first of the electron theories, that is one in which the forces on a moving electric charge depend on its velocity as well as its position.[14] It could also be used to obtain the formulas for induced currents. Others were also involved in furthering the study of electricity and magnetism, for example Bernhard Riemann and Hermann Helmholtz. In Britain, Kelvin applied vectors to the theory of magnetism (1851) and established the two magnetic vectors that we know as **B** and **H**. He also introduced the terms susceptibility and permeability into magnetic science and worked on the energy involved in magnetic and electrical phenomena (1853).

However, Maxwell's was the major contribution.[10,12,14] His theory evolved through three long papers published between 1855 and 1864, the first of which has already been briefly mentioned. The theory was published again in a more definitive version in 1873 in his *Treatise on Electricity and Magnetism*, a famous book that became known as the Electrician's Bible.

In the first paper Maxwell did two things. First, he considered a comparison between the lines of force and the lines of flow of an incompressible fluid, a hydrodynamic model; and second, he represented the electrotonic state mathematically. It was here that he gave his first version of Ampère's law, which we would write as curl $\mathbf{H} = \mathbf{J}$ (though Maxwell used the component form rather than the vector-calculus form for the expression). $\mathbf{J}$ was the conduction current only; the displacement current had not yet been thought of. In all Maxwell arrived at a total of six laws through which he gave mathematical support to Faraday's lines of force, but he saw these laws only as 'a temporary instrument of research' and not as a physical theory. "I do not think that it contains even the shadow of a true physical theory," was his comment on the idea expressed by these laws. It was meant to be a mathematical description of what happened, not a physical explanation. For those who really wanted a physical theory of electrodynamics Maxwell pointed to the work of Wilhelm Weber, on which he later commented, "The value of his researches, both experimental and theoretical, renders the study of his theory necessary to every electrician." Why then did Maxwell go to the trouble of writing his own paper? Because, he answered, "it is a good thing to have two ways of looking at a subject, and to admit that there are two ways of looking at it."

Maxwell also expressed his hope of discovering a 'mechanical conception' of Faraday's electrotonic state, a hope that was fulfilled in his second paper, 'On physical lines of force,' published in four parts in 1861–62, a mere 18 000 words or so. In this second paper Maxwell set for himself the task of investigating the mechanical results of states of tension and motion in a medium, and then comparing them with the observed phenomena of magnetism and electricity. Recall that Faraday's lines of force were believed to delineate lines of strain in a medium or in space. What Maxwell was now doing was to seek a mechanical (as well as mathematical) representation or model to describe electrical and magnetic phenomena.

The outcome was a mechanical conception based on molecular vortices in a magnetic medium, which has become known as the vortex model. It was applied in detail to magnetic phenomena, electric currents, and static electricity, and to the action of magnetism on polarized light. Each vortex (a sort of rotating cylinder) had its axis aligned with a line of magnetic force. Since the vortex revolved in the same direction as its neighbours, it became necessary to place particles like idle wheels (which represented electricity) between them so that the edges would not clash. The kinetic energy of the vortex motion represented the magnetic energy; the drift of the idle-wheel particles was the electric current (Fig. 4.5). Other electromagnetic pheno-

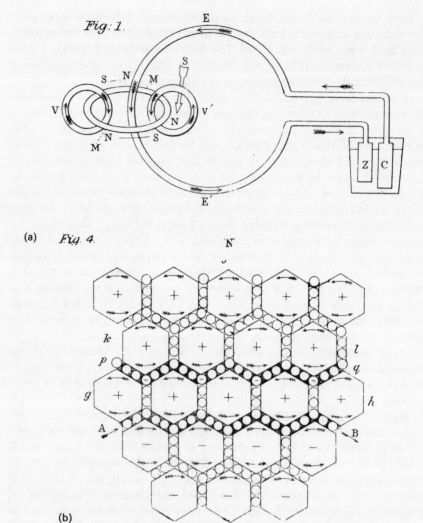

(a)

(b)

Figure 4.5 Maxwell's mechanical model of the ether. (a) Relationship between electric
current and magnetic vortices. Starting a current E E' puts vortices V V' into
motion and creates lines of magnetic force M, or vice versa. Analogy with
toothed wheels driven by toothed rack. (b) AB are idle-wheel particles of
electricity located between vortices (shown hexagonal) and moving from A
to B. Lines of magnetic force enter and leave plane of paper. This model can
be used to illustrate how by starting or stopping primary current AB a
secondary current (row pq) can be induced (after Ref. 12)

mena, such as tension and electromotive force, could also be represented; a
summary is given in Table 4.1. It was a very complicated mechanical model of
the ether and also a very useful one, which still holds a fascination for the
modern reader.

Table 4.1 Maxwell's Vortex Model, 1861–1862 (sources: Refs. 10, 12, 14)

Electromagnetic Phenomenon	Mechanical Representation
Magnetic field	A region of space containing molecular vortices (cylinders) rotating axially in the same direction
Line of magnetic force	Axis of a vortex
Uniform magnetic field	Several parallel vortices
Magnetic energy	Kinetic energy of vortices
Electricity	Idle-wheel particles between vortices, very small in size and mass when compared to the vortices
Electric current	Sideways movement of the idle-wheel particles
Electromotive force	Tangential force exerted by vortex cells on idle-wheel particles
Electric potential	Pressure of particles on each other

In the application of the model to electrostatics it was assumed that the particles of electricity could be displaced from their equilibrium position by an external electric field, and that the elasticity of the vortex cell would return them to their equilibrium position when the field was removed. A steady displacement was interpreted as an electrostatic field but a change of the displacement was a current, the famous 'displacement current' which then had to be included in Maxwell's second version of Ampère's law.

The analysis of his model led Maxwell not only to displacement current and his equations, but also to the demonstration that the electric and magnetic vectors are at right angles to each other and are propagated, in air or vacuum, with a velocity numerically almost equal to the known velocity of light. He concluded: "We can scarcely avoid the inference that light consists in the transverse undulations of the same medium which is the cause of electric and magnetic phenomena."[12]

It has been said that there is nothing new under the sun. Sir Edmund Whittaker has pointed out[14] that there are similarities between Maxwell's model of the ether and that suggested by Johann Bernoulli (the younger) 125 years earlier, with which he won the prize of the French Academy in 1736. In his theory of light Bernoulli mistakenly used longitudinal waves, not the transverse type used by Maxwell, yet Sir Isaac Newton had already stated a major objection to longitudinal waves. Otherwise the similarities between Bernoulli's ether and Maxwell's led Whittaker to comment, "One feels that perhaps no man ever so narrowly missed a great discovery." Perhaps he should be placed equal first with his compatriot Colladon.

Table 4.2 Maxwell's Equations (symbols have their usual meanings)

Vector Form	Integral Form	Physical Description
$\text{curl } \mathbf{E} = -\dfrac{\partial \mathbf{B}}{\partial t}$	$\oint_C \mathbf{E}.\mathbf{dl} = -\dfrac{\partial}{\partial t} \int\!\!\int_S \mathbf{B}.\mathbf{dS}$	Faraday's law. A changing magnetic field induces an electric field proportional to the rate of change
$\text{curl } \mathbf{H} = \mathbf{J} + \dfrac{\partial \mathbf{D}}{\partial t}$	$\oint_C \mathbf{H}.\mathbf{dl} = \int_S \mathbf{J}.\mathbf{dS} + \dfrac{\partial}{\partial t} \int_S \mathbf{D}.\mathbf{dS}$	Ampère's law. A current produces a magnetic field proportional to the total current, conduction plus displacement
$\text{div } \mathbf{D} = \rho$	$\oint_S \mathbf{D}.\mathbf{dS} = \int_V \rho \, dV$	Gauss's law. The total electric flux density from a closed surface equals the total charge enclosed. Or, electric lines of force start and stop on positive and negative charges
$\text{div } \mathbf{B} = 0$	$\oint_S \mathbf{B}.\mathbf{dS} = 0$	The net magnetic flux density out of a closed surface is zero. Or, magnetic lines of force form closed loops, starting and stopping nowhere

The third paper was published in 1864 and in it Maxwell presented his theory of the ether and its relationship with electric and magnetic fields. Now his work had the status of a theory. Gone was the mechanical scaffolding with which it had been built. Through Maxwell's own later work, and through that of others, the long list of mathematical equations now presented would be reduced to the four equations we know today (Table 4.2). This paper, 'A dynamical theory of the electromagnetic field,' added a further 20 000 or more words to Maxwell's publications and introduced the term electromagnetic field, which was defined as the space that contains and surrounds bodies in electric or magnetic conditions.

Part VI of this lengthy epistle was boldly entitled the Electromagnetic Theory of Light. Maxwell showed that a plane wave was propagated through the field with a velocity equal to the number of electrostatic units in one electromagnetic unit. When the experimental value for this ratio was compared with experimental values for the velocity of light, Maxwell remarked that they agreed "sufficiently well." He noted that no use had been made of electricity or magnetism in the measurements of the velocity of light. And, referring to the only available measurement of the ratio of the two systems of units, he made the beautiful comment: "The only use made of light in the experiment was to see the instruments."[12] Already his case rested on very strong supports. Others would be provided in due course.

The constant we denote by the letter c, the velocity of light, which had been knocking on the door of mathematical physics for so long, was finally brought in from the cold. It was but a short step to the conclusion that light was an electromagnetic wave and to the toppling of visible light from the pedestal on which mankind had put it, to be relegated to its new status as merely one of the many members of the family that make up the electromagnetic spectrum. What had begun as Faraday's lines of force had become Maxwell's electromagnetic theory of light.

Though his life was a short one (he died at forty-eight, possibly of cancer) Maxwell, like nearly all the great scientists, contributed much more to our knowledge of nature than the theory for which he is so well known. His first paper was about 'Oval curves' and was published when he was only fourteen. It was read to the Royal Society of Edinburgh by a Professor J. D. Forbes because, according to one biographer,[15] "it was thought somewhat undignified in those days for a mere school-boy to be allowed to address directly the members of the Society."

Only in those days? What would be the reaction today, one wonders, if a teenager tried to present a paper on, say, nuclear physics to one of the learned societies?

In 1857, at a more 'dignified' age, Maxwell won the Adams Prize at Cambridge for a paper on the structure of Saturn's rings, a subject on which he later wrote a book. He also made contributions to kinetic theory and to statistical mechanics; students who study semiconductor theory today still make use of Maxwell–Boltzmann statistics. The theory of colour vision also came under his scrutiny and he was probably the first to project a colour photograph, using three black-and-white slides, one exposed for each of the primary colours. He was a member of a team set up by a committee of the British Association for the Advancement of Science to look into electrical measurements, and in the Royal Society's Catalogue of Scientific Papers he is credited with sixty-eight publications, only one of which he shared as joint author. Perhaps as a diversion from his scientific work he also had a love for poetry and tried his hand at writing some himself.

Unlike Faraday Maxwell came from a moderately well-to-do family, part of the landed gentry of Scotland. He was born in Edinburgh and spent part of his career in Scotland and part in England. He entered the University of Edinburgh in 1847 and moved to Cambridge in 1850, graduating from Trinity College as Second Wrangler in January 1854. Two years later he was back in Scotland as professor of natural philosophy at Marischal College, Aberdeen. After three years the position was abolished in a merger between two local colleges to form the University of Aberdeen. (Not even a genius, it would seem, is safe from redundancy.) Scotland's loss was England's gain as Maxwell next took a chair at King's College in London, where his major work was performed and where he had much closer contacts with other physicists, especially Faraday. Five years later he was back in Scotland in temporary retirement from teaching duties and living at the family seat at Glenlair, Kirkcudbright. After a severe illness he settled down to work there; among other things he wrote the *Treatise on Electricity and Magnetism*. In 1871 he was lured back to England to take a new chair at Cambridge, where the decision had been made to do more to encourage the teaching of physics (especially heat, electricity, and magnetism) and to endow a new research unit, the Cavendish Laboratory, of which Maxwell became the first director. It was also while he was at Cambridge, in his last spell of work, that he edited and published the century-old work on the electrical researches of Henry Cavendish. His death there in 1879 brought the career of one of the greatest of physical scientists to an untimely end. On the centenary of his birth, in 1931, memorial tablets to both Maxwell and Faraday were unveiled in Westminster Abbey.

Acceptance of Maxwell's Theory

Electromagnetic theory, with Faraday's and Maxwell's work at its core, has long been of central importance to electrical engineering science. That position was won only slowly. In the early days it had its rivals. Later, even as it became accepted, it still had eminent opponents. It is not uncommon for great scientists, who in their younger days boldly propelled progress through the barriers of the scientific establishment, to become obstacles to progress themselves in their old age. Objections to new theories are not always overcome by careful argument; sometimes they simply die away with the passing of a generation.

Maxwell's theory was not the only electromagnetic theory of light. In 1867, three years after Maxwell, Ludwig Lorenz of Copenhagen published his own independent theory. According to Whittaker[14] it "lacks the rich physical suggestiveness of Maxwell's," although it is important to physicists for its discussion of retarded potentials (to account for the delay between an electromagnetic disturbance and its perception at a distant point). Others besides Maxwell were also attracted to the idea of providing mechanical models of the ether to represent the phenomena of electricity and magnetism and, some hoped, gravity as well. Included in what could be a long list were Kirchhoff, Helmholtz, C. A. Bjerknes, and many more. But Maxwell's theory was particularly authoritative and stood the test of experimental verification, even though some gaps remained to be filled by others. Maxwell had not conquered the phenomena of reflection and refraction; they were left to Helmholtz (1870) and H. A. Lorentz (1877). Lorentz (a Dutchman, not to be confounded with the Danish Lorenz) in particular distinguished himself by advancing the theory of electromagnetism. Another item missing from Maxwell's theory was a theorem for the energy flow in an electromagnetic field. This was provided in 1884 by J. H. Poynting and independently by Oliver Heaviside a year later.

On the experimental side further verification for Maxwell's theory began to accumulate. In 1875 the aforementioned effect of an electric field on light, previously sought by Faraday, was discovered by John Kerr. A year later an American, H. A. Rowland, provided experimental proof that rapidly moving electrically charged matter produces a magnetic field, as does an electric current. That had been more or less assumed by Maxwell and others. The Rowland effect became even more significant after J. J. Thomson's discovery of the previously hypothetical electron in 1897. One story told about Rowland is that once when testifying at a trial he gave his own name in answer to a question about who was the greatest living American physicist. Usually a modest man, he could only explain afterwards, "What could I do? I was under oath."[16]

On the theoretical side G. F. FitzGerald, an Irish physicist, in the early 1880s published a series of papers pointing to the possibility of radiating electromagnetic waves into space. He proposed a 'magnetic oscillator' as a suitable device but offered no means of detecting the waves produced. Originally he intended to write on the 'impossibility' of the idea, but he changed it to 'possibility' before publication. Somehow it seems unfair to the Irish, the butt of so many jokes, that it is an Irish physicist, and a good one at that, who is singled out to be remembered for such an abrupt change of mind. FitzGerald's other main claim to fame, the suggestion that the length of a material object depends on its velocity, might also seem 'a bit Irish' as the saying goes, despite the fact that it became an integral part of relativity theory. But we are getting ahead of the story.

If Maxwell was correct, FitzGerald reasoned, then energy need not remain within an electrodynamic system. "It seems highly probable," he wrote, "that the energy of varying currents is in part radiated into space and so lost to us." That was a fair statement of the theoretical possibility of a radio transmitter. Yet the experimental demonstration of both transmitter and receiver came from Heinrich Hertz in Germany, another brilliant physicist who, like Maxwell, died young.

In 1884 Hertz, a protégé of Helmholtz, examined the connection between Maxwell's equations and other, more classical ideas of electrodynamics and was led to propose a principle of unity of electric force, that is that the electric force produced by a changing magnetic field was the same thing as the electric force experienced in electrostatics. He went on to conclude that a varying magnetic field must induce an electric field in the surrounding space and would therefore exert forces on electrostatic charges. From his ideas he was able to derive Maxwell's equations without the help of Maxwell's mechanical model. Two years later he made an experimental observation which, though it had been made by several practical experimenters before him, when taken together with his deep grasp of Maxwell's theory opened the way forward to the transmission and reception of electromagnetic waves. The observation was that when an open circuit is formed from a circular piece of wire so as to leave only a small gap between the ends, a spark can be made to strike across the gap whenever a spark discharge is produced by a nearby induction coil. Hertz found that the secondary spark occurred even when there was no physical connection with the primary circuit, provided the resonant frequencies of the two circuits were similar. With this discovery he had provided himself with the most rudimentary transmitter and detector with which he could examine electromagnetic waves. Even though the actual observation was not new, what was new was that this knowledge was now in the mind of a man who not only knew of Maxwell's theory but had studied it thoroughly and understood it and, what is more, was a good experimenter as well as a good theoretician. In a relatively short time Hertz verified Maxwell's prediction that electromagnetic waves could be propagated through air with a finite velocity.

Before examining Hertz's experimental work let us take a brief look at how others had toyed with the generation and detection of electromagnetic waves without being in a position to relate their work to Maxwell's. Some prominent names were involved, Henry and T. A. Edison among them.

Ever since men had first learned to produce electric sparks, which are oscillatory discharges, they had been also in a primitive way radiating electromagnetic energy into space. The problem lay in detecting these weak waves. The usual method, if it can be called that, employed a crude antenna, some form of earthing (grounding), and some type of detector. Antennas came in all shapes and sizes; metal plates, wires, overhead pipes, tin roofs, and so on. The earth return was made via any convenient system, such as a buried metal plate or water pipe. And detectors, the really critical element, were based on loose metallic contacts or, more often, on small gaps between conductors. Not one of the experimenters really understood what he had achieved. Henry magnetized steel needles many meters away from his source spark and was led to be 'disposed to adopt the hypothesis of an electrical plenum.' That was twenty-two years before Maxwell's theory. Edison, eleven years after Maxwell's theory, thought he had found a new force of nature 'as distinct from electricity as light or heat,' a statement which revealed he knew nothing of Maxwell's work. Hughes, the inventor of the microphone, obtained reliable results and discovered the standing waves that result when a transmitted wave interacts with a reflected wave. However, he was persuaded that conduction was taking place through the air. Henry, Edison, and Hughes were not the only ones to transmit and receive electromagnetic waves before Hertz, but they will suffice to highlight the difference between Hertz and his predecessors. (A more detailed account of their work, and that of others, is given in Chapter 8.[17])

Hertz was the first to have a theoretical scientific grasp of what he was seeking to achieve experimentally. The others, by and large, were playing with a baffling phenomenon without any understanding of it.

Heinrich Rudolf Hertz began in 1886 by discovering that electric waves propagated with a finite velocity along a wire.[18] Oliver Lodge performed similar work at about the same time and both men were anticipated by Wilhelm von Bezold in 1870. But it was Hertz who discovered that the wire was not necessary: a spark could be produced at the detector without a metallic connection to the transmitter, particularly if the dimensions of the detecting circuit corresponded to the wavelength of the transmitter's oscillations.

In the first of a series of experiments conducted in 1887–1888 and published from 1888 to 1890, Hertz investigated the effects of placing various dielectrics between the transmitter and receiver: wood, sulfur, paraffin, and asphalt. His results confirmed one of the basic principles of Maxwell's theory, the polarization of a dielectric by electromagnetic forces. Next Hertz compared the velocity of propagation of waves via wire with that through air and found that the velocity through air was of the same order of magnitude as the velocity of light. The two velocities were not quite the same, which was not predicted by Maxwell's theory. This erroneous result has been ascribed to experimental problems caused by a wavelength too large for the size of the room. For a time doubt was thrown on all of Hertz's work, but later experiments by Ernst Lecher proved the two velocities to be equal.[18]

The next step was to give definite proof that electromagnetic radiation consisted of waves just as light did. Hertz reflected the radiation from walls covered with zinc sheeting and obtained standing waves by interference. Such results left no room for doubt. Then came a theoretical paper in which his transmitter was analyzed by Maxwell's theory and finally the experimental *pièce de resistance* performed in December 1888.[19] For this work the equipment was modified to achieve a higher frequency. Hertz himself estimated the wavelength at about 66 cm (455 MHz), probably the shortest waves he used; we cannot know for sure, especially since he modified his equipment several times and so changed the frequency. It has been tentatively concluded that at various times he operated from 50 to 500 MHz, in what are now the VHF and UHF bands.[20] (The difficulty of reliable frequency measurements continued into the early days of radio; experimenters often had only the vaguest ideas of just where they were in the electromagnetic spectrum.)

The transmitter, or primary conductor as Hertz called it, used in the final set of experiments consisted of an adjustable spark gap, fixed at 3 mm, in the middle of a 26 cm-long brass dipole; the poles of the spark gap were formed by two spheres (Fig. 4.6). It was fed by a small induction coil. A parabolic reflector was made from a square zinc sheet 2 m on a side. This simple and elegant device was held together with paper, wood, sealing wax, and rubber bands, and could be dismantled quickly for the frequently needed repolishing of the pole surfaces. One wonders what the reaction of some of today's engineers and students would be to wood-and-sealing wax equipment being used to roll back the frontiers of knowledge.

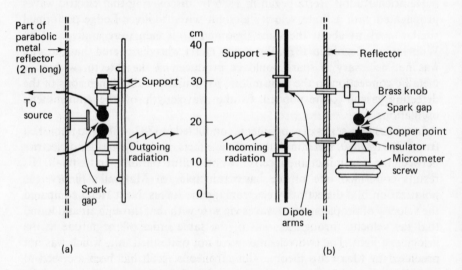

Figure 4.6 Hertz's transmitter and receiver: (a) transmitter (b) receiver (after Ref. 19). The scale is approximate

The receiver or secondary conductor also had a dipole antenna: each arm was 50 cm long. Two wires connected these arms to a tiny spark gap between a brass sphere and a fine copper point. The materials were chosen so that the soft point would not dig into the hard sphere. The spark gap was adjustable by means of a watch spring and micrometer screw that controlled the position of the point. Reception was achieved up to a distance of 16 m, the width of the available room.

With this simple equipment Hertz further demonstrated the validity of the laws of optics for electromagnetic waves. He had already shown that their velocity was finite and that they could be reflected to produce standing waves. Now he refracted them with a 100 kg prism made from pitch and even polarized them by reflection and by use of a wire screen. With the transmitter in the vertical plane, he found the electric field oscillated in the vertical plane and the magnetic in the horizontal plane. Also noted were the rectilinear (ray-like), properties of the waves: a complete shadow was cast by tin foil or gold paper. Their equivalence with light could hardly be doubted.

Hertz's work brought him fame not only in scientific circles but with the public as well through publicity in newspapers, magazines, and public lectures.[20] That he had obtained verification of important points in electromagnetic theory was what mattered to Hertz and other scientists; that he had demonstrated the basis of communication across space without wires was what caught the imagination of Marconi and others. These two consequences of his work were to grow ever more divergent as radio communication was developed, with a few notable exceptions, by cut-and-try inventors rather than by scientists.

When he did this work Hertz was professor of physics at the Technische Hochschule in Karlsruhe. The quick recognition brought offers of several positions; he accepted one as successor to R. J. E. Clausius at the University of Bonn. It was there that he completed his papers on electrical theory, which were collected into a book published in 1892 and translated into English the following year. It joined Maxwell's own great treatise on electromagnetism as a fundamental text of the first rank. It was a great loss to electrical science that both of them died at early ages: Hertz at 36, Maxwell at 48. Hertz's name is commemorated in the unit of frequency, despite a pathetic attempt by the Nazis to abolish its use because he was half Jewish.

Support for Maxwell's theory also came from Oliver Heaviside in Britain, who published a long series of complex papers starting in 1882 and continuing for a decade.[21] In these papers he simplified, explained, interpreted, and used Maxwell's theory in the solution of practical engineering problems. In so doing he presented the theory in the form in which it is usually used today, based on the vector forms of the fundamental electric and magnetic fields E and H rather than the mathematical concepts of vector and scalar potentials, a style perhaps more useful to engineers than Maxwell's physicist's approach. Heaviside also suggested new terminology in his approach to magnetism and so gave us our modern terms of resistivity, self-inductance, mutual inductance, permittivity, impedance, reluctance, and distortion. He was also an early advocate of rationalized units to rid science of the 'disease' of 4π, a constant that was for ever cropping up in the use of the old electrostatic and electromagnetic systems of units. In common with Faraday, but in contrast to Maxwell and Hertz, Heaviside had little formal education; yet when the University of Göttingen awarded him an honorary degree in 1905 the citation described him as 'among the propagators of the Maxwellian science, easily the first.' Among his achievements based on his studies of Maxwell were his investigation of the skin effect (the tendency of high-frequency currents to flow mostly near the surface of conductors) and his examination of the role played by inductance in telecommunications, to which we shall return in the next chapter. He is also remembered for his 1902 suggestion of a reflecting layer in the upper atmosphere to explain radio communication beyond the horizon, also suggested the same year by the British-American engineer A. E. Kennelly. Heaviside also produced a lengthy treatise on electromagnetic theory, the third important tome on the subject.

In addition to being the first to use the now common vector-algebra versions of Maxwell's equations, Heaviside is thought to have been the first to refer to the symmetry of the equations, a point sometimes stressed by teachers. A logical extension would be the introduction of a magnetic conduction term to match the electrical conduction term, and what may be called a magnetic pole density equivalent to the electric charge density. In other words, it would be necessary for north and south poles to exist independently of each other, a phenomenon which has not been observed to date. In the 1930s Paul Dirac calculated that the attractive force between such hypothetical monopoles would be 4500 times stronger than that between an electron and a proton, which might explain the absence of such monopoles. Few physicists today accept magnetic monopoles even as an hypothesis, though speculation on the subject continues.

Even the work of Hertz and Heaviside did not dispel all doubts about Maxwell's theory. In particular, Kelvin had lifelong doubts. In the early days he is said to have described it as "a failure, the hiding of ignorance under cover of a formula."[10] Later, in 1888, he referred to displacement current as a "curious and ingenious, but not wholly tenable hypothesis." FitzGerald in 1896, and Lodge in 1898, wrote to Heaviside relating, respectively, Kelvin's disagreement concerning Maxwell's theory of the propagation of electro-magnetic waves and the prediction of radiation pressure (the mechanical pressure experienced by an irradiated object). However, radiation pressure was confirmed only a year later by P. N. Lebedev in Russia, one more confirmation of a Maxwellian prediction.[22]

Perhaps a more commonly held view was that expressed by Henri Poincaré in 1894. "There still remains, therefore, much to be done; the identity of light and electricity is from today something more than a seducing hypothesis: it is a probable truth, but it is not as yet a proved truth."[23]

Relativity and Quanta

Most of us tend to compartmentalize our work by erecting false boundaries around subjects. Electrical engineers have tended to lose sight of the intimate links between electromagnetism, a 19th century science, and relativity and quantum theories, which are too readily seen as 20th century innovations. Both in fact had their roots firmly in electromagnetism.

Maxwell's theory was mainly concerned with the laws of the electromagnetic field and not (despite the original mechanical model) with the actions of matter itself. Towards the end of the 19th century this situation was changed by Lorentz, who based electromagnetic theory on the existence of electrons that acted on each other and were at rest, or moving, in a stationary ether. At the time electrons had not yet been discovered but theories of the existence of a fixed minimum electric charge were prevalent. G. J. Stoney, in Ireland, suggested the name electron for this fixed charge in 1891, and J. J. Thomson discovered the particle which eventually took the name six years later.

Lorentz's interest in electromagnetism dated back to 1875, when he wrote his doctor's thesis on the theory of the reflection and refraction of light, a problem that Maxwell had left unsolved because he could not satisfy himself as to the correct boundary conditions. Lorentz also pointed to other problems that remained to be studied: chromatic dispersion, the rotation of the plane of polarization, the influence on light of the movement of the medium, emission, absorption, and radiant heat. And he suggested that if light and radiant heat really did consist of electrical vibrations then it would be natural to assume that the molecules of the bodies that were the source of the vibrations were "the seats of electrical oscillations." In other words, light was produced by electrical charges oscillating on an atomic scale. Maxwell's theory, he commented, was "far from having attained its final form."[24]

The Lorentz theory was put forward in two papers published in 1892 and 1895. Electrical effects were now to be explained on the assumption of the existence of material particles with a definite mass and charge and moving through a stationary ether. But movements of bodies through the ether were known to yield experimental problems. If two reference systems, one for the ether and one for the electrons, were moving with uniform motion relative to each other then it seemed, from classical mechanics, that Maxwell's equations could only be valid for one of those frames of reference, the absolute one of the ether, not both. If that were true then it ought to be possible to detect the existence of this absolute system, the Newtonian ether. However, experiments to do just that, the very sensitive ether-drift experiment by A. A. Michelson and E. W. Morley for example, gave null results. Something was wrong, either with Maxwell's electromagnetic theory or with classical mechanics. In 1892 in an effort to obtain an explanation, FitzGerald and Lorentz independently suggested that all bodies moving through the ether contracted in the direction of their motion. This seemingly strange idea came from Lorentz's extension of electromagnetic theory and showed that Maxwell's equations were valid in both frames of reference after all. It was classical mechanics, not electromagnetism, that was wrong. The idea of an absolute frame of reference had taken a bad knock. As 19th century scientists were putting the finishing touches on their classical picture of the universe, the beautiful picture they were painting began to fall apart.

The Lorentz transformation became part of the theory of relativity. The

Lorentz force also dates from the same time, 1895. Ten years later, when Albert Einstein published his paper on special relativity, it was entitled, 'On the electrodynamics of moving bodies.' It is a title that surprises many electrical engineers, who sometimes tend to feel that relativity has nothing to do with them. It is said that Einstein's teachers had told him that he would never amount to anything and that his indifference was demoralizing to both his teachers and to other students. Years later Einstein himself stated that theoretical science would be better pursued by a plumber who would not have to justify himself with publications and instead could concentrate on really important problems. In 1905 when he published three fundamental papers, one of which later earned him a Nobel prize, he was employed, not as a research scientist or university lecturer, but as a clerk in the Swiss patent office. Maybe he was then practicing what he was later to preach.

In his relativity paper Einstein introduced two postulates. One was the constancy of the velocity of light; the other, that the laws of optics and electrodynamics obeyed the principle of relativity. With that the classical Galilean transformations and the idea of an absolute frame of reference were abandoned in favour of the Lorentz transformations. In Einstein's own words, "The special theory of relativity owes its origin to Maxwell's equations of the electromagnetic field."[25]

Possibly the greatest success of the Lorentz electromagnetic theory was its explanation of the effect of a magnetic field on spectra, the Zeeman effect discovered in 1896. The discovery and explanation taken together earned Zeeman and Lorentz the 1902 Nobel prize. The corresponding effect produced by an electric field, the Stark effect, was discovered by Johannes Stark in 1913 and, like the anomalous Zeeman effect, is explained with the aid of quantum theory.

Quantum theory, the second half of the revolution which occurred in physics at the start of the 20th century, also solved two of the problems in electromagnetism to which Lorentz had pointed in 1875, the explanation of the emission and absorption of radiation. The quantum theory has been of far more importance to electrical engineers than has relativity, and has led to an understanding of semiconductors and to their widespread use in modern electronics. A brief discussion of the development of quantum theory will be found in Chapter 9; for the moment we shall be content to take a glimpse at how it arose out of problems with electromagnetic theory.

Classical physics, including electromagnetic theory, was quite unable to explain certain effects that loomed over it around the turn of the century; questions concerning blackbody radiation, the specific heat of elements, the photoelectric effect, and so on. Any problem concerned with light or radiation was also concerned with electromagnetism.

The problem about the emission of radiation from a black body was that classical theory predicted an increase of intensity with decreasing wavelength, whereas experiment showed that at short wavelengths exactly the opposite happened. In an otherwise quite successful theory, that was rather disconcerting. In 1900 Max Planck, a professor at the University of Berlin who like Hertz had studied under Helmholtz and Kirchhoff (and in fact replaced Kirchhoff in Berlin when he died), offered an explanation of the experimental results by postulating that energy is emitted in the form of electromagnetic radiation only in discrete units and not on a continuous basis. The discrete units he called quanta, hence quantum theory. The beginning of quantum theory was not warmly welcomed, not even by Planck himself.

A second problem area was the photoelectric effect, the generation of electrons by incident radiation. Hertz is usually given the credit for its discovery in 1887 in connection with his experiments on electromagnetic radiation, though he did not realize it; he shares credit for it with W. Hallwachs. Again the problem was that the theory did not explain what happened in practice. The emission was found to vary with the frequency of the incident radiation, but below a threshold frequency there was no emission at all, no matter how intense the incident radiation was. Below the threshold it was like trying to get blood out of a stone. Above the threshold one had a perfectly happy blood donor, but classical theory could not say why. Einstein used the quantum theory in 1905 to offer an explanation, but in so doing he claimed that the condition that forces the radiation to be in units or quanta was not a condition of physical matter as Planck had it, but a condition of the field or of light itself. Light, an electromagnetic wave, had according to Einstein a particle nature as well. Not many people were very happy with that suggestion. Lorentz saw it as scattering the undulatory theory and all its triumphs to the winds.[26] The patent office clerk, it seemed, had thrown the baby out with the bath water, yet in the course of time experimental evidence supported Einstein.

Conclusion

In this chapter we have seen a bare outline of how the science of electromagnetism developed from the first fumblings for electromagnetic induction to the spawning of quantum and relativity theories. Someone born about 1820 could have lived through the entire period of this scientific evolution and seen the development of equally dramatic engineering applications. John Adams, the second U.S. president, was an old man when it began; Dwight Eisenhower was a boy when it ended. Queen Victoria lived through virtually the whole period. She was a baby when Oersted burst his news on the scientific centres of Europe and she died as Planck introduced quantum theory. When she was in her teens the electromagnetic telegraph was launched and later she exchanged telegrams with President Andrew Johnson when the Atlantic cable opened. She was in her late fifties when Edison began to capture the headlines and she saw the rise of engineers such as Werner von Siemens, George Westinghouse, Sebastian de Ferranti, and even Guglielmo Marconi. Electromagnetic machines generated electricity to be brought into the home for lighting and later for other uses too. And maybe the Queen was amused by the telephone, another electromagnetic device, and Marconi's exploits with radio.

There is a story from the early days of electromagnetism that Faraday was once asked what use it was. He answered by asking, what was the use of a baby? That particular baby proved to be very useful indeed.

References and Notes

1. S. Ross, *Notes and Records of the Royal Society* 20: 184, 1965. This excellent account is the basis for the first part of this chapter; it gives a detailed account of the search for electromagnetic induction.
2. W. A. Atherton, *Am. J. Phys.* 48: 781, 1980.
3. F. Cajori, *A History of Physics*, Macmillan, London, 1929. (Reprint by Dover, New York, 1962.)
4. A. Einstein and L. Infeld, *The Evolution of Physics*, Cambridge University Press, London, 1961.
5. Both Ross (Ref. 1) and Williams (Ref. 8) have given details of some of Faraday's attempts prior to the actual discovery and have discussed the correspondence between him and Ampère.
6. M. Faraday, *Faraday's Diary*, Bell, London, vol. 1, 1932.
7. T. Coulson, Chapter 3, Ref. 8.
8. L. P. Williams, Chapter 3, Ref. 2.
9. *Catalogue of Scientific Papers*, (1800–1863), vol. 2, 1868; vol. 4, 1870; (1864–1873), vol. 8, 1879, Royal Society, London.
10. B. Gee, *Phys. Educ.* 13: 287, 1978.
11. L. P. Williams, *Nature* 187: 730, 1960.
12. W. D. Niven, Ed., *The Scientific Papers of James Clerk Maxwell*, Cambridge University Press, Cambridge, vol. 1, 1890. (Reprinted by Dover.)
13. L. Campbell and W. Garnett, *The Life of James Clerk Maxwell*, Macmillan, London, 1882.
14. E. T. Whittaker, Chapter 2, Ref. 4.
15. D. K. C. MacDonald, *Faraday, Maxwell and Kelvin*, Heinemann, London, 1965.
16. I. Azimov, Chapter 3, Ref. 9.
17. C. Süsskind, see Chapter 8, Ref. 1.
18. C. Süsskind, *IEEE Spectrum* 5(12): 57, 1968. The present account is also based on Refs. 14, 19, and 20.
19. W. F. Magie, Chapter 2, Ref. 5.
20. H. G. J. Aitken, *Syntony and Spark: The Origins of Radio*, Wiley, New York, 1976.
21. W. Jackson, *Heaviside Centenary Volume*, IEE, London, 1950. This volume gives an account of Heaviside's work.
22. For a discussion of Kelvin's reservations see Ref. 14. Also, A. M. Bork, *Am. J. Phys.* 31: 854, 1963.

23. H. Poincaré, *Nature* 50: 8, 1894.

24. R. Taton, Chapter 3, Ref. 7.

25. P. A. Schilpp, Ed., *Albert Einstein: Philosopher-scientist*, Tudor, New York, 1949.

26. J. H. Jeans, *The Growth of Physical Science*, Cambridge University Press, London, 1951.

27. Since this chapter was written it has been reported that William Sturgeon (who made the first electromagnet) has a claim to having discovered electromagnetic induction before Faraday. A strong argument has been put forward that Sturgeon made a limited and hesitant assertion to the effect that motion of a conductor through a magnetic field would produce an electric current. See: J. O. Marsh and G. L. Hodkinson, "Faraday, Sturgeon and the Discovery of Electromagnetic Induction," IEE Weekend Meeting on the History of Electrical Engineering, Brighton, 2–4 July 1982.

5 TELECOMMUNICATIONS

Leaflets published around 1843 advertised a new commercial venture, a 'galvanic and magneto electric telegraph.' This, the world's first commercial electric telegraph, went into operation in 1839 on the Great Western Railway in England. A telegram, or rather a telegraphic despatch, cost a shilling (5p), and could be sent thanks to the work of William Cooke and Charles Wheatstone. Members of the public were allowed to view the equipment, again on payment of a shilling, and the *Morning Post* recommended the visit to all who loved to see the wonders of science.

The telegraph was the first important contribution electrical engineering made to society and, though telegrams are less common now than they were, the telegraph is the ancestor of our present system of telecommunications. A telex is simply a modern business telegram.

The electric telegraph evolved to a commercial enterprise over a period of some ninety years, and it is impossible to point to any one person as being the original inventor. Dozens of inventors devised telegraph systems, some better than others, but it was not until after the discovery of electromagnetism in 1820 that a commercial electric telegraph became practical. Then the important names began to stand out; P. L. Schilling in Russia, Gauss and Weber in Germany, Cooke and Wheatstone in Britain, and Morse and A. Vail in America.

Telegraphy was not something new to the 19th century, though large-scale telegraphy certainly was. Fire, smoke, and light had all been used to send messages before the idea of using electricity first occurred to anyone. From the 16th century, impractical systems based on the magnet were suggested. They were generally called sympathetic telegraphs and leaned more on the ideas of magic than engineering. However, once it had been realized that some materials would conduct electricity over reasonable distances the idea of using it as a means of communication was inevitable, though any practical system would have to prove itself in competition with successful mechanical systems such as Claude Chappe's semaphore, which was used in France from 1792 and for which the word telegraph was coined. By the time it was replaced by an electric one, around 1852, Chappe's network extended about 3000 miles (5000 km). In Britain, a shutter telegraph was used by the Admiralty from 1795 to 1816, largely because of the fear of a Napoleonic invasion. This system eventually linked London with the harbour towns of Deal, Portsmouth, Plymouth, and Yarmouth. Both of these mechanical systems did useful work during the Napoleonic Wars.

It is generally accepted that the first suggestion of what could have been a workable, though clumsy, electric telegraph was made by 'C.M.' in an anonymous letter published in *Scot's Magazine* in 1753, just 24 years after Gray's demonstration of the transmission of electric charge.[1] The proposal was to use static electricity from a friction generator to charge a wire, and detect the charge at the far end of the wire by watching for its electrostatic effects on a small piece of paper. One wire was proposed for each letter of the alphabet. When a letter was signalled, the appropriate piece of paper would be attracted to its wire. Though C.M.'s telegraph was not actually built, there are claims that the first working model of an electric telegraph was built in Paris in 1787, with a pithball electrometer as the detector.

In this early period the main activity, all performed by inventive amateurs, centred on devising better receivers and transmitters and, whenever possible, reducing—and protecting—the wires used. Essentially these tasks are similar to those facing today's telecommunications engineer. Pithball electrometers were popular as detectors and, after 1800, electrochemical effects were used. In 1809 S. T. von Sömmering built a receiver in Munich that exploited the fact that a weak current can decompose water. The bubbles of hydrogen and oxygen so produced were used to indicate the letters (Fig. 5.1). Another electrochemical system was the recording telegraph suggested by H. G. Dyar in America in 1826–1827, in which litmus paper was turned red by the action of a spark at the receiver. Though he built a successful experimental telegraph at Long Island, and proposed a line to Philadelphia, Dyar was deterred from progressing further by the threat of legal action for conspiracy to conduct "secret communication from city to city."[2]

At the transmission end, the invention of the primary battery in 1800 offered a more convenient power source than the friction generators and, after Faraday's discovery of electromagnetic induction in 1831, hand-operated magneto-electric generators came into use.

One wire for every letter was certainly a cumbersome way of building a telegraph. One way of improving things would be to insulate the wires with pitch-impregnated paper and form them into cables. Such a scheme was suggested by the Spanish physician Don Francisco Salvá, who is also remembered as an early supporter of vaccination. The cables could then be buried in the ground or even laid on the sea bed. A single-wire system provided obvious savings. Some reports claim that Salvá devised a single-wire system in 1798, though little is known of it and the reports may not be reliable. However, a single-wire electrostatic system was built by Francis Ronalds in England in 1816, in two parts, one clockwork and one electrical. Two synchronized clocks

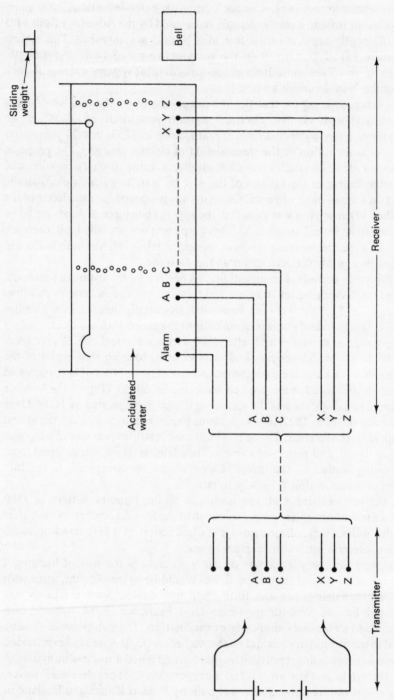

Figure 5.1 Principle of Sömmering's telegraph

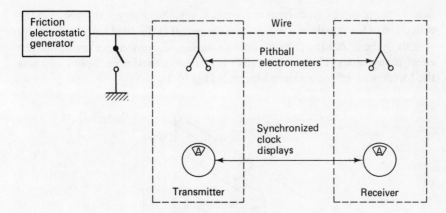

Figure 5.2 Principle of Ronald's telegraph

were used to display letters one at a time, and the single wire was used to signal to a pithball electrometer when the desired letter was in the display (Fig. 5.2). Correct functioning depended on maintaining synchronization and on the assumption that the transmission of electricity along a wire was virtually instantaneous. Ronalds was able to satisfy himself on both counts. He also understood that if a long wire was insulated and buried in a glass tube, following one of his ideas, then this configuration "might destroy the suddenness of a discharge."[1] It is quite remarkable that he so clearly foresaw this capacitance effect, which later bedevilled the Atlantic telegraph cable and which was eventually studied by Faraday, Kelvin, Heaviside, and others.

When Ronalds offered his telegraph to the British Admiralty it was rejected with the pompous reply that telegraphs were no longer necessary. Two years earlier, Ralph Wedgwood, a member of the famous pottery family and also the inventor of carbon paper, had received a similar rebuff to his own offer of a telegraph.

Practical systems faced other problems besides electrical ones and official indifference. Both Salvá and Ronalds feared wilful damage, but Ronalds at least had a pragmatic answer to vandals: "Hang them if you can catch them, damn them if you cannot, and mend it immediately in both cases."[1]

Ingenious though many of them were, none of these electrostatic systems produced a commercial telegraph. Possibly they would have done so had their death knell not been sounded by the discovery of electromagnetism. Even in its simplest applications electromagnetism offered much better detectors than any pithball electrometer or electrochemical contrivance could achieve.

European Electromagnetic Telegraphs

After Oersted's discovery of electromagnetism, it was suggested that a magnetic compass needle could be used as a telegraph detector since it would detect the presence of a current in a wire. J. S. C. Schweigger showed that the deflection of the needle was increased if the wire was doubled back under it. By repeating the technique so as to obtain a loose coil, Schweigger multiplied the effect many times. The resulting instrument was called a multiplier and was the forerunner of the modern ammeter (Fig. 5.3).

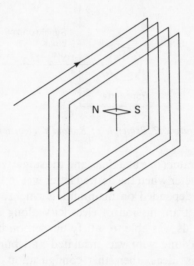

Figure 5.3 Principle of Schweigger's multiplier

Earlier, in Munich, P. L. Schilling, a member of the Russian diplomatic corps, had met Sömmering and had become interested in his electrochemical telegraph. After returning to Russia Schilling continued his work and planted the seeds from which grew the future electromagnetic needle telegraphs, in which the detector consisted of a compass needle that reacted to the energization of a nearby electromagnet. With the multiplier and a compass needle as a detector he produced various experimental systems that depended on one or more wires and employed codes. However, no serious attempt was made to put the telegraph to commercial use until about 1836. Schilling died in 1837, before the plans were carried out.

In 1833 two scientists, Gauss and Weber, operated an electromagnetic telegraph at Göttingen in Germany, mainly for scientific experiments on the transmission of electricity. This system, which covered a distance of over a mile (2 km), operated until 1838 and modifications and improvements were made to it in that time. Apart from Schilling, whose first electromagnetic telegraph dated back to around 1825, they were the first to operate such a system; but since they could not afford the time to develop their telegraph, they invited someone else to do that for them. The result was C. A. Steinheil's telegraph of 1837. It used copper conductors, a magnetoelectric generator, an alarm to alert the clerk at the receiver, and a receiver that printed the results onto paper by deflecting magnets which were connected by capillary tubes to cups of ink. Steinheil is also usually credited with the important invention of the earth or ground return (which did away with one of the wires), although there are other claimants and some experimenters had used it much earlier, with or without knowing it.

Meanwhile one of Schilling's telegraphs was seen in Heidelberg by William Fothergill Cooke, a visiting British physician, whose imagination was caught. On his return home he built his own equipment, in which magnetic needles were deflected so as to point towards letters. Cooke embraced telegraphy wholeheartedly but quickly ran into problems over operating his alarm at long distances. Long-distance telegraphy was then regarded by the British scientific community as impractical, largely as a consequence of experiments conducted by Peter Barlow, a respected physicist, who had carefully examined the conduction of current through wire with a view to settling the rivalry between the one-fluid and two-fluid theories of electricity. One of the questions posed was whether the current dropped as the length of wire was increased. Barlow concluded that it did, moreover to such an extent that electromagnetic telegraphy was impractical. His results became well known and were widely accepted on both sides of the Atlantic. They had been published two years before the announcement of Ohm's law.

Cooke sought expert help and was eventually directed to Professor Charles Wheatstone of King's College in London. Wheatstone had also been experimenting with telegraphy and it was with him that Cooke formed a successful, though at times acrimonious, partnership in 1837, the year that Queen Victoria ascended the throne. Their first British patent, the world's first for electrical communication, was sealed on 12 June 1837.

The solution of the problem that caused Cooke so much trouble appears simple today with our easy acceptance of the concepts of voltage, current, resistance, and Ohm's law. But to Cooke, and to Wheatstone, it was very difficult. Ohm's law was not yet properly known in Britain. The fairly high resistance of the long wires that were used in telegraphy meant that the current would be small at the far end. The obvious remedy was a voltage increase, but an improvement could also be made at the detector: an increase in the number of turns on the coil to strengthen the magnetic field that deflected the needle. The importance of this modification had to be explained to Wheatstone by Joseph Henry when he visited London in 1837. Henry stressed that many turns of fine wire were needed. The increased resistance of the coil was trivial compared with the resistance of the line and the high internal resistance of the battery. However, such understanding of the role played by resistance only came about after Ohm's law of 1827 became established in Britain in the early 1840s.

Cooke, who was the entrepreneur as well as the original motivator of the partnership, next sought a market. He turned to the recently formed and rapidly expanding railway companies. He was convinced that they would need telegraphs for signalling. In 1836 he wrote a pamphlet telling them so, which was not generally published until it appeared in a book in 1856. Another book by Cooke, *Telegraphic Railways*, was published in 1842.[2] In 1837 the London and Birmingham Railway Company arranged a trial at which the railway pioneer Robert Stephenson was present. Cooke later remarked that Stephenson had played with the instruments more than anyone else. A successful experimental line was built but no further orders were received.[2,3]

Cooke then turned to the Great Western Railway. Another trial section was built, extending almost 13 miles (20 km) from Paddington station in London to West Drayton. Five-needle instruments were used (Fig. 5.4). Five copper wires plus one return wire were covered with hemp and bound into a cable, or 'telegraph rope,' and buried in an iron pipe alongside the track; care was taken to exclude water and to allow for testing facilities. It was successfully put into operation in 1839 and later extended to Slough, a total of nearly 18 miles (30 km). But the expense of the line, £165 ($800) per mile, retarded further exploitation.[3]

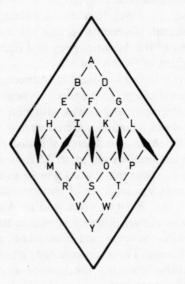

Figure 5.4 Display of Cooke and Wheatstone 5-needle telegraph

With the aid of codes the number of wires was reduced. Eventually only one wire was used with an earth return. As the costs came down, the telegraph began to spread.

The public's imagination was caught by the application of telegraphy in law enforcement on at least two occasions: when a railway pickpocket was caught with the help of the telegraph and, more important, when an accused murderer was arrested thanks to the use of the Slough–Paddington telegraph. John Tawell had escaped from the scene of the crime in Slough by train, but his description was telegraphed to Paddington, where he was arrested. Later he was hanged. (A similar episode marked the beginnings of radiotelegraphy: in 1910 the notorious murderer Dr. Crippen was arrested with the aid of the new ship-to-shore radiotelegraph.)

One might expect that the railway companies would have been quick to apply the telegraph to railway signalling. Apparently they were not. In a *Daily Telegraph* interview[4] published in 1898 the chairman of one railway company commented that electrical signalling had been applied on railways in the Netherlands before it had come to Britain, where the railway was born. He further stated that in the early days, "the wire was not employed for railway work at all, and it was looked upon as a toy." Toy or no toy, other lines were built. One measure of their rapid growth was the royalty paid to Wheatstone in two successive years: £444 in 1844 and £2775 in 1845.[5]

Cooke had continued to act as the commercial manager and contractor, but in 1845 The Electric Telegraph Company was formed and bought Cooke and Wheatstone's patents. Other companies and telegraph lines soon followed. London and Dover were linked in 1846 and thoughts soon turned to crossing the English Channel. By 1852 about 4000 miles (6000 km) of wire were in use in Britain; by 1868 there were several companies with more than 16 000 miles (25 000 km) among them and Britons were sending nearly six million telegrams a year. (In America the 16 000-mile figure had been reached considerably earlier, in 1852.)

Because most of the telegraphs followed railway lines, many towns were not served and agitation began for the government to purchase the private companies. An inquiry was held to determine whether the Post Office could work the telegraph system successfully. It was decided that it could and the system was nationalized in 1870, at a cost to the State of nearly £8 ($40) million.[6] In the years of uncertainty before nationalization little new investment was made by the private companies and the State had to spend at least another £2 ($10) million on improvements, far more than officially estimated. It would seem that escalating government costs are nothing new to our present age.

Although the early telegraphs in Britain were of the needle variety and employed skilled operators using special codes, Wheatstone in particular considered that a simple 'ABC' telegraph was needed for private unskilled operation. He produced at least two models, patented in 1839 and 1840, which were later improved. In one a clockwork escapement, triggered by an electromagnet, rotated a dial on which the alphabet was printed; hence the name 'ABC'. The transmitter had a similar dial, which was rotated by finger (as was to be done in telephones later) and sent pulses from a battery down the line. The 1840 patent included a transmitter that depended on induced current. The ABC telegraphs were slower than the coded needles but after about 1860 they found a market in private use.

Among others in Britain who also attempted to build telegraphs, one man in particular nearly became a very serious rival to Cooke and Wheatstone. Edward Davy of London submitted a description of his telegraph in 1837, the same year that Cooke and Wheatstone got their patent. Cooke and Wheatstone claimed infringement but Davy was also granted a patent and began negotiations with railway companies. He was close to floating a company with a successful telegraph when he emigrated to Australia, leaving the field free for Cooke and Wheatstone. Davy's telegraph, which depended on Daniell cells, needles, and alarms, also used an 'electrical renewer' and 'relays of metallic circuits'. Possibly this was the first relay. He is credited with inventing the first relay to use a galvanometer needle "to bring into contact two metallic surfaces so as to establish a new circuit, dependent on a local battery; and so on ad infinitum." Others, notably Henry, Morse, Wheatstone, and perhaps Cooke, also independently invented relays around this period.

By the mid-1840s commercial telegraphy was a reality. It contributed to and benefited from the social trend to better communications in an age when travel became swifter and easier both on land and at sea. One example is the way the telegraph helped to establish Greenwich Mean Time as standard throughout Britain, by transmitting time signals from London so that railway time tables could be corrected from local time to 'railway time' or 'London time.'

By 1847 two networks were in operation in Britain, one for the north and one for the south.[7] They were joined on 14 November 1847 and the stockmarket quotations for the day were telegraphed from London to Manchester. Sending a telegram could be expensive. The tariff for twenty words was a penny (2¢) a mile for the first fifty miles, a halfpenny a mile for the next fifty, and a farthing ($\frac{1}{2}$¢) a mile beyond one hundred miles. Rival companies forced down the rates and in March 1850 ten shillings ($2.50) was the maximum charge for any distance. By the end of the 1850s most inland telegrams cost one or two shillings, similar to the one shilling per message charged by the original Slough–Paddington line.

The Slough–Paddington line was back in the news again on 6 August 1844 when the first press telegram in Britain was sent to *The Times* from Windsor Castle to announce the birth of a son to Queen Victoria. Special public greetings telegrams were introduced much later, in 1935. In the following February some 50 000 lovers took advantage of the special St. Valentine's Day telegram. One young man who had spent 8s 9d ($1.75) is said to have concluded with, "And now I've asked you to be mine—By gosh! it's cost me eight-and-nine!"[7]

Telegraphy in America

Samuel Finley Breeze Morse, the American Leonardo according to one of his biographers,[8] would still be remembered even if he had never had his dream of an electric telegraph. He is recalled as one of America's foremost artists, a founder of the National Academy of Design, and one of its first daguerreotypists, the forerunners of today's photographers.

Unlike Cooke, who had been introduced to electric telegraphy via

Schilling's working model in Heidelberg, Morse conceived the idea for himself and for a long time found it difficult to believe that anyone could have beaten him to it. While on a trip to Europe he saw and studied Chappe's mechanical telegraph. On his return voyage to America in 1832 he thought out his early ideas for electric telegraphy. His career as an artist was increasingly relegated to second place.

By the end of the voyage his notebook was crammed with sketches and ideas, many of which were dropped before he built his first commercial line. From the beginning he wanted a receiver that would give a permanent record of the message. He devised a system in which words were coded into groups of numbers, and began work on a code book. Each number was further coded into dots and dashes for transmission, but this was not the famous Morse code, which came later. Speedy transmission was to be achieved by the assembly of lead types of the numerals on a long bar, which could then be quickly drawn under contacts so as to switch a battery in and out of a two-wire circuit. At the receiver the dots and dashes were to be embossed onto moving paper tape. Evidently Morse's proposals owed little to the rush of ideas prevalent in Europe from about 1825 to 1837, most of which concentrated on magnetic needles.

After landing at New York he tinkered with telegraphy until reality forced itself upon him. He had a family to support, paintings to finish, and very little money. America was then no great benefactor to artists, even good ones.

By the end of 1835 Morse was at New York University. Apart from his teaching and painting, he was once again working on his telegraph. One of his friends who saw this early telegraph was L. D. Gale, a professor of science, and in him Morse found a partner. As Cooke sought technical assistance from Wheatstone, so Morse learned from Gale, especially of Henry's achievements. Joseph Henry in particular understood the advantages of using a large number of turns around an electromagnet, and of using an 'intensity' rather than a 'quantity' battery, that is one with a high electromotive force (emf, or voltage) rather than one able to sustain a large current. Henry had pointed the way to telegraphy in 1831 when he used his knowledge to signal through more than a mile of wire so as to energize an electromagnet, which caused a bell to be struck by a pivoted permanent-magnet armature—the first electric bell.

A third man, Alfred Vail, who had been still a student only a year before, joined the partnership in September 1837, bringing with him financial backing as well as mechanical skill. Vail was a good mechanic and made many improvements to the telegraph including, according to his own claim, the introduction of the Morse key. Some have claimed for him the honour of devising the Morse code itself. F. O. J. Smith, a fourth partner, joined in March 1838. As a lawyer his job was to steer the telegraph through the labyrinths of Washington. Unfortunately his character was somewhat questionable. He was a Congressman who tended to abuse his position to further the telegraph, of which he was now part owner.[8] At times, for various reasons, Morse was close to despair.

In 1838, with improved equipment, public demonstrations were given for the first time. The dots and dashes of the first version of the now famous Morse code were embossed onto paper and gave a transmission speed of ten words per minute. Six years later the code was improved to become what is known today as American Morse, and with further modifications as International Morse.

Meanwhile in Britain and Germany telegraphs were being built and put into operation. Morse went ahead with his delayed application for a patent and received it on 20 June 1840, eight days after Cooke and Wheatstone secured their own U.S. patent.

Finally, in 1843, after long and frustrating delays that once more brought Morse to the brink of financial ruin, Congress granted $30 000 (£6000) for an experimental line to link Washington with Baltimore, about 40 miles (60 km) away. The original plan was to bury the iron wires in lead pipes, but that scheme failed because of defective insulation. The wire was recovered and strung up on chestnut poles, with glass doorknobs used as insulators. As the two-wire line reached out from Washington it was regularly checked by messages sent both ways. Morse's telegraph was at last opened to the public on 24 May 1844 with the famous first message, 'What hath God wrought?' But it was that year's Democratic convention in Baltimore that finally blasted the Morse telegraph into America's consciousness. The vote for the Presidential nomination ran to nine ballots; the result of each was telegraphed immediately to Washington, where Morse had established his office in the Chamber of the Supreme Court. As the excitement grew, Senators flocked to Morse's room in such numbers that the Senate was adjourned. The Morse telegraph had arrived at last, and with a storm of good publicity.

By 1845 Morse had extended the line to New York and Boston, using one wire and an earth return. The original embossers were eventually replaced by inkers, and they in turn gave way to the sounder, a device made famous in our own time by western movies. The sounder arose from the operators' skill at following the clicks of their receivers. The typewriter was first pressed into service in 1878, one herald of the future teleprinters and the rest of our own telecommunications systems. It had gone a long way beyond the day when Washingtonians had enquired what it would cost to send a parcel to Baltimore by telegraph.

Rarely can anything so novel have caught on so quickly. After only four years there were 6000 miles (10 000 km) of wire in use in America, and after eight years the figure was over 16 000 miles (25 000 km), about 70 per cent of the total world figure of 23 000 miles (37 000 km). Licences were granted and independent companies formed and merged. The first big merger took place in 1851 and produced the Mississippi Valley Printing Telegraph Company, which later became famous as Western Union.

One example of how the Morse system spread is the story of its introduction into Germany, the home of Sömmering, Gauss, Weber, and Steinheil. In 1846, Werner Siemens, then a 29-year-old artillery officer, saw one of Wheatstone's

ABC telegraphs in Berlin. These dial systems were easy to use and became very popular for private and metropolitan use; speeds of 8 to 15 words per minute could be achieved. Siemens saw some possibilities for improvement and soon designed his own system, which he put into production after he had established a partnership with the mechanic Johann Georg Halske in 1847. This was the beginning of the famous firm of Siemens & Halske (Fig. 5.5). In the same year an early example of the Morse system arrived in Europe. In 1848 it found itself transmitting, in competition against the Siemens equipment, the speech from the throne of the King of Prussia at the opening of the Diet in Berlin. The Morse equipment sent the whole speech in $1\frac{1}{2}$ hours, whereas the Siemens dial took $7\frac{1}{2}$ hours. As a result, Siemens & Halske were invited to start manufacturing the Morse equipment.[9]

Figure 5.5 Siemens needle telegraph, 1847

Social Impact

Everywhere it went the telegraph brought social changes, the first of the long line of social changes caused by electrical and electronic engineering. Until the advent of the telephone around 1880 it was the standard metropolitan communications system used by individuals and businesses for local, distant, and foreign communications. It ended the isolation of the police and fire services in cities. It was used by stockbrokers and newsmen. Reporters no longer had to rely on the mail; news was received while it was still fresh. A press wire service was started in 1849 by J. Reuter in England, who supplemented the incomplete European lines with carrier pigeons.

Warfare has always demanded the best communications. The telegraph made its military debut in the Crimean War in 1854 and became of major importance in the American Civil War. The line from New York to San Francisco was completed in 1861, just in time to be pressed into service. Electrical engineering received a baptism of fire as it was dragged into the war machine.

The Crimean campaign also brought what was probably the first long-term maintenance contract in electrical engineering. Siemens & Halske, who were to build a colossal network stretching from the Gulf of Finland to the Black Sea, were pressed against their better judgment by the Russian government to accept a condition that they should keep the lines operational for 12 years for a fixed annual sum of 250 000 rubles. As the lines proved to be quite reliable, the small German firm netted nearly two million rubles, a sum that helped to start it on its way to becoming one of the world's first major electrical manufacturers.

As well as being useful to one's own military, telegraph lines were also useful to the enemy and so became potential targets. In a later war, the Spanish–American War, submarine cables were destroyed in 1898. And in the 1904–1905 Russian–Japanese War, radiotelegraphy made its debut into active service.

But telegraphy had peaceful uses too. The improvement of safety on railways, the advancement of meteorology, the improvement of time standards, the measurement of longitude, and the transmission of stock exchange information are all early examples of its impact.

Technically the telegraph was continually improving. One writer referred to telegraphs with the comment that "any ingenuous clockmaker could produce modifications of them and the ink on the receipt of purchase of one would scarcely be dry, before another, perhaps better and cleverer, would be offered from the same fertile source."[10] Even so, improvisation reigned; Americans in particular seemed to be developing to a high pitch the ancient art of muddling through, supposedly a British talent. As a result C. F. Varley in 1867 accepted an invitation to go from England to help bring the American apparatus up to European standards and to recommend standards for current, voltage, and resistance.

But while the early telegraphs grappled their way across cities to link business houses, and across land to link cities, imaginative minds were already casting their thoughts across the seas.

Submarine Telegraphy

Submarine telegraphy was one of the great technical adventures of the 19th century, something akin to the exploration of space today.

The idea of underwater cables was not new. Salvá had toyed with it, and Sömmering and Schilling had carried out trials across a river in 1811. From about 1838 onwards many people experimented with underwater cables, including Wheatstone, Morse, Ezra Cornell, and even Samuel Colt, the inventor of the revolver. In 1840 a House of Commons committee held an inquiry into the feasibility of linking Dover with Calais. Then, in 1842, a significant event took place when guttapercha, the gum of a Malayan tree, was introduced into Britain. Among those who recognized its potential as a water sealant and insulator were Faraday, Wheatstone, and Wilhelm (later Sir William) Siemens, the London representative of the Siemens concern. (They probably did not recognize its potential for three other uses eventually found for it; in golf balls, in chewing gum, and as a filler for tooth cavities.)

Guttapercha soon became the important dielectric for submarine cables and was used in the first really significant attempt at laying an undersea cable in 1850, when the brothers Jacob and John Watkins Brett laid a 0.085 in. (2 mm) copper conductor covered by a 0.5 in. (1.3 cm) layer of guttapercha across the English Channel from Dover to Cap Griz Nez. Telegraph signals were exchanged the same day by use of a Cooke and Wheatstone needle telegraph, since a printing telegraph had failed to respond. *The Illustrated London News* informed its readers of the "first interchange between France and England." However, success was short lived. According to the often repeated tale a Boulogne fisherman trawled it up and cut it. One version says he wanted the 'gold' at its centre; others say the brave fellow decapitated a sea monster.

The following year, the year of the Great Exhibition at the Crystal Palace in London, a second attempt was made, this time with a core of four copper wires each 0.065 in. (1.7 mm) in diameter. The cable, manufactured by R. S. Newall and Co. of Gateshead, again had guttapercha as the insulator, now accompanied by tarred hemp. It was finished off with ten 0.3 in. (7.6 mm) iron wires as armour. At seven tons per nautical mile it proved difficult to handle, particularly as the art of cable laying was being learned on the job. The cable ran out short of the French shore. More was added on and, on 13 November 1851, the first successful undersea cable went into public service, a service that continued for 24 years. Now customers began to flock to it, in numbers sufficient to justify a second Channel cable, which opened two years later to link Dover with Ostend.

Other submarine cables followed as the shorter and shallower waters were spanned, but not always at the first attempt. Much expensive cable was lost. London's Thameside found itself with a new industry as cable manufacture got under way. For a time, submarine cables remained a British monopoly. Scotland was linked with Ireland, England with Holland; the Mediterranean and Black seas were crossed, the Mississippi River and the Gulf of St. Lawrence spanned. As thoughts turned to a transatlantic link, Ireland and Newfoundland were tied to their respective continents.

The Atlantic Telegraph Company was formed in 1856. Charles Bright, John Brett, and Kelvin in England, and Morse and Cyrus Field in America, were among those involved. Meanwhile Britain celebrated the end of the Crimean War amid fears of a new one with the United States. The break-up of Spanish America and the Monroe Doctrine of no European colonies in the Americas clashed with Britain's stand over her territories in Central America. In Washington the British minister was dismissed. The powderkeg was ready to blow, according to one British report, but the political crisis subsided and plans continued for the Atlantic cable.

Experiments were conducted to ensure that communication was possible over such a distance without intermediate stations, and in 1857 the first attempt was made, only to end in failure when the cable broke after a mere 380 miles (610 km) had been laid.

Lessons were learned and a second attempt made the following year. The cable was carried by two ships and spliced in mid-ocean, and each ship then set off for home while paying out the cable. Again it broke. But seven weeks later they tried again and, on 6 August 1858, the Atlantic was spanned. Congratulatory messages were exchanged and celebrations began, one of which nearly set the New York City Hall on fire.[11] But after success came failure. Quite falsely, operators decided that induction coils would give a faster working speed than batteries and so high-voltage induction coils (up to 2 kV has been quoted[12]) were used. Probably as a result the insulation began to fail. For a time the very sensitive Thomson (Kelvin) mirror galvanometer prolonged the cable's life. But after a few weeks, in October, it failed. The failure, together with the following year's failure of the Red Sea cable to India, Britain's most glittering colony, led to a government enquiry that lasted nine months. One fact to emerge was that only just over 3000 miles (5000 km) of cable was working out of 11 364 miles (18 200 km) laid.

Meanwhile, in America, Western Union was planning to reach Europe via British Columbia, Alaska, the Bering Straits, and Siberia. Although the attempt was started it was not completed, but one side effect was to encourage the United States to purchase Alaska from the Russians.

Britain was now looking both eastward and westward. To the east an indirect and rather shaky overland link had been established with India, with a line which one writer summed up with the comments that "anyone cabling to India needed to be lucky."[9] A map published by the *Illustrated London News* in 1865 showed the line continuing to Rangoon, with extensions proposed to Singapore, Java, Australia, Hong Kong, and Shanghai (Fig. 5.6). To the west the Atlantic still beckoned.

The British government's committee reported its findings in 1861 and blamed the previous failures on poor design, manufacturing, and handling of the cable. The committee had consulted many of the big names and it helped to consolidate the work of many British scientists. Much was learned especially about the theory of electrical transmission, the effects of impurities on the conductivity of copper, the design and manufacture of cables, and the techniques for laying them. With the results of such a thorough investigation available the prospects for a new attempt were encouraging.

A new cable was designed in which seven strands of high-quality copper wire were covered with four coats of guttapercha, surrounded by hemp, and armoured with ten iron wires. The overall diameter was 1.127 in. (2.8 cm) and the breaking tension was nearly 8 tons. Extra protection was given to the shore ends. The entire cable was packed into the hold of the *Great Eastern*, at 22 500

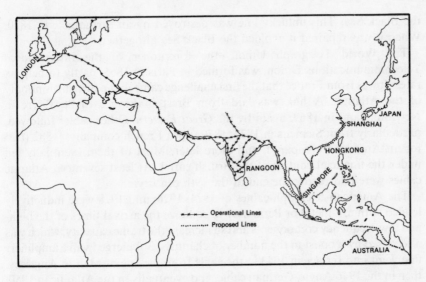

Figure 5.6 Telegraph map, London–Asia (after 'Illustrated London News', 8 July 1865)

tons the largest ship of the era, and laying began in July 1865, after the Civil War in America had come to an end. The press watched eagerly, but with only 600 miles (1000 km) to go the "wire that is to make thought simultaneous in the two hemispheres"[13] parted.

Success came the next year. While Europe was preoccupied with Otto von Bismarck's invasion of mighty Austria, the *Great Eastern* not only laid a new cable but grappled up the last one, spliced fresh cable onto it, and completed that too. By the end of July 1866 two cables linked Europe and America, from Valentia Bay in Ireland to Trinity Bay in Newfoundland. They operated for 6 and 11 years, respectively.

The telegrams were soon flowing, though at a cost of £20 ($100) or more not many individuals could have made use of the new facility. *The Illustrated London News* carried the congratulatory telegrams between Queen Victoria and President Andrew Johnson, and among the first genuine items of news conveyed by the telegraph and published in the same issue were, "Grant has been created a full General and Sherman a lieutenant-general," and "The cholera is spreading in New York, Brooklyn and the neighbourhood."[14]

The rate fell quickly. It was down to $1.575 per word by 1868 and 40 cents a word in 1885. Five letters counted as a word. Some rate cutting followed as rival companies fought for business but agreement was reached at 25 cents a word in 1888. (The competition by the Marconi Wireless Company in 1916 started more rate cutting, until agreement was reached in 1923 at 20 cents per word for both radio and cable.)

The *Great Eastern* went on to lay a new Red Sea cable and complete the new direct telegraph to India. Siemens completed its own direct link to India via the Black Sea. This unlucky line was destroyed by an earthquake in 1870. When it was repaired it avoided the Black Sea altogether.

The World Telegraph Union, the forerunner of the International Telecommunications Union, was formed in Paris in 1865, partly to help fix rates. It was from France that the first challenge came to the British monopoly of the Atlantic. A link was laid from Brest in France to St. Pierre in Newfoundland in 1869, again by the *Great Eastern*. Other cables followed, particularly from Siemens in 1874, 1879 (for a French company), 1882 (two for an American company), and more later. Most of them eventually fell under the financial control of the British cartel. At least seventeen Atlantic cables were laid before the end of the 19th century.

The American telegraph cables of 1924, 1926, and 1928 were inductively loaded with mu-metal or Permalloy to equalize the arrival times of the high- and low-frequency components and so increase the traffic capacity, which was also raised by a boost in the number of channels. Submerged valve amplifiers first went into service in 1943 in the cable from the Isle of Man to Anglesey, then in the 1946 Anglo-German cable, and eventually in the Atlantic in 1950.

In the 1940s there were twenty Atlantic telegraph cables, but 1956 saw the beginning of the end when the first telephone cable was laid and offered the frequency band of one telephone circuit for telegraph use. That one band gave eighteen telegraph channels. If the whole cable had been used for telegraph channels only it would have offered about forty times the capacity of all the previous transatlantic cables.[15] Ten years later, in 1966, the last of the exclusively telegraphic cables was abandoned, bringing to an end a system that had served for a century and whose place has now been taken by telephone cables and communications satellites.

Technical Improvements

Many and varied were the technical improvements that formed the bridge between the first commercial international telegraphs and today's telex, telephones, and telemetry.

Even before the first Atlantic cable was completed R. E. House, Bakewell, and J. G. Gintl had shown the shape of things to come with, respectively, a very early printing telegraph, a copying or facsimile telegraph, and duplex telegraphy. But the first important development was the Hughes printing telegraph of 1854, a robustly built American machine that gained a big market in Europe after its introduction via France (Fig. 5.7). About forty words per minute could be achieved, a higher speed than with the Morse sounder.[16] D. E. Hughes was encouraged in his work by the Associated Press wire service, which saw in it a means of breaking the American Telegraph Company's monopoly of the Morse system and so reducing the telegram rates.

Instead of using the slow step-by-step motion, Hughes employed synchronized free-running type wheels. When one of the keys on a piano-style

Figure 5.7 Hughes printing telegraph

keyboard was pressed the corresponding pin, from a set of pins arranged in a circle, was raised and grazed by a horizontally rotating arm. The momentary contact was used to transmit a pulse whose timing depended on the position of the rotating arm, and hence on the particular pin raised. The basic idea was something like that used in a present-day car distributor. The pulse was used at both the transmitter and receiver to press a gummed paper strip against a type wheel and so record the character. The printed message was then glued to a message form. The two type-wheels had to be synchronized with one another and with the rotating arm. In 1911 it was estimated that 3000 of these machines were handling the bulk of telegraph traffic in continental Europe.[17]

Meanwhile Wheatstone had introduced his automatic Morse transmitter, a machine for which he was knighted and which, like the Hughes equipment, stood the test of time and was used for over half a century. Punched paper tape was used to store the message for transmission and to feed the data to the telegraph key. This automatic system was used almost exclusively by the British; perhaps its greatest success was in the London-Teheran section of the line to India, which contained ten automatic repeaters.

Where sensitivity was the criterion for the receiver, as in the Atlantic cables, it was Thomson's mirror galvanometer that held the day. A magnetized steel needle about 0.4 in. (1 cm) long was glued to a tiny mirror and suspended by a silk fibre in a bobbin of wire; the bobbin was wound in four sections that could be connected in various ways to adjust the sensitivity. The pointer was a light beam reflected from the mirror onto a screen 3 feet (1 m) away, a technique that had been used earlier by Gauss and Weber. Latimer Clark gave a convincing demonstration of the mirror galvanometer's sensitivity when, using the two Atlantic cables in series, he received a signal from a battery consisting of a small zinc rod immersed in dilute acid held in a silver sewing thimble. The tiny cell sent a signal around 4000 miles (6400 km) to the galvanometer beside it. It also demonstrated the absurdity of the argument that had led to the use of high-voltage induction coils in 1858.

William Thomson, who had taken an active interest in submarine telegraphy from its earliest days, invented another high-sensitivity receiver in 1867, the syphon recorder, which had the advantage of producing a permanent record. It also meant that one man could operate the receiver instead of two. Thomson was knighted for his scientific work on submarine telegraphy in 1866 and was made a baron in 1892, when he became Lord Kelvin.

Duplex telegraphy, the art of sending two simultaneous signals along the

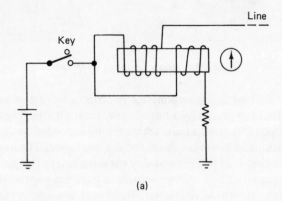

(a)

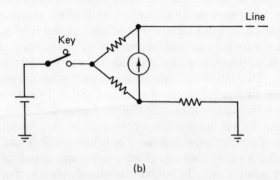

(b)

Figure 5.8 Principles of duplex telegraphy: (a) differential (magnetic effects of coils cancel on transmit, sum on receive); (b) bridge (current flows through meter only on receive)

same wire in opposite directions, doubled the capacity of any telegraph line. Two techniques evolved (Fig. 5.8). One used a differential relay in which two coils were used in the electromagnet, with half the current going through each. At the transmitter the magnetic effects of the currents cancelled; at the receiver, they were summed. The second method used bridges to obtain the same results. The bridge technique had evolved via Wheatstone from S. H. Christie. Then in 1862 Varley suggested the use of capacitors at each end of the line to sharpen the signal.

J. G. Gintl of Vienna is usually credited with the first idea of how to achieve duplex using the differential system as early as 1853. Karl Frischen of Hanover perfected it a year later and it was immediately adopted by Siemens & Halske.

J. B. Stearns in 1872 patented the use of his artificial line with a differential duplex system, which became the most popular for aerial land lines. Earlier, from about 1867, a duplex system had been used on American land lines without any capacitors at all, probably a bridge system. In Britain the Wheatstone automatic system proved fast enough for the available traffic and duplexing was not carried out until after nationalization in 1870, when the differential system came into use.

Submarine cables, with their much higher capacitance distributed throughout the cable, presented a problem different from that of land lines, and the Stearns system had to make way for that of Herbert Taylor and Alexander Muirhead of 1875. In this method the capacitive and resistive components were combined in one unit, a much closer analogy to the actual cable than the alternative of using separate (and alternate) capacitive and resistive components. The Muirhead 'artificial' line, along with the bridge duplex, made long-distance submarine duplexing possible and became dominant by the turn of the century.

Two systems therefore evolved: Stearns's plus differential duplex for land lines, and Muirhead's plus bridge duplex for submarine cables. Duplexing had almost doubled the traffic on many cables and land lines, but attempts at further increasing the traffic density by quadruplex, sextuplex, and even octuplex, met with little success. Attempts at sending ever more signals along a single wire were leading nowhere. A new idea was needed. Why not try dividing the time available into periods and subdivide each period into very short time slots? Each time slot could then be allocated to a given signal. Provided the total time taken for one period was not too long the idea should work, or at least that is what Émile Baudot, a French telegraph clerk, thought. The result was a new technique, which, loosely speaking, may be classed as time-division multiplex (TDM).

Others before Baudot had experimented with multiplex systems and another Frenchman, Bernard Mayer, had put one into practice. His system had some success in France until it was displaced by Baudot's around 1874. Baudot used a rotating switch, called a distributor, which divided the time of one period between two, three, four, or more transmitting and receiving machines. Because of the time division a new code was needed in which each pulse had the same duration, unlike Morse's dots and dashes. Baudot used what would now be called a five-bit code giving thirty-two combinations, and so each transmitter had five keys. This was a radical change from Morse. Now every letter was represented by five bits instead of some having only one, others two, and still others having three or more. At the receiver the letters were printed onto gummed paper tape that could then be cut and glued to message forms.

The Baudot system met with great success, partly owing to its excellent construction, and became the standard system in France. Later it spread throughout Europe and appeared in South America, India, and Ceylon. In 1916 it was reported that the international circuits were operated almost entirely on the Baudot system.[18]

In Britain it was the Murray TDM system which slowly came into use. In America the Buckingham system was developed by Western Union. Donald Murray's system was said to operate faster than Baudot's, 40 to 45 wpm as against 20 to 30 wpm, and it is to Murray, a New Zealander, that we owe the present 'Baudot' code.

The Telephone

Mechanical telephones, like mechanical telegraphs, had been around for some time before the 'real thing' appeared, though they must have been little more than toys. The string telephone is said to date from 1667 and speaking tubes probably appeared early in the 19th century. In 1861 a German schoolteacher, Philipp Reis, published a description of a 'telephon' that could be credited with transmitting single-frequency sound, although it is not usually credited with transmitting speech. Earlier, in 1857, Antonio Meucci, an Italian-American, developed a primitive electrical telephone but was unable to get financial backing. His patent expired three years before Bell filed for his own telephone patent. Another American, Elisha Gray, came very close to beating Bell to the invention.

The real telephone was the invention of Alexander Graham Bell, a statement that has stood the test of much litigation in court.

Bell inherited a family tradition of studying speech and elocution; one aim was to teach the deaf. Born in Edinburgh in 1847, he later moved to London where he studied at University College before emigrating to Canada in 1870, and onward to the United States, where he opened the Boston School for the Deaf. Bell was also interested in multiple telegraphy and worked on a device he called a harmonic telegraph. A vibrating tuning fork would generate an intermittent current in a line; at the other end another tuning fork, of the same resonant frequency, would be set in motion. Several pairs of tuning forks at various frequencies, each with its own Morse key, would comprise a one-line multiple-channel telegraph that used one frequency per channel. This work, together with his work on the mechanical voice recorders he used as aids in teaching the deaf, led to the telephone.

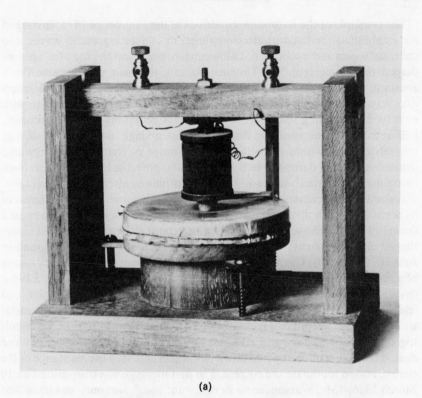

(a)

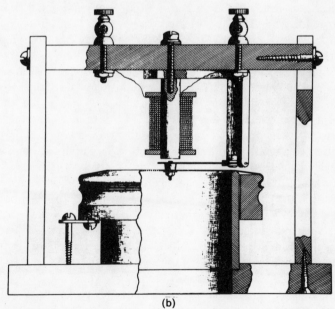

(b)

Figure 5.9 (a) Bell's first telephone (Gallows telephone). (b) Cut-away diagram

The principle of the telephone, conceived in the summer of 1874, envisaged the established methods of electromagnetism converting audio waves into electrical waves and back again. The important point was the use of a "speech-shaped electric current," as Thomas Watson, the man who became Bell's assistant, later put it.[19] In time the tuning forks of the harmonic telegraph gave way to several tuned reeds across the pole face of an electromagnet, with each reed responding to a specific frequency. Continuing their experiments, Bell and Watson accidentally discovered that a single damped and slightly magnetized reed would respond to a wide range of frequencies. They also realized that the damping had accidentally prevented the current from becoming intermittent; instead, it was a continuous alternating current.

A stretched parchment membrane, with one end of a single reed fastened to its centre, was then arranged over a pole of an electromagnet. Speech caused the membrane (and hence the reed) to vibrate and generate the desired voice-shaped electric current. This instrument transmitted some muffled sound in June 1875 (Fig. 5.9). The next year articulate speech was transmitted by means of a damped reed receiver and a new type of liquid transmitter (Fig. 5.10), a device previously invented by Elisha Gray in his own telegraphy and telephony work. In this apparatus a diaphragm was used to position a metal wire in dilute acid held in a metal cup, all of which formed part of the circuit together with a battery and receiver. Speech vibrated the diaphragm and caused the wire to move up and down in the acid and vary the resistance of the circuit, and so modulate the current. The first message was transmitted on 10 March 1876, "Mr. Watson, come here. I want you." Not only was it the first

Figure 5.10 Above: Bell's liquid transmitter and its receiver. Opposite: cut-away diagram

message, it was also the first emergency call: Bell had spilled acid over his clothes.

The liquid transmitter and the reed receiver were soon replaced by an improved membrane transmitter and what became known as the iron-box receiver. With more tests, demonstrations, and lectures, other improvements followed. The telephone was exhibited at the Centennial Exhibition in 1876 at Philadelphia and among those who saw it were Kelvin and Elisha Gray.

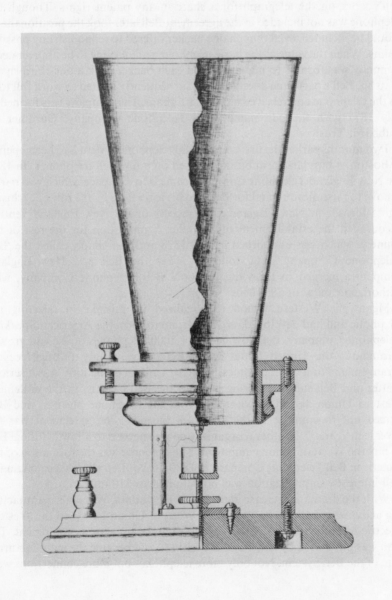

Kelvin was impressed, but Gray must have been rather disconcerted, for he had filed for a caveat on the same day that Bell filed for a patent. Bell's patent, number 174 465, is the most lucrative ever issued, a record that Western Union must have eyed with remorse after it had rejected an offer to exploit it for $100 000. Western Union was not alone in mistaking the potential of the new invention. One far-sighted newspaper reporter in 1875 wrote of the telephone, "It is an interesting toy . . . but it can never be of any practical value."[30]

The early corporate history of the companies formed to exploit the telephone is interesting.[19] Early in 1875 two friends formally agreed to finance Bell's work on the telegraph for a share in any patent rights. Though the telephone was not included in the agreement Bell later took the position that it should be included, even though one backer offered to relinquish any possible claims. When the value of the basic telephone patent came to be appreciated a company was formed to manage it and each backer took a one-third share. Watson, Bell's part-time assistant, was subsequently invited to work full time on the telephone and received a one-tenth share. The company was formed in July 1877 with the odd name, 'Bell Telephone Company, Gardiner G. Hubbard, Trustee.'

Two months earlier the first commercial telephone system had been opened in Boston, a burglar-alarm business based on a few Bell telephones. In 1878 the New England Telephone Company came into existence, which was partly a move to raise more capital for Bell and his associates by the sale of exclusive rights for use in New England. The success of the New England venture encouraged the establishment of a similar organization for the rest of the country, which was established in July 1878 and was simply called the 'Bell Telephone Company.' The following year the Bell and New England companies merged to form the National Bell Telephone Company, with authorized capital of $850 000.

Meanwhile Western Union had realized its mistake in rejecting the telephone and had developed its own system to form the American Speaking Telephone Company. Bell now had over 3000 telephones in use and growth throughout the United States was spectacular, mainly through leasing arrangements to local companies. Western Union competition was a serious matter and Bell sued for infringement of the Bell patents. In the settlement Western Union agreed to withdraw from the telephone business and Bell agreed not to compete in the telegraph business; the agreement was for seventeen years.[19] Another reorganization was needed to allow National Bell to buy the Western Union equipment. The outcome was the formation of the American Bell Telephone Company in 1880. It acquired all rights belonging to Bell interests. Capitalization was now limited to $10 million.

With the dramatic increase in the use of telephones, Watson's manufacturing activities proved to be too small. Western Union had used the Western Electric Manufacturing Co. of Chicago to make some of its equipment. The firm was purchased by American Bell in 1882 and became the manufacturing arm of the Bell System. Not only more telephones but more connections were needed between exchanges. The cost of long intercity lines was too much for American Bell and so yet another company was formed to construct and operate lines throughout North America. This company, the American Telephone and Telegraph Co. (AT&T), was formed in February 1885, nine years after Bell filed for his patent. By 1899 AT&T had become the parent organization of the Bell System.

Research and development were important from the very beginning. In 1883 an experimental shop was formed by American Bell and was the first of a line of organizations that eventually led to the formation of the Bell Telephone Laboratories (now called simply Bell Laboratories) in 1924. The laboratories had a dual responsibility both to AT&T and to Western Electric. They remain one of the world's prime organizations for research and development.

Bell's early telephone was extremely simple and demanded no new knowledge to understand it. The only source of power was the human voice and anyone with an understanding of induction, explained by Faraday over forty years earlier, could have understood it. Indeed Maxwell expressed disappointment. He had expected something so far removed from, say, Kelvin's syphon recorder, as that was from a common electric bell. He commented that "the disappointment arising from its humble appearance was only partially relieved on finding that it was really able to talk."[20] But if Maxwell was disappointed the public was not. It captured its imagination even more than the telegraph had done. By the end of the first year 778 telephones were in service, and by 1880 Americans were making about 240 000 phone calls a day, a figure vastly greater than the 80 000 telegrams a day. The rise in use of the telephone was phenomenal, especially up to the early 1900s (Fig. 5.11). The number of telephones in the world today is measured in hundreds of millions.

After Bell, Edison produced a carbon transmitter and D. E. Hughes, already mentioned for his invention of a printing telegraph, invented an instrument which he called a microphone. In 1878 he recounted how the use of a light bar of graphite, mounted with sensitive loose contacts between two other blocks of graphite, would produce a loud sound in a telephone receiver.

Edison's, Gray's and Hughes's microphones modulated a battery current rather than using the human voice to induce a current. The output was no longer limited to the power of the human voice. A favourite experiment has been described as trapping a house fly in a match box and placing it near the microphone; the fly's footsteps were said to sound like the tramp of an elephant.[16] More likely the sounds were caused by movement of the loose contacts, the same loose variable-resistance contacts that led on to the further development of microphones.

The Electrician reported the microphone as follows: "A child's half-penny wooden money-box for a resonator, on which was fixed by means of sealing wax a short glass tube, filled with a mixture of tin and zinc, the ends being stopped by two pieces of charcoal to which were attached wires, having a

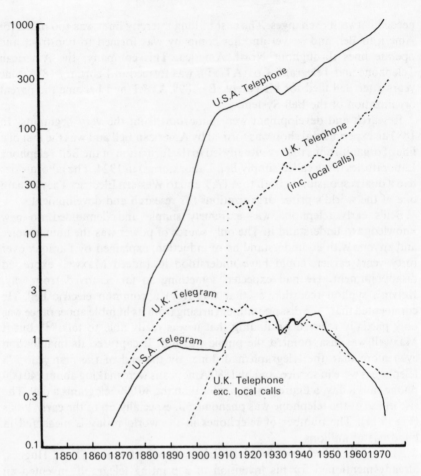

Figure 5.11 Annual per capita number of telegrams and telephone calls for Britain and USA (sources: Government statistics, Refs. 31–34)

battery of three Daniell cells—consisting of three small jam pots—in circuit."[21]

As the market for telephones grew the problem of switching between them became important. The concept evolved of having local switching centres for local telephones and then interconnecting these local exchanges. At first all switching was performed manually; the telephone, like the typewriter, has been hailed for helping in the emancipation of women by providing respectable jobs for ladies in which even a Victorian father could find no fault.

A. B. Strowger, a Kansas City undertaker, patented the first widely used automatic switching system in 1889 and advertised it as the "girl-less, cuss-less, out-of-order-less, wait-less telephone."[22] Electromagnets energized by pulses received from the caller's telephone operated a pawl-and-ratchet mechanism to move a wiper over a bank of switches. Strowger's switch was unique in providing two directions of motion. By 1895 it had been developed into the basic system that has survived in many places to this day. To call a number such as 34, the caller pressed one key three times and another key four times, which led to many errors. In 1896 Strowger's engineers improved things by inventing the dial technique, which went unchallenged until 1961. Strowger's company, Automatic Electric, continued to develop automatic switching; the Bell system preferred manual switching until about the 1920s.

Throughout the 20th century the improvement of exchange switching has been one of the main aims of telephone engineers. Probably it always will be; it has been both their biggest headache and their biggest triumph. With the aid of automatic exchanges many telephone subscribers can now dial their way around the world. Strowger's system laid some of the basic concepts. From about 1906 linear switches, as well as rotary ones, were developed and led to Bell's 'panel' system in which a small amount of 'common control' was used. Common control was to be a fundamental concept of later exchange systems; it enabled some parts of the switching system to be used only briefly by a caller and then released for use by another customer. In this way parts of the exchange were not tied up by a single caller. Common control is used extensively in crossbar switching, perhaps the most important of the electromechanical switching systems. The first proposal for a crossbar switch has been claimed for J. N. Reynolds of the Bell System (1913) but credit for the invention is usually given to the Swedish engineer G. A. Betulander, who invented the crossbar switch in 1919.[19] For the next 20 years its successors were used in Sweden. Instead of using a sliding or wiping action for the switches the crossbar system employs what are basically relay-type switches to interconnect bars arranged as a mechanical matrix. Circuits are employed to receive and memorize each called number, select a route through the switching system, and seek alternate routes if the primary path is busy. Speed and reliability are better than those achieved by previous systems though not as good as those of the later electronic exchanges. The crossbar system was adopted and developed by AT&T in the 1930s for their large-city exchanges, but in Britain the Post Office continued to use the Strowger system with its step-by-step connections.

The present and future of telephone exchanges lies with neither of these systems but with electronic switching, which is inherently faster and more reliable than any electromechanical switching. Electronic exchanges are also much more versatile, as they are virtually program-controlled special-purpose electronic computers. Special features such as the interception of calls to a given number can be obtained by the insertion of a new block of instructions into the control program, which may be done remotely over the telephone line. In an electromechanical exchange expensive physical changes would be necessary to achieve the same result. Electronic exchanges have depended on advances in transistor and integrated-circuit technologies. The transistor itself was a product of Bell Laboratories concern for improving switching systems.

The first production program-controlled electronic exchange appears to have been Bell's No 1 Electronic Switching System (ESS), which was first installed in 1965. Since then, especially in the 1970s, electronic exchanges have spread.

One measure of the impact of automatic switching on the telephone system is the number of operators employed by the system. Figures for the Bell system show that in 1925 and in 1965 roughly the same number were employed (around 150 000), whereas the number of telephones had risen from about 12 million to about 75 million. The cost, in terms of both manpower and dollars, would have been far too great for manual exchanges to have supported 75 million telephones in 1965. About one million operators would have been required.[19]

Advancement of Theory

The next important development in telecommunications was the use of carrier waves and frequency-division multiplexing. Both had appeared by the end of World War I, accelerated by the development of radio and deeper theoretical understanding of electrical transmission, particularly of the roles played by capacitance and inductance in transmission lines.

Ohm had pinpointed the role of resistance in the late 1820s (though many of his contemporaries had taken some convincing of its significance) and Kirchhoff had stated his law in 1844. The use of relays had enabled land lines to stretch great distances before the Atlantic cable finally forced telegraph engineers to examine the role played by capacitance, a problem Ronalds had foreseen in 1823. Lord Kelvin first examined this role in submarine cables by treating the transmission of a pulse as essentially the same problem as charging up a very long and thin Leyden jar capacitor. The guttapercha was the dielectric and the wire and the sea formed the two 'plates' of the capacitor; the cable was a resistance with a capacitance to earth distributed along it. The time constant involved in charging the cable to send a pulse was relatively long, which greatly impeded transmission since the cable had to be discharged again before the next pulse was sent. Little wonder the operating speed was low. Kelvin's mathematics treated the problem as one of diffusion and was based on Fourier's treatment of heat diffusion, a treatise that had also helped Ohm towards his famous law. The pulse diffused along the cable from one end to the other. The operating speed could be increased by an increase in the sensitivity of the receiver, as evidenced by the success of Kelvin's mirror galvanometer, or by a reduction in the product of the cable's capacitance and resistance, Kelvin's KR law.

Once this capacitance problem was understood something could be done about it. The capacitance–resistance product could be minimized. Positive and negative pulses could be used so that one helped discharge the other. Submarine telegraphers learned to cope, and on land lines where air was the dielectric the problem was not as acute. Yet though the telegraph engineers might be content, the telephone engineers were not. Telephoning over hundreds of miles of wire was a daydream. The signals used were of course weak and electronic amplifiers were undreamt of. Also crosstalk (interference) from telegraphs was a real problem. Better microphones and receivers were not the real answer, and use of resonators and microphones at intervals along the line as acoustic amplifiers did not solve the problem either.

The problem seemed to be a new one. The telephone was an AC instrument and different frequencies were propagated at different velocities. Speech became unintelligible. Something was missing, yet that something had been around for a long time. One of the men who eventually helped solve the problem, Columbia University professor M. I. Pupin, wrote in 1934 about the 19th century engineer: "There was one word in the vocabulary of his language that he refused to learn. That word was 'inductance.' The telegraph engineer of those days had a holy horror of the so-called 'choak [sic] coils' in the telegraph line; the telephone engineer inherited that fear, and hence he paid small attention to the apostles of the inductance doctrine. The foremost among those apostles was the late Oliver Heaviside."[23]

Heaviside, a nephew by marriage of Charles Wheatstone, had been trying with little success to tell the world about the effects of inductance and the great need for it in communications. Born in 1850 in London, he received little formal education and was mainly self-taught. Though one of Britain's greatest mathematical physicists, he had considerable difficulty for a long time in getting his papers into print. He did not follow the accepted Cambridge mathematical doctrine; he preferred vectors to quaternions; he evolved his own operational calculus; and his methods were said to have shocked the mathematicians. It was those mathematicians, competent as they were, who had difficulty in understanding his work. When they refereed his papers for publication they turned them down seeking clarification, something that Heaviside, living the life of a recluse in Torquay, found difficult to forgive. Later in life he was led to caustic gibes such as, "Whether good mathematicians when they die, go to Cambridge, I do not know."

His papers were eventually published in the weekly *Electrician*, though few could understand them. Eventually, the truth would out and his message was heard: "It is the very essence of good long distance telephony that inductance should *not* be negligible."[24] Inductance, previously a nuisance, was to take its rightful place in the loading of cables.

Heaviside's approach lay through understanding and extending Faraday's and Maxwell's work on electromagnetism. We have already seen that it was he who "cleared away the debris of Maxwell's battle" and, like Hertz, presented Maxwell's theory in the form in which we know it. He saw no difference in principle between the new radiotelegraphy through free space and the older type guided along wires. If inductance, a word which he coined, was used properly, it would raise the role of the neglected magnetic field until it was equal to that of the previously dominant electric field. If the two were of equal importance, the receiver would 'see' every feature of the transmitted wave and

we would have a distortionless circuit, something that can only be approximated in practice. Resistance and capacitance were distributed along the
wire and inductance should be too, in order to compensate for the effects of
capacitance. It could be continuous or lumped; if the latter, inductance coils
must be deliberately inserted at intervals along the wire.

Heaviside did much more than merely suggest the use of inductively loaded
cables. His mathematics, although not as rigorous as the mathematicians
might have wished, introduced such sophisticated techniques as complex
variables and Laplace transformations into electrical engineering, and a
whole new nomenclature: inductance, attenuation, reluctance, and reactance,
among others.

Experiments with loading coils were made by S. P. Thompson, G. A.
Campbell, C. J. Creed, and others, but it was Pupin who patented the criteria
for loading coils in 1899, and C. E. Krarup who produced practical
distributed loading by forming a closed spiral of iron wire around the
conductor in 1902. Campbell made the major contribution to the Bell
telephone system. Loading coils were used on land lines but the problems of
installing and maintaining coils at sea led to the use of distributed loading on
submarine cables.

Correct loading prevented signals from being scrambled, but some
problems remained. Attenuation was overcome with the help of valve
(thermionic-tube) amplifiers from about 1913 onwards and the abandonment
of the earth return in favour of twisted pairs reduced crosstalk. The British
General Post Office experimented with a 660 mile (1100 km) telephone link in
1915 but submarine telephony was still restricted to short distances. For a long
time radiotelephony, complete with fade outs, was the rule for transoceanic
telephony.

The other new technique that was evolving, frequency-division multiplexing
(FDM), was largely a byproduct of radio and filter work performed before
and during World War I. FDM carrier systems took their place in both
telephones and telegraphs, with a 4 kHz bandwidth allowed for each
telephone channel. New magnetic materials such as permalloy made inductive
loading easier. The New York–Azores cable of 1925 could be worked at 1900
letters per minute, about four times the usual speed of such a cable, thanks to
the replacement of soft iron by permalloy. It was so fast that it outstripped the
speed of the standard equipment used with it and new equipment and methods
had to be worked out.

Other advances also took place. The simplex circuit of the 1880s enabled the
telegraph to use the telephone wires. A telephone cable linked England and
France in 1891, one year before the Bell Telephone Co. introduced a big step
forward, the solid-back carbon transmitter. Its higher efficiency enabled
longer lines to carry satisfactory conversations. In America private enterprise
reigned supreme; in Britain, the telephone was added to the GPO monopoly of
the telegraph in 1896.

The Modern Era

The era of complete dominance of telecommunications by wire was nearing an end. Marconi had been transmitting and receiving electromagnetic waves over increasingly long distances and in 1901 he bridged the Atlantic. In 1906 the American engineer Reginald Fessenden inserted a microphone into his radio transmitter and asked if anyone could hear him. Radio was coming; so was the aeroplane. With flights eventually crossing the Atlantic and reaching out to Australia, the telegraph, already reeling from attacks by telephones and radio, saw a third, though junior, partner join the attack against it—the airmail letter. By the 1920s the decline of the telegraph's fortunes was only too evident, especially in Europe (Fig. 5.12).

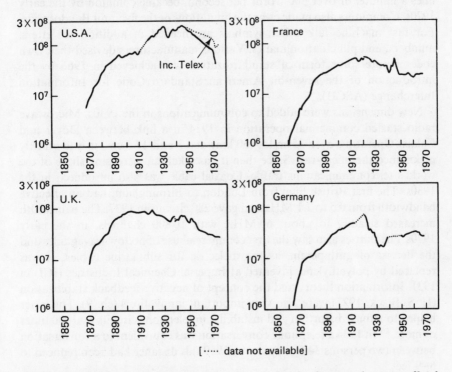

Figure 5.12 Telegrams per annum for USA, Britain, France, Germany (sources: official statistics, Refs. 31–34)

In 1925 D. M. Murray[25] in Britain was writing of the "new telegraphy," the start-stop printer or teleprinter. He pointed warning fingers at the advantages of the rivals: the telephone, radiotelegraphy, and air mail. But Murray was not despondent. "The telegraph is not, but should be, at every businessman's elbow like the telephone," he wrote . . . "We must 'teletype' as well as 'teletalk'."

And soon teletyping, or teleprinting, was a reality. The start-stop teleprinter was already under development and by the early 1930s it had reached the market. The trade name Teletype is well known. The early version was the backbone of the American manual teletypewriter exchange (TWX) started in 1931 by AT&T; 65 000 had been manufactured by 1946. The operating speed was slowly increased and automatic switching started in the USA in 1962. The European system, Telex, originated in Germany and spread rapidly after World War II. Telephones and Telex grew to have complementary roles in business.

With the arrival of electronic computers the telegraph, in the form of the teleprinter, received a huge and unexpected market. Previously undreamt of writing speeds came with the ever faster peripheral lineprinters; a thousand lines a minute, or over 300 words per second, became common by the early 1970s. Computers also made clear the limitations of the five-unit Baudot code. For fast machine talk other symbols are needed in addition to letters, numbers, and punctuation marks. Various manufacturers devised their own codes before some form of standardization was achieved in 1966 by the introduction of the seven-bit American Standard Code for Information Interchange (ASCII).

New dimensions were added to communications in the 1930s. Microwave radio started commercial operation in 1934 in a link between Dover and Calais. Operating over 35 miles (60 km) at 17 and 17.5 cm, it was quickly nicknamed the micro-ray. Since then it has taken an increasing share of the world's telecommunications traffic. Coaxial cable was also introduced in the 1930s. The first British line, from London to Birmingham, had an effective bandwidth from 0.5 to 2.1 MHz and gave 280 circuits in 1937. The bandwidth increased steadily to about 60 MHz, with 10 800 channels, in the early 1970s.The thirties also saw the first commercial use of hollow waveguides and the demise of guttapercha as the dielectric for submarine cables. It was replaced by polyethylene, invented at Imperial Chemical Industries (ICI) in 1931. Information theory and the concept of negative-feedback stabilization (H. S. Black, 1927) were also very important innovations (ch. 9). The latter helped to reduce distortion and instability in telephone circuits by 1000 times or more.[26] In 1934 a telephone conversation was equivalent to a conversation between two persons 34 feet apart; by 1959 this distance had been reduced to 6–9 feet.

As mentioned, a telephone cable across the Atlantic was at last completed in 1956. Two cables were in fact used, one for east–west and the other for west–east communications. They were laid between Oban in Scotland and Clarenville in Newfoundland and gave 35 telephone circuits; the 36th was used for 18 telegraph channels. Fifty-one repeaters were used each way with three thermionic valves each. Reliability and long life were of the utmost importance. Cables and repeaters were designed for a lifetime of at least 25 years. Transistors made their debut in submarine cable repeaters in 1964 in an already existing cable linking Britain and Belgium.

The Pacific was first spanned by a telegraph cable in 1902 and is now the home of the world's longest submarine telephone cable, the 9000-mile Commonwealth Pacific. It was opened in December 1963, cost £35 ($85) million, and stretched from Australia to Canada via New Zealand and Hawaii.

As the radio spectrum became increasingly crowded, and as the market for international communications grew, so submarine cable laying took on new life, particularly with the advent of coaxial cable. Cable laying continues to be a lucrative business and an average of six or seven major cables have been laid every year since World War II.

Long-distance telecommunications took another step forward in 1962 with the launching of the orbiting communications satellite Telstar, owned by AT&T. Earlier experiments had been made with passive, aluminium-coated reflecting baloons (Echo 1 and 2), and small active satellites (Score and Courier). Telstar was followed by a variety of military and civilian satellites, both American and Russian. The Intelsat series of geostationary satellites provided an almost global communications system. Intelsat 1, or Early Bird, was launched in 1965 and gave 240 two-way telephone circuits. Prime-time colour television was charged at the rate of $22 350 (£10 000) per hour.[27] By 1977 eight Intelsat 4 and 4A satellites offered 40 000 telephone circuits to six continents and prime time colour television cost only $5100 (£2500) per hour. Intelsat 5 satellites offer 12 000 telephone circuits each, quite a contrast to Early Bird's 240 of 15 years earlier.

Pulse code modulation (PCM), invented by A. H. Reeves in France in the 1930s, was not used commercially until the 1960s. Since then the enormous advances made in electronics have led to a gradual change from analogue to digital techniques. Electronic switching at the exchanges also encourages this move to digital electronics. By the end of the 1970s digital telephony, spurred on by advances in semiconductor technology, was almost an equal partner with analogue, bringing with it better means of controlling noise, stability, and accuracy.

And so the quest continues, searching for ever wider bandwidths, ever higher frequencies, ever lower distortion. The progression has been from electrostatic pithball telegraphs, which can be traced back to Stephen Gray's experiments with moistened hemp in 1729, to the use of wires, cables, coax, waveguides, microwaves, and satellites. In the search for greater bandwidth coherent lightwaves conducted along optical fibres have been installed in installations in Europe, USA, and Japan since 1977. Fibre-optic waveguides offer other advantages besides wide bandwidth, not least their small size, light weight, insensitivity to electromagnetic interference, and low transmission losses.

And what of the future? We are now moving into what has already been dubbed the Information Society where information, and access to it, is rated as a vital part of a nation's resources. Digital electronics, one way or another, is the key to that society. One thing we can be sure of: whatever system we get in the future someone, somewhere, will find that it is still not quite good enough.

References

1. J. J. Fahie, Chapter 2, Ref. 6.
2. E. A. Marland, *Early Electrical Communication*, Abelard–Schuman, London, 1964.
3. C. M. Jarvis, *JIEE* 2: 130–137, 584–592, 1956.
4. *Daily Telegraph*, p. 10, 30 September 1898.
5. C. Singer *et al.*, Eds., *A History of Technology*, Oxford University Press, Oxford, vol. 4, 1958.
6. *Encyclopaedia Britannica*, 11th edition, 1910–1911.
7. P. Robertson, *The Shell Book of Firsts*, Ebury Press and Michael Joseph, London, 1975.
8. C. Mabee, *The American Leonardo: The Life of Samuel F. B. Morse*, Knopf, New York, 1959.
9. G. Siemens, *History of the House of Siemens, 1847–1914*, Karl Alber, Munich, vol. 1, 1957.
10. H. Sharlin, *Elec. Eng.* 80: 54–58, 1961.
11. C. Bright, *Submarine Telegraphs*, Crosby Lockwood and Son, London, 1898.
12. F. Scowen, *Electronics and Power* 23: 204–206, 1977.
13. *Illustrated London News*, pp. 82–83, 29 July 1865.
14. *Illustrated London News*, p. 102, 4 August 1866.
15. M. J. Kelly *et al.*, *Elec. Eng.* 74: 192–197, 1955.
16. J. A. Fleming, *Fifty Years of Electricity*, Wireless Press, London, 1921.
17. D. Murray, *JIEE* 47: 450–529, 1911.
18. H. H. Harrison, *Electrician* 77: 798–800, 1916.
19. M. D. Fagen, Ed., *A History of Engineering and Science in the Bell System*, vol. 1 (1875–1925), Bell Laboratories, New York, 1975.
20. J. C. Maxwell, Rede Lecture, 1878, *Nature* 18: 159–163, 1878.
21. *Electrician* 34: 395, 1895; also *Nature* 55: 496–497, 1897.
22. D. Barnes, *Electronic Design* 18: 42–51, September 1977.
23. M. I. Pupin, *Elec. Eng.* 53: 691, 1934.
24. F. Gill, *Bell Syst. Tech. J.* 4: 349–354, 1925.
25. D. M. Murray, *JIEE* 63: 245–280, 1925.
26. E. I. Green, *Elec. Eng.* 78: 470–480, 1959.
27. B. I. Edelson, *Scientific American* 236: 58–73, February 1977.
28. P. Dunsheath, *A History of Electrical Engineering*, Faber, London, 1962.
29. D. G. Tucker, *J. Inst. Electronics and Telecom. Engrs.* 22 (No. 3): 101–106, 1976.
30. J. A. Brady, *Telephone Eng. and Management* 84 (No. 5): 101, 1980.
31. U. S. Bureau of the Census, *Statistical Abstract of the United States 1974*, (95th edition), Washington, D.C., 1974.

32. Central Statistical Office, *Annual Abstract of Statistics 1976*, HM Stationery Office, London, 1976.
33. U.S. Bureau of the Census, *Historical Statistics of the United States, Colonial Times to 1957*, Washington, D.C., 1960.
34. B. R. Mitchell, *European Historical Statistics 1750–1970*, Macmillan, London, 1975.

6 ELECTRIC LIGHTING AND ITS CONSEQUENCES

The electrical engineering industry really began life as an industry to provide electric lighting, first by means of arc lamps and then by incandescent lamps. The industrial applications of electricity before the commercial exploitation of electric lighting were trivial compared with what came after. The telegraph, telephone, and electroplating had raised small industries to develop, install, maintain, and run those services, and the communications industry has grown into an enveloping giant in its own right. But it was lighting that first demanded central power stations for the efficient mass generation of electric current and then placed that current in the home, office, factory, and street for use in lighting systems, and later for other applications as well. In this way, and others, lighting had a profound effect on the technical and commercial development of the industry, including even electronics. Further, it was the profits from electric lighting that supported the electrical industry in its formative years and enabled some of the early companies (for example General Electric in America and Philips in the Netherlands) to grow into large industrial concerns operating internationally in most of the major areas of electrical engineering. Even today the profits from light bulbs are important to major firms.

The prehistory of electric lighting was fairly short and served to indicate the possibilities. A few interesting scientific experiments on glow discharges had been made in the 18th century by Hauksbee in England, Nollet in France, and others. Early in the 19th century Davy produced a brilliant light using a large battery to maintain an arc between two charcoal electrodes, and wires were raised to incandescence by the passage of a current through them from Leyden jars in the 18th century and from chemical batteries later. Such experiments and demonstrations proved that light could be produced by electrical means: by gas discharge, by incandescence, and by continuous arcs. It was the task of later generations to develop these principles into useful, reliable, and commercially feasible electric lamps that could successfully compete with other commercial systems for the production of artificial light.

The Rivals

A generation that can turn night into day at the flick of a switch may easily forget that candles and oil lamps have been man's traditional sources of artificial light. They are still with us today. Many households keep a couple of candles handy for 'when the electricity goes off'' and decorative candles remain popular; a romantic dinner for two would hardly be the same under the blaze of a quartz–halogen lamp. This tradition of oil and candle was broken when gas was first piped out from a central source early in the last century. There were around 40 000 gas lamps in the streets of London by 1823, for example.[1]

Gas lighting was developed from experiments conducted in Britain, France, and Germany towards the end of the 18th century. Improvements were slowly made to produce purer gases that gave less smoke and soot, and to devise more efficient burners.[2] Karl Auer von Welsbach, a Viennese chemist and appropriately a former pupil of Bunsen, produced an incandescent gas mantle in 1886 that was particularly efficient and gave a light approximating to daylight. This device had been developed in part in response to the growing challenge from electric incandescent lighting, with which it successfully competed for decades.

Gas lighting became increasingly important commercially from about the middle of the 19th century. In the United States the capital invested in gas companies in 1850 was about $6.7 (£1.35) million and it increased by a factor of ten over the following 20 years. By the time Thomas Edison launched his electric lamp on the market late in 1880 about $150 (£30) million had been invested in gas lighting. Gas shares dropped on news of his success. By comparison the capital of General Electric, the company that resulted from Edison's work, had reached $35 (£7) million thirteen years after the invention of Edison's successful lamp.

However, we are getting ahead of the story. Before the incandescent lamp there was the arc lamp.

Arc Lamps Make a False Start

Though arc lighting was eventually eclipsed by other forms of lighting it did perform some important and lasting functions: it helped to establish some early electrical manufacturing companies, it provided experience for engineers, particularly in the design of improved generating equipment, and it established electrical engineering as a useful engineering discipline outside electrochemistry and telecommunications.

After Davy's demonstrations of 1800–1802 of a continuous electric arc maintained between two carbon electrodes, the production of a carbon arc lamp was a possibility awaiting development. Three problems had to be dealt with to achieve success: a means of producing carbon in a form that would minimize the burning away of the electrodes, a means of regulating the arc to compensate for changes caused as the electrodes burned away, and a relatively cheap and reliable source of current.[2] Even if success could be achieved the light would be too brilliant for use in the home or office; the potential market lay in street lighting, lighthouses, illumination of large spaces, floodlighting, and so forth.

The trigger for the development of the electric arc into an early form of arc lamp was probably the improvement that took place between 1836 and 1842 in the manufacture of batteries. Certainly arc lamps were exhibited from about the mid-1840s. The Daniell cell made its appearance in 1836 and was quickly followed by a variety of other improved chemical batteries. Carbon was also improved. In 1844 in France, Léon Foucault, famous for his measurement of the velocity of light and his pendulum experiment, devised a hand-regulated arc lamp with electrodes made of retort carbon, a hard deposit of fairly pure carbon produced during the manufacture of coal gas. Over the next couple of years two patents were granted in Britain for the purification of carbon. One of the patentees was W. E. Staite, who had been experimenting with arc lamps since 1834. In 1836 he had shown that a lamp could be regulated by clockwork. By the mid 1840s, then, the three problems might appear to have been solved.

Staite continued to improve his lamp with the help of W. Petrie and they gave many demonstrations; for example, the portico of the National Gallery in London was floodlit on 28 November 1848. *The Illustrated London News* (ILN) reported this demonstration and stressed that the arc lamp was nothing new: "Year after year it has been exhibited at every course of philosophical lectures since the time of Sir Humphry Davy." ILN called attention to unanswered questions relating to the cost of the lamp and to the almost continual attention arc lights needed to keep them operating. The magazine warned that in providing an arc lighting system, "if there is a serious defect upon one point, ruin would be entailed upon all who enter the undertaking."[3] Staite learned that lesson the hard way. When the Patent Electric Light Co. failed after only a few years in about 1850 he lost most of his own money.[4]

The problems of improving the carbon and regulating the light were solved by many people, but the limitations imposed by the primary batteries defeated everyone. Bright has tabulated most of the important developments in the early evolution of the arc lamp and listed 23 lamps developed between 1844 and 1859 (14 English, 8 French, 1 American).[2] Many ingenious solutions were found for the problem of regulation: electromagnetic devices, gravity, floats, and clockwork and other mechanical devices. Carbon electrodes were tried in the shape of rods, discs, wheels and plates, with or without additives such as tar, sugar, pitch, powdered coke, and china clay. Carbon rods seem to have been the most successful in the end. However, the limitations of the batteries caused most inventors to abandon the development of the arc lamp by 1860. "For a dozen years," says Bright, "no improvements on existing lamps were patented."

The battery problem can be illustrated by an example. Staite's lamp is said to have consumed only half an inch of carbon per hour, probably a good enough solution to the electrode problem. However, the battery used a third of a pound of zinc each hour for a 100-candlepower light. The arc lamp at the South Foreland lighthouse (1858), powered by a steam-driven generator, was rated at 1000 candlepower.[5] Such a battery-powered lighthouse beacon would have consumed over 3 lb of zinc per hour, hardly an acceptable situation.

Obtaining plentiful cheap electric current remained an unsolved problem until the early 1870s, when efficient electromechanical methods of current generation became more widely available. Until then the widespread use of arc lighting was to remain a pipedream.

Generators: Pixii to Gramme

The development of mechanical methods of generating electricity began in 1832, soon after Faraday's announcement of the discovery of electromagnetic induction, when the French instrument maker Hippolyte Pixii exhibited a 'magneto-electric machine' that produced a somewhat discontinuous alternating current. Ampère suggested a commutator to convert the output to a direct but undulating current, a waveform that was to be around for a long time. Over the years such simple hand-driven machines were developed into large, power-driven generators to supply lighting systems.

Pixii had used a horseshoe magnet which was hand cranked so that its poles rotated under a pair of coils (Fig. 6.1). In London, Joseph Saxton, and later E. M. Clarke, rotated the lighter coils and left the heavy magnet in a fixed position, a technique that was widely adopted. Clarke made what were probably the first commercial generators; they were used in laboratories and in electrotherapy, for the relief of rheumatism and other ailments. When Saxton accused him of pirating his design Clarke replied that he had made modifications and had received Saxton machines for repair.

Table 6.1 and Fig. 6.1 outline some of the developments. Designers tried various positions for the coils with respect to the magnetic poles. From the early 1840s the number of magnets and coils used in machines began to increase. Floris Nollet of Brussels was the first to attempt to build a large power driven magneto-electric generator, on which he obtained a British patent in 1850. His idea was to use limelight in lighthouses, which he would obtain by heating lime to incandescence in an oxy-hydrogen burner and using his generator to produce oxygen and hydrogen by electrolysis. Nollet died in 1853 but an Anglo-French group formed the Compagnie de l'Alliance to develop his ideas. Though this attempt failed, one of the engineers, F. H. Holmes, returned to England convinced that he could make arc lighting into a successful lighthouse venture. Arc lighting had failed previously because there was no satisfactory source of current. With Holmes's development work on magneto-generators, arc lighting again became a commercial possibility.

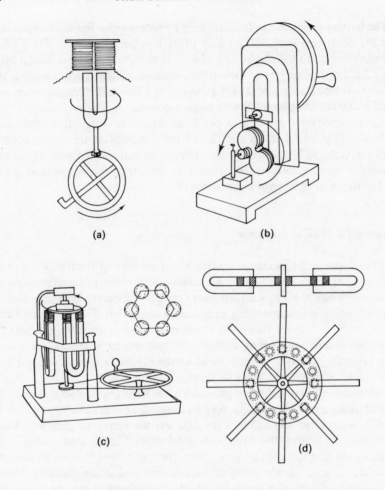

*Figure 6.1 Early generators: (a) Pixii, 1832 (Paris), 2 fixed coils, 1 magnet;
(b) Clarke, 1834 (London), 2 rotating coils, 1 magnet; (c) Stöhrer, 1843–
1844 (Leipzig), 6 rotating coils, 3 magnets; (d) Millward, 1851
(Birmingham), 16 rotating coils, 8 magnets*

Holmes approached Trinity House, the British lighthouse authority, and a trial was arranged. Faraday acted as the judge and was delighted with the results.

For further trials two larger machines were constructed in which two wheels rotated at 90 rpm between three vertical frames. Each wheel carried 80 coils and each frame supported 20 magnets, a total of 160 coils and 60 magnets. The generators weighed over 5 tons, absorbed 2.75 hp, and were belt driven by a noncondensing steam engine.[6] Their output has been estimated at around 1.7 kW.[4] The arc lamp was manufactured by Jules Duboscq of Paris, a leading designer, and had an automatic feed for the carbons. Trials proved to be

Table 6.1 The Early Development of Generators (sources: Refs. 4, 6, and 7)

1831	Faraday (GB)/Henry (USA)—discovery of electromagnetic induction
1832	Pixii (Fr.)—hand driven, 1 magnet, 2 coils, magnet rotates
1833	Saxton (GB)—hand driven, 1 magnet, 2 coils, coils rotate
1834	Clarke (GB)—commercial production of hand-driven magnetos
1844	Stöhrer (Ger.)—3 magnets, 6 coils, hand driven
1849	Nollet (Bleg.)—proposal for power-driven machine
1851	Millward (GB)—power-driven magneto, 8 magnets, 16 coils
1856	Siemens (Ger.)—'H' armature instead of coils
1857	Holmes (GB)—power-driven magneto, 36 magnets, 120 coils
1858	Holmes (GB)—power-driven magneto, 60 magnets, 160 coils
1860	Pacinotti (It.)—ring armature instead of coils
1863	Wilde (GB)—patents dynamo-electric machine with magneto exciter
1866– 1867	Wilde (GB), Siemens (Ger.), Varley (GB), Wheatstone (GB), Farmer (USA), Ladd (GB), self-excited dynamo
1870	Gramme (Belg.)—dynamo ring armature
1872	Hefner-Alteneck (Ger.)—drum armature
1880	De Meritens (Fr.)—distributed rotor winding in magnetos

successful and in 1862 a system was put into operation at Dungeness lighthouse. Though initially the light had many failures the Dungeness system operated for 13 years. Several other British lighthouses were electrified as confidence grew, and arc lamps became an established form for a few lighthouses. One 1867 design by Holmes for Souter Point in northeast England remained in service until 1900.

Meanwhile in France the Alliance company had been refloated and was producing machines for use with arc lamps in French lighthouses. Arc lamp searchlights were also produced in France and some were used by the French Army during the siege of Paris in the Franco-Prussian War of 1870–1871.[7] Yet arc lamps were still few and far between. Even in 1880 there were only ten electric lighthouses in the whole world.[6]

Though there was now a use for arc lamps, major improvements in electric generators and new lamp designs were needed before large-scale power generation could become feasible. The main move was from magnetos to dynamos. Meanwhile, magnetos still had room for improvement. A distributed rotor winding replaced the standard coils from about 1881 and gave a much improved and more continuous output. Still, the future lay with dynamos. It came in three stages: the replacement of permanent magnets by electromagnets, the self-excitation of the dynamo, and better designs for armatures.

In April 1866 Faraday read to the Royal Society a paper by a Dr. Henry Wilde of Manchester. Wilde had invented what became known as a dynamo-electric generator, that is a generator in which electromagnets are used to produce the magnetic field. The current to excite the electromagnets came from a magneto-electric machine (one using permanent magnets) mounted on the same drive shaft. Wilde had discovered, as he put it, that "an indefinitely small amount of dynamic electricity or of magnetism is capable of evolving an indefinitely large amount of dynamic electricity." The next step was to abandon the magneto and have only the dynamo itself, with the small amount of residual magnetism in the field coils used to generate a small current that could then be fed back to the field coils to increase the magnetic field and generate a larger current. The result was a self-excited dynamo. Priority for its invention is confused as results of similar experiments were communicated to several bodies simultaneously during 1866–1867 by various investigators, including Wilde, Wheatstone, and S. A. Varley in England; Werner Siemens in Germany; and M. G. Farmer in America. All may have been stimulated by Wilde's magneto-excited dynamo. Priority was still being debated in 1900.[8]

The efficiency of a generator depends on the design of the armature, or coil, as well as on the method of producing the magnetic field. Werner Siemens produced the first big improvement in 1856, the shuttle or 'H' armature. It was an iron cylinder with two deep longitudinal slots in which insulated wire was wound, with the ends of the wire terminating at the two curved plates of a simple commutator. Its main advantage was its small diameter, which meant it was suitable for operation at high speeds and with smaller magnets when used in magnetos. It was also used in dynamos. The early dynamos had solid cores and heating became a problem. Operation was limited to about three hours and water cooling was used on some machines.

Although the 'H' armature enjoyed a period of popularity, it still produced a pulsating direct current. The final step that made large-scale commercial DC generation possible was the invention of the ring armature, first invented by Antonio Pacinotti in Italy in 1860 and reinvented in an improved form in 1870 by the Belgian engineer Z. T. Gramme, an employee of the Alliance company. For the first time a truly continuous current was produced from a machine, 70 years after Volta and almost 40 years after the discovery of electromagnetic induction.

Gramme's armature consisted of a continuous wire wound on the same principle as Pacinotti's and tapped at intervals for connection to a multi-segment commutator. The more segments there were, the smoother was the current. The armature was mounted on a ring core of iron wire coated with a bituminous compound to insulate the individual strands and so reduce eddy currents. Gramme's generator was exhibited in Paris in 1871 and patented. Within a short time it was being manufactured in Paris and abroad under license. A new era in the generation and use of electricity had begun.

Questions of cost and efficiency became more important as commercial production began. Improvements were made. Variations on the Gramme machine were produced in several countries; by Emil Bürgin in Switzerland, R. E. B. Crompton in England, Charles Brush and Farmer in America. Friedrich von Hefner-Alteneck, the chief designer at Siemens & Halske, modified the shuttle and ring designs to produce the drum armature in 1872, in which copper was saved and preformed coils could be used. It remained a standard for a quarter century.[7]

Power-driven self-excited dynamos with ring or drum armatures solved the problem of plentiful and continuous production of direct current. After a few years of design consolidation, electrical engineering moved out of its infancy into adolescence and the world was never the same again.

Commercial Arc Lighting

By the late 1870s the potential market for arc lighting was well developed. What was needed to meet it was a dynamo-lamp system that was simple, reliable, and capable of providing a better light than gas for about the same price. Arc lighting was about to become a successful venture on both sides of the Atlantic.

One of the best remembered European systems was the Jablochkoff candle, invented by the Russian army officer Paul Jablochkoff in Paris in 1876. It consisted of two parallel carbon electrodes separated by a layer of kaolin. An alternating current supply allowed the carbons to burn evenly and so maintain the arc at the tip, a neat solution to the regulation problem that still troubled street arc lights. A British patent was granted in 1876 and trials were held in Paris, London, and elsewhere. In one form four candles were used in one unit, with the current switching automatically to the next candle when needed. Though Jablochkoff's candle enjoyed success and helped establish street arc lighting, its defects and expense eventually led to its disuse.

Americans played only a minor role in the early development of arc lighting at first but from about 1877 their role grew rapidly. Most of the American inventors were young men and some of the firms they founded to exploit arc lighting are still important today. The Thomson-Houston Electric Company became one of the co-founders, with the Edison companies, of the giant General Electric Co. (GE); its British offshoot, British Thomson-Houston (BTH), was one of the co-founders, with the British branch of Westinghouse, of Associated Electrical Industries (AEI), which later became part of the British conglomerate the General Electric Company (GEC). To avoid confusion between these two general electric giants they will be referred to as GE and GEC, respectively.

In America the chemist C. F. Brush built his first dynamo in his mid-20s and obtained financial backing from a friend, the vice president of the Cleveland Telegraph Supply Company.[9] It was agreed that the company should make and sell Brush dynamos, arc lamps, and anything else Brush should invent. So successful was this association that the company changed its name in 1880 to the Brush Electric Company; Brush had become the American pioneer of arc lighting (Fig. 6.2).

Figure 6.2 Brush arc lamp, Los Angeles, about 1885

In 1877 the Franklin Institute of Philadelphia invited all manufacturers to submit dynamos for a competitive trial. Only three machines actually competed, those of Gramme and Brush, and another American machine, the Wallace–Farmer dynamo, made in Connecticut. Brush won. In the years that followed he further improved the dynamo, lamps, and carbons, and his simple and reliable system became a commercial success in both America and

Europe. In 1879 the world's first central power station, two Brush dynamos supplying twenty-two arc lamps, opened in San Francisco. It was a financial success and after six months was powering fifty arc lamps. Other stations followed. By 1885 the Brush plant could produce 1500 arc lamps per month. It was producing over 20 million carbons per annum by 1890 and had diversified into incandescent lighting also.

Among those attracted to arc lighting by Brush's pioneering success was the English-born Elihu Thomson, an instructor at the famed Central High School in Philadelphia.[9] With an older colleague, E. J. Houston, he was invited to be on the committee set up to conduct the Franklin Institute's 1877 trial. Thomson and Houston later designed their own dynamo and arc lamp and received limited local financial backing. When a new company was proposed to finance and manufacture equipment based on their patents, Thomson became its electrical engineer and Houston remained at his post as a teacher. When the American Electric Company went into business, Thomson and Houston together received 30 per cent of the stock plus $6000 (£1200) in cash. By the end of 1881 a system was ready that was believed to be better than its rivals. However, the company was sluggish at marketing. Worse, the most enthusiastic backer committed suicide. Thomson was on the point of withdrawing himself and his patents when a group of Massachusetts businessmen, led by C. A. Coffin, a former shoe salesman, took over. Coffin and Thomson merged salesmanship and organizational ability with sound technical judgment and inventiveness. In 1883 the company's name was changed to the Thomson-Houston Electric Company, a name that was to become famous. Late in 1889 they took over Brush Electric, part of Coffin's merger policy that took Thomson-Houston to a dominant position in the American arc-lighting industry. When GE was formed in 1892, Coffin became head of a firm that for a long time dominated electrical engineering everywhere. Apart from the communications industry, Westinghouse was the only significant American general electrical manufacturer to evade GE's takeover or merger moves—a rivalry that continues.

In Britain one of the most important firms was Crompton and Company. Colonel R. E. B. Crompton, the founder, led an interesting life.[10] At the age of 11 he visited his brother in the trenches of the Crimean War, came under fire, and was decorated. His interests included military and civilian motor transport as well as electrical engineering and he was a founder member of the Royal Automobile Club. He died during World War II at the age of 94.

Colonel Crompton was in his 30s when he left the army and turned his attention to arc lighting. He began by supplying Gramme dynamos bought in France but by the end of 1880 his Chelmsford works in Essex had manufactured the first of a new machine designed by Crompton and by Emil Bürgin, a Swiss engineer. Over 400 were made before the design was abandoned about 1886. The Glasgow Post Office was Crompton's first large indoor lighting installation: 180 gas jets were replaced by two arc lights in the sorting office in 1880. Large railway stations, with high roofs ideal for arc lighting, such as King's Cross in London, were also early customers. Business changed with the rise of the incandescent lamp and later innovations but the company continued to thrive. In 1927 the company merged with F. and A. Parkinson Ltd to become Crompton Parkinson Ltd. It later became part of the Hawker-Siddeley group and the lighting division continued to be an important part of the business.

In Germany Siemens & Halske also turned to arc lighting. Hefner-Alteneck produced his own arc lamp in 1878 and four years later proposed a unit of light to facilitate measurements. This German firm must have had a penchant for units. Werner Siemens had earlier suggested a unit of resistance, the siemens, which was widely used until replaced by the ohm. Arc lighting was supplied to railway stations, factories, and the German houses of parliament.[7]

Though arc lighting enjoyed a period of success it was almost trivial compared with the eventual success of incandescent lighting. Two technical developments helped delay the end. Staite, in 1846, had shown that enclosing the arc in glass prolonged the life of the carbons; unfortunately, the glass became blackened with carbon. Purer carbon led to a revival of the idea and commercial exploitation began around 1893. The second improvement came from Germany in 1899 and was called the flaming arc. The Jablochkoff candle was its ancestor. Nonconducting salts were added to the carbon; they evaporated into the arc and made it brighter. However, carbon life was short. Cheaper labour costs in Europe led to the development of the flaming arc. In America, where electricity was cheap, the enclosed arc became popular.

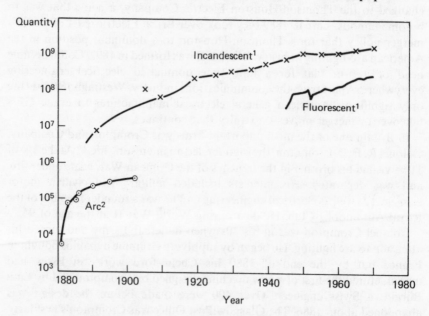

Figure 6.3 Approximate number of electric lights, USA, 1880–1970: 1-Production, 2-In Service

Figure 6.3 clearly shows the levelling off in use of arc lamps in the USA. Their market declined after about 1910 as metal filament lamps came into use and, with the advent of high-power filament lamps, their fate was sealed before the end of World War I.

The Incandescent Lamp

Few people give even a passing thought to the humble light bulb. It has long since lost its glamour. Yet it was this glass bulb with its little filament that placed the electrical industry on a sure footing (Fig. 6.4).

The brilliance of the arc lamp made it unsuitable for widespread domestic use. Hardly anyone wanted a lighthouse in their living room. Methods were sought to 'subdivide' the electric light so as to attack gas lighting in the pursuit of the enormous potential market for small domestic-type artificial lighting. Several successful lamps were made towards the end of the 1870s; those by Swan in England and Edison in America are the most famous.

It has been pointed out before that almost every invention is preceded by a series of other inventions and discoveries on which it depends.[11] Yet it is the way of the world to shower the praise, honour, and sometimes wealth as well on one individual, who is then often regarded as the sole originator. Things are simpler that way. That the selected individual, such as Morse with the telegraph or Edison with the light bulb, deserves the praise need not be questioned. That they were the sole inventors, or even the first inventors, is plainly not true. As Swan put it, "There are no inventions without a pedigree."[12]

After the improvements in batteries around 1840 there were many attempts to make an electric incandescent lamp. The first patent for such a lamp was granted in 1841; another followed in 1845. Both were issued in Britain but these were not the only inventions. Bright listed twenty such inventions in what he termed the precommercial period of the incandescent lamp.[2] The nationalities of the inventors included British, French, German, Belgian, Russian, and American. The materials used as the incandescent element included carbon, platinum, and iridium. Some were enclosed in glass, others not; some in vacuum, others in air or nitrogen. None could be commercially successful because of their short life, caused by oxidation of the element, and the expense of battery current.

These two problems were effectively solved by two inventions. A German chemist in England, Herman Sprengel, invented a mercury vacuum pump in 1865 and ten years later William Crookes perfected a method of using it to evacuate glass bulbs. This was the key to solving the problem of oxidation. As with the arc lamp, the other problem was solved by the dynamo. By 1875 incandescent lighting was again a possibility. The main question to be answered concerned the composition and form of the incandescent element.

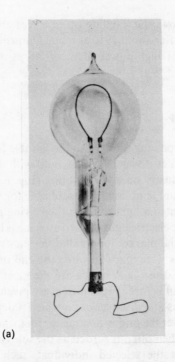

(a)

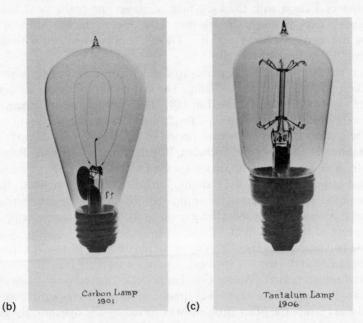

(b)   (c)

Carbon Lamp
1901

Tantalum Lamp
1906

*Figure 6.4 Early incandescent lamp filaments: (a) Edison's carbonized cardboard
filament, c. 1880; (b) carbon, 1901; (c) tantalum, 1906*

THE FAMILY ALBUM—CHANGING BULBS - - - - By GLUYAS WILLIAMS

FINDS UPPER HALL LIGHT WON'T LIGHT. BULB MUST HAVE BURNED OUT

UNSCREWS BULB

GOES DOWN CELLAR FOR NEW BULB. FINDS BOX OF NEW BULBS IS EMPTY

DECIDES NOBODY EVER GOES INTO ATTIC STORE-ROOM, IT'LL BE ALL RIGHT TO TAKE BULB FROM THERE

SCREWS OTHER BULB BACK IN STORE-ROOM SOCKET AND THE STORE-ROOM BULB INTO HALL SOCKET

IT STILL WON'T LIGHT. FINDS HE GOT BULBS MIXED AND PUT THE GOOD BULB BACK IN THE STORE-ROOM

SIGHS WEARILY, AND GOES UP TO STORE-ROOM TO CHANGE THEM ROUND AGAIN

GETS HALL LIGHT TO LIGHT AT LAST, JUST AS WIFE CALLS CAN HE COME FIX THIS STORE-ROOM LIGHT. SHE'S GOT TO GET SOMETHING OUT OF THE TRUNK

(Copyright, 1928, by The Bell Syndicate, Inc.)

Gluyas Williams's 1928 cartoon on the frustrations of burned-out light bulbs

Edison

Thomas Alva Edison, the wizard of Menlo Park, was the last of the six most important inventors to start work on the problem. After buying a Wallace-Farmer dynamo he commented, "Now that I have a machine to make the electricity, I can experiment as much as I please. I think . . . there is where I can beat the other inventors, as I have so many facilities here for trying experiments."[2]

His advantage was his laboratory, a prototype of the modern industrial laboratory except that it developed mostly the ideas of one man, Edison, rather than those of many. Today's industrial research laboratory is said to have had its origins in the German organic-chemical industry. According to Drucker[13] the turning point came with the synthesis of aspirin by Adolf von Baeyer in 1899. Nevertheless, as far as electrical engineering is concerned,

Edison's Menlo Park, N. J., laboratory of the late 1870s was a forerunner of the modern research unit.

Despite having been described as 'addled' by his teacher, Edison already had several financially successful inventions behind him, particularly in telegraphy, before he turned to serious work on the incandescent lamp in 1878. The first stage was to examine the gas lighting industry. Ninety per cent of its revenues came from home or office lighting and that was where he would challenge them. To do it he would need a lamp with luminescence similar to that of the gas jet and a marketing policy based on that of the gas companies. So closely did Edison achieve these goals that he even referred to his lamps as burners and sent monthly bills expressed in light-hours, as did the gas companies. The message was put across. He was selling that familiar stuff called light, not that unfamiliar stuff, electricity.

His first idea for the incandescent element was carbon. He soon abandoned it as a failure, although it later became the basis of his first commercial lamp. Even as he turned away from it, other inventors turned to it. By this time it was apparent that more funds than Edison alone could provide would be needed. His reputation was such that funding presented no serious problem and in October 1878 the Edison Electric Light Company was founded with a capital of $300 000 (£60 000). New York's leading investment banker, Drexel, Morgan and Company, was one of the backers. This was strong support indeed, particularly as Edison was being financed to invent a completely new lighting system, not to develop an existing one. Edison himself received 2500 of the 3000 shares of the new company plus $30 000 (£6000) in cash, most of which he eventually spent on experiments.

One of the problems in choosing the incandescent material was the high temperature required to achieve incandescence. Obviously it had to be less than the melting point. Carbon, platinum, and iridium were popular choices with inventors. Platinum was Edison's next choice and some initial success led to a slight panic in gas shares. Consideration of the whole lighting system probably led to the critical decision to use a high-resistance filament (100 ohms), which would keep the current low and make the use of copper conductors of small cross section for the mains supply possible. Some 'experts' believed in low-resistance filaments but Edison thought they would make the cost of the mains prohibitive. The lamp could not be considered by itself; the entire system had to be efficient and commercially viable.

A thin high-resistance platinum wire gave better results than any other tried previously. Many other materials were tried, including boron, silicon, iridium, rhodium, chromium, zirconium, zirconium oxide, titanium oxide, and osmium. Thin wires of tungsten, osmium, and tantalum, so successful in later light bulbs, were not then available. Edison's research technique is illustrated by a quotation attributed to him: "I've tried everything. I have not failed. I've just found 10 000 ways that won't work." But eventually he and his team found a way that would work. In October 1879 they carbonized a piece of cotton sewing thread by heating it in an oxygen-free atmosphere. When used as a filament in an evacuated glass bulb it burned for nearly two days. Carbon was the answer. Carbonized bristol board lasted even longer, up to 170 hours, and carbonized paper filaments were used in Edison's first commercial lamps.

Still the experiments continued. Bamboo was found to be especially good and "Jules Verne type explorers"[12] were despatched to seek even better vegetable fibres. Bamboo became the standard filament material in Edison lamps for the next few years.

Another technical problem that threatened the experimental lamps was solved by Edison and Swan at about the same time. At the high operating temperatures of the lamps, the gas and water vapour absorbed in the filament, glass, and stem were released and so impaired the vacuum and lowered the effectiveness of the lamps. This problem was overcome by heating the glass and filament during manufacture while they were still connected to the vacuum pump, which removed the offending gases.

Edison tried to protect all his important discoveries or inventions by applying for patents. Late in 1879 he applied in several countries for patents for his cotton-thread filament lamp, a lamp that never saw commercial development. The U.S. patent (1880) was particularly important, as it was broader: it protected structural vegetable fibre filaments and thus gave some protection to the later bamboo filaments even though they were protected by their own patent of 1881. After a long legal battle the paper filament patent was awarded to two other Americans, William Sawyer and Albon Man, for their low-resistance filament. It took six years for that decision to be handed down.

Swan

Joseph Swan, a chemist from Newcastle in the northeast of England, is remembered for a variety of work other than his electric lamp. His inventions of bromide photographic printing paper and artificial cellulose thread, a prototype man-made fibre, are just two that have had long-term effects.

Swan played a part in both the early and the commercial periods of the filament lamp. In the 1860s he achieved incandescence with carbonized strips of paper and cardboard but the problem of obtaining a good vacuum deterred further work. This problem was solved by the new vacuum pump, but not until 1877, and two years later the concomitant problem of outgassing was also solved. Swan's experimental lamp of 1878–1879 employed a slender carbon rod, about 1 mm in diameter, as the incandescent element. It was exhibited in December 1878 but its lifetime was not sufficient for commercial exploitation.

As Swan and Edison were perfecting their lamps at almost the same time the question has arisen of who produced the first practical lamp. Yet that is really too naive a question since the early designs changed so quickly and because (particularly with Edison) the whole lighting system should be considered.

The dates of patents do not present a simple picture either, since Edison applied for his patent as quickly as possible, whereas Swan initially believed the broad features of the lamp to be unpatentable.

Swan gave the first public demonstrations from December 1878 onwards of a working incandescent lamp that used a carbon rod as filament but was not put into production. In November 1879 Edison patented his carbonized cotton-thread filament lamp, which also did not go into commercial production. By the end of the year he had given his first public demonstration using carbonized bristol board, which did go into production. Early in 1880 Swan introduced a parchmentized cotton filament that he patented and used in his first commercial lamps. Whether this feature was inspired by Edison's patent is open to question. Both used cotton but the treatment was different. Edison's first commercial lamps of late 1880 again used bristol board, which was in turn replaced by bamboo. Swan's first commercial lamps of 1881 used his parchmentized cotton thread.

Swan formed a company, with Crompton as chief engineer, to manufacture lamps and make lighting installations. One of the first installations was at Kelvin's house, a clear indication that Swan was not then thinking along Edison's grand lines of central stations powering city blocks.

For a while parchmentized or carbonized thread was the universal basis of lamp filaments, but in 1883 Swan patented a technique for producing artificial thread. Nitrocellulose dissolved in acetic acid was squirted through a die into a coagulating agent such as alcohol. The resulting thread could be carbonized to produce a filament more uniform than natural fibres and therefore less subject to excessive local heating. Another inventor devised an alternative technique at about the same time and the two collaborated. The squirted-cellulose filament remained the standard until carbon filaments became obsolete. As a sideline, Swan prepared some particularly fine thread which his wife crocheted into lace mats and doilies that were exhibited in 1885 as 'artificial silk,' the first man-made fibre.

In Britain the Swan and Edison companies merged in 1883 to form the Edison and Swan Electric Light Company, or Ediswan, which merged with a Philips affiliate about 1920.[2] In 1882 the Electric Lighting Act had been passed and limited the tenure of electricity supply companies to twenty-one years. It has been described as "one of the most short-sighted and retrogressive pieces of legislation to reach the Statute Book during the later 19th century."[14] Designed to protect the public by encouraging caution and preventing monopolies in the exploitation of an untested field, it instead stifled commercial development. In the six years that passed before it was amended British industry fell behind its rivals. One possible benefit was that it may have encouraged the Edison-Swan merger, which gave the new company an effective patent monopoly in Britain. A flaw remained to be eradicated. Swan's earlier work might be construed as 'prior art,' which would have invalidated the patents and opened the door to competition. To avoid that risk, a filament was defined to be less than 1 mm in diameter, so that Swan's work of 1878–1879 with a 1 mm carbon rod could not be considered as prior work on the filament lamp. It is perhaps ironical that it was Swan's parchmentized cotton thread, and later his squirted cellulose thread, that became the standards.

Other Inventors

Swan and Edison were not alone in inventing incandescent lamps. M. G. Farmer has already been noted for his work on dynamos. He also worked on the telegraph and the incandescent lamp. In 1879 he patented a lamp consisting of a horizontal carbon rod in a nitrogen atmosphere or vacuum. Like Edison he advocated the parallel connection for the lamps and a voltage-regulated dynamo in preference to the constant-current series connections used with arc lamps. His lamp patents were of only minor commercial importance. Another pioneer was H. S. Maxim, also remembered for his machine gun. After unsuccessful attempts he produced a commercial lamp in 1880 after, it is said, Edison had personally explained the whole process to him. Two other Americans, W. E. Sawyer and Albon Man, as noted earlier, beat Edison to the patent for the carbonized paper filament. They used a low-resistance filament in a nitrogen filled bulb. A company was formed in New York in 1878 to develop and manufacture their lamp.

Apart from Americans another Englishman was also on the scene. St. George Lane-Fox patented his first lamp in 1878 using loops of high-resistance platinum-iridium wire in an air or nitrogen atmosphere. When others paved the way he also turned to carbon in a vacuum but developed his own carbonizing technique. The Anglo-American Brush Electric Light Corporation Ltd., formed in 1880, sold his system in Britain.

Commercial Development

Edison's was the best of the early lamps in nearly all important aspects and it won several awards in Europe and America. Edison also had the foresight to design a complete lighting system rather than just a lamp. It was a system far superior to any conceived by his competitors. It included a central power station with improved dynamos, a distribution system, lamp fittings, switches, cables, fuses, meters, and everything else that was needed to provide electric lighting for an area of a city.

Arc lighting had used fairly high-voltage constant-current dynamos with efficiencies around 50 per cent. The arc lamps themselves were arranged in series; when one went out, all the others did too. Edison and others opted for parallel connections for incandescent lamps so that they could be switched on and off individually. For the parallel connection a low constant voltage rather than a constant current was needed. Edison's dynamo was much more efficient than its rivals, and eventually he settled on a 110 V supply. Competitors with more easily made, bulkier, lower-resistance filaments generally chose voltages of 40 to 70 V, though as skills increased and the designers realized that higher voltages (and lower currents) meant lower copper losses, 110 V became fairly standard in America.

After demonstrations of experimental power stations at Menlo Park (the site of Edison's laboratory), Pearl Street in New York City was selected for the first commercial station for incandescent lighting. Coverage of the Wall Street financial district also afforded a practical demonstration of an economical lighting system to the financial community. However, it was in London in the spring of 1882 that an Edison station was first operated commercially. Edison's new Jumbo dynamos could supply about 1200 lights each, about four times more than previous dynamos (Fig. 6.5). The Pearl Street station officially opened on 4 September 1882; Edison himself switched on the lights in the offices of his chief financial backer. The total cost of bringing the system to its commercial stage over the period 1878–1884 was nearly $500 000 (£100 000), far more than originally estimated.[9] Evidently underestimating the costs of large projects is nothing new to our own age.

Figure 6.5 Edison's Jumbo dynamo, 1881

The Edison Electric Light Company had been founded in October 1878 mainly to provide funds for the research work and to hold the resulting patents. For manufacturing, other companies were formed with Edison Electric Light as a 'parent.' Examples include the Edison Lamp Company (1880) for the lamps, the Edison Tube Company (1881) to manufacture the street mains, and the Edison Machine Works (1884) to make the dynamos. S. Bergmann and Company, founded by a former Edison employee who had gone into business for himself, manufactured the screw-type lamp bases (said to have been inspired by the screw top of a kerosene can) and the sockets, switches, meters, fuses, and other components. After the success of Pearl Street other local Edison companies were founded to provide lighting from central power stations. About a dozen existed by the end of 1883, and nearly sixty by 1886. The Edison Company for Isolated Lighting (1882) provided lighting where a central station was not needed. Over 330 000 Edison lamps had been sold by 1886, about three-quarters of all the lamps made in America at that time.

From about 1884 onwards Edison gradually withdrew from the commercial side to concentrate on further inventions. The many companies came under the control of lawyers and financiers and were consolidated into one concern with capital of $12 (£2.4) million. After the 1892 merger with Thomson-Houston the new company, General Electric, had capital of $35 million, sales of $12 million, and around 10 000 employees (Fig. 6.6). This giant dominated the American electrical industry; Charles Coffin of Thomson-Houston became its first president.

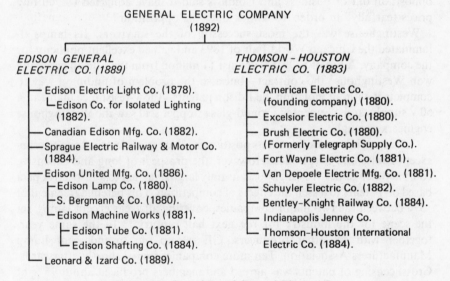

Figure 6.6 Genealogy of General Electric Co.

Meanwhile others in the USA were manufacturing and selling patented incandescent lamps. The United States Electric Lighting Co. in 1882 controlled the Weston, Farmer, and Maxim patents. Swan had licensed the Swan Lamp Manufacturing Co. to produce his system, Brush Electric had the American rights to the Lane-Fox patents, and Thomson-Houston had secured the Sawyer-Man patents. William Stanley had patented a carbonized silk filament and a self-regulating dynamo, which were acquired by George Westinghouse's Union Switch and Signal Company; its electrical department was so successful that it was incorporated as a separate company in 1886, Westinghouse Electric. It became GE's biggest rival.

With so many competing patents and a lucrative, growing market exploited by all, victory in the ensuing legal battles became vital. Most of the victories went to GE, particularly as the courts upheld Edison's basic lamp patent in the first major patent battle, which was initiated in 1885 and finally settled in 1892. Between 1885 and 1901 the Edison companies and GE spent an estimated $2 million (£400 000) defending their patents in court, about four times the amount originally spent to develop the system.[15]

Edison's basic patent was upheld six months after the formation of GE but only two years were left in which to exploit it. The Edison companies had spent a great deal of money to develop their lighting system only (as they saw it) to watch rivals imitate their product and steal some of the market. GE now felt it had a legal and moral right to all the profits as a return on its investment, and it could now offer protection to its licensees who were facing unlicensed competition. A largely successful policy was launched of applying for court injunctions against competing lamp manufacturers. The aim was to put the opposition out of business and Coffin is said to have admitted to "cutting prices fearfully" in order to "knock out" the opposition.[15]

Westinghouse was the most successful of the survivors. Its lamps illuminated the Chicago World Fair of 1893 and gained excellent publicity for the company. A bid of $399 000 against $1 million from the Edison company won Westinghouse the contract—but also the problem of finding a way to complete it without violating the Edison patents. A low-resistance filament, a 50 V supply, and a cemented ground-glass stopper seal saw the Westinghouse engineers through.

GE meanwhile consolidated its position in all areas of electrical engineering except cables and telephones. However, the prospect of long and expensive court action against Westinghouse, its only large rival, led to a restrictive pact based on patents to lessen the effects of competition. This patent agreement of 1896 between the two major companies, covering all areas except lamps, set the scene for the industry for the next half century.[15] In the same year, together with six other producers, GE set up the Incandescent Lamp Manufacturers Association. Ten more companies joined shortly afterwards. Crosslicensing of patents was agreed and members produced about 95 % of American lamps. They fixed prices and allocated markets. Though GE's basic patent had expired, the company still dominated this first American lamp cartel by virtue of size. In 1911 antitrust proceedings were brought against the leading firms but little change resulted. By then GE's share of the American trade was about 80 per cent.

Such cartels were popular in Germany even if frowned upon in America. The first lamp ring was formed in Germany in 1894, led by Siemens & Halske and AEG, the descendant of the German Edison Company. By 1903 this cartel had grown into an international agreement including the major manufacturers in Germany, Austria-Hungary, Holland, Switzerland, and Italy. In Britain, Ediswan enjoyed a virtual monopoly until 1893 when its patent expired. Then, faced with new competition from the Continent, the price for a lamp fell by 73 per cent, fair comment on the role a monopoly can play. In 1905 the British manufacturers also formed a cartel, the British Carbon Lamp Association.[15]

Metal Filament Lamps

The period from 1897 to 1912 was one of great change in the electric lighting business. Domestic leaders had emerged by 1897. The carbon lamp was firmly established and mass production had begun, but it was not yet supreme. Improved gas and arc lighting were still strong competitors; to defeat them decisively, radical improvements were needed, which meant a change of filament material.

Carbon melts at about 3800°C but an operating temperature above 1600°C led to blackening of the bulb. At 1600°C the output was about 3.4 lumens per watt and about 98 per cent of the electrical energy was wasted as heat. A material was required that could be operated successfully at a temperature well above 1600°C and new scientific knowledge of metals and rare earths had to be used in the search for it.

One of the first successes, and still used in special applications, came from H. W. Nernst of the University of Göttingen in Germany. He took refractory metal oxides, including those of magnesium, calcium, and rare earths; he mixed them into a paste, squirted them through a die, and dried them to make a rod. When cold the rods were electrical insulators but when heated, with an alcohol burner or match, they became conductors and a current produced a strong white light. In Europe the patents were taken up by AEG of Germany, and in America by Westinghouse.

Others pursued the quest for metal filaments, particularly of osmium, tantalum, and tungsten, but there were no known methods of pressing or drawing these metals into wires. In 1898 the same Austrian inventor who had staved off the decline of gas lighting, K. A. von Welsbach, helped the great rival by devising a method of making an osmium filament. He made a brittle wire by mixing powdered osmium into a paste with a cellulose binder. It was then squirted through a die, sintered to fuse the metal particles, and heated to vapourize out the binder. In this way was born the metal filament lamp, very fragile and so expensive that the metal was recovered from burned-out lamps. It was made in small quantities from 1902, the same year that Siemens & Halske developed a better lamp made from tantalum. The German firm managed to produce purer tantalum than had been previously available and that could be drawn into wires. Bright reports that Siemens then secured all world sources of this expensive metal and became the only manufacturer of tantalum filaments, though others were licensed to assemble the Siemens filaments into lamps.[2] In America, GE and National Electric (largely owned by GE) paid $250 000 (£50 000) plus royalties for the exclusive privilege to this successful lamp. It was sold in the USA from 1906 to 1913.

Tungsten is the metal used today. It had been known for a century or so and was easily available from about 1890, although filaments could not be made until 1904. Many companies and inventors were working on it when success came to Alexander Just and Franz Hanaman, of the Technische Hochschule in Vienna. They used an atmosphere of tungsten oxychloride and hydrogen to deposit almost pure tungsten onto a slender carbon filament. The carbon was dissolved away to leave a tiny tungsten tube. Another technique they used was akin to that developed by von Welsbach for osmium filaments. Others were also successful and all inventors sought patents.

At the Austrian Welsbach company it was discovered that the osmium lamp could be improved by the addition of tungsten to the osmium. The new lamp was given a name that has since become famous: Osram (1906). The first letters came from osmium, the last three from Wolfram (German for tungsten). It was quickly found that the less osmium the better the filament, and the Osram lamp was soon made from tungsten only. The early tungsten lamps represented a big improvement over carbon lamps. The efficiency was at least doubled and, unlike with carbon, was largely maintained throughout their life.

Britain and America, the homes of the carbon lamp, were now dependent on licensing from Germany and Austria. In Britain GEC bought rights to the Osram lamp and, with a German company, set up the jointly owned Osram Lamp Works. In America, Westinghouse purchased the Austrian Welsbach Co. and GE obtained U.S. patent rights from the German Welsbach Co. Half a million or more tungsten lamps were sold in the first year on the American

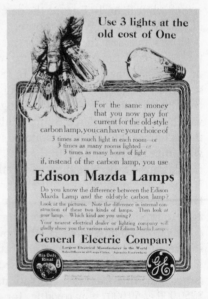

Figure 6.7 Advertisements for Mazda lamps in U.S.A., 1909–1911

market. In 1909 GE introduced a new trademark, Mazda, taken from the Persian god of light, as a symbol of lamps that comprised the latest technical advances from both home and abroad (Fig. 6.7). The Mazda trademark was introduced to Britain by British Thomson-Houston (BTH), then controlled by GE. Later it passed to AEI and on to Thorn-EMI.

The early tungsten lamps were successful even though the metal was brittle and could not be drawn into a strong wire. The next stage marked the start of the manufacture of our present-day lamp, based on drawn tungsten wire. Its development also illustrates the shift of emphasis from Europe to America, and the further movement of research into large laboratories where teamwork was encouraged. Both moves are illustrative of the change from 19th to 20th century styles of research. The individual inventor working with his friends and scientific advisers, such as Swan, Cooke, and Morse, had largely given way to trained teams of scientist-engineers.

In the period 1908–1910 a team led by W. D. Coolidge at GE's research laboratory at Schenectady, N. Y., developed a method for making ductile tungsten and drawing it into a wire by repeated heating and hot swaging of the metal, followed by a hot drawing process. In 1910 about a third of the laboratory's funds were spent on tungsten research. A patent was granted in 1913. The new lamp was introduced to America in 1911 as the Mazda B and was sold in several sizes: 25, 40, 60, 100 and 150 W. It reached Europe via GE patent agreements with AEG and BTH.

It was the GE laboratory that also produced the next improvement. In 1913 Irving Langmuir proved that the blackening of the glass bulb was caused solely by evaporation of the tungsten. A reduction was achieved when the lamp was manufactured with an atmosphere of nitrogen at almost atmospheric pressure, rather than the low pressure used by some of the pioneer inventors. Argon was used when it became available commercially about 1918. It was also Langmuir who discovered that coiling the filament reduced the energy losses caused by convection and conduction. The coiled-coil filament came much later, in 1934, and further increased the efficiency by up to 20 per cent. Longer filaments could also be used; a modern 100 W bulb has a coiled-coil filament which if unwound would be over 1 m long.

Several minor improvements were also made. Lamp bases were standardized in America to the Edison screw around 1900, about the same time that production was mechanized. The bayonet cap used in Britain traces back to the Edison and Swan Co. in 1886. Early lamps had a glass tip seal at the round end of the bulb. The seal could be eliminated beginning about 1900, but the extra step was expensive and tipless bulbs only became standard after World War I. Another improvement was the introduction of inside frosting, brought about by a two-stage acid wash. Satisfactory frosted or 'pearl' lamps date from 1925.

The carbon lamp manufacturers had formed cartels for mutual benefit, and did so again for tungsten. Loose national cartels were formed and an informal international ring existed before World War I. GE held a strong position through its holdings in overseas companies such as BTH, AEG, and Tokyo Electric. Both the Japanese and the Dutch (Philips) industries grew stronger during the war since the major suppliers were among the belligerents and were too involved with the war to meet overseas markets. After the war a new company was set up to extend GE's influence, International General Electric. In 1924 a gigantic world cartel, as it was later described in an antitrust suit, was set up with GE as 'the hub.'[15] It was called Phoebus, a sort of lamp men's Organization of Petroleum Exporting Countries (OPEC). GE was the only major lamp producer not to sign the Phoebus agreement. Instead, it exercised control of Phoebus through its shareholdings in other companies. Between 80 and 90 per cent of European production was controlled by the cartel when it came to grief, not through the action of lawyers but by the outbreak of World War II. The exact effect of cartels on the price paid by the consumer for his light bulb is probably impossible to evaluate but at least one pair of authors maintain that "it was without doubt the policy of the members of Phoebus to maintain relatively high prices."[15] In Sweden in 1928 a new company was formed in protest against the high prices charged by Phoebus members. Though members dropped their price by 27 per cent the new company was still able to undersell them and make a profit.[15]

Discharge, Fluorescent, and Other Lamps

The coiled-coil gas-filled incandescent lamp of 1934 has continued as the basic general-purpose light bulb to the present, although it is not the only source of electric light. Discharge and fluorescent lamps are widely used in outdoor and indoor lighting, respectively. Discharge lighting had some success in the early 20th century but the major advances did not come until the 1930s.

Discharge lighting goes back to 18th century experiments with frictional electricity and poorly evacuated glass globes. Suggestions were made for applications as early as the 1860s and patents were granted but the first commercial discharge tube did not appear until 1895. That was the Moore tube, made by an ex-GE employee who left to develop his own ideas. In 1912 his companies were absorbed by GE and the ex-employee was once again an employee. D. M. Moore used discharges in nitrogen, carbon dioxide, and air. His tubes enjoyed some commercial success despite their extreme length and kilovolt supply. Peter Cooper-Hewitt, another individual American inventor, used a mercury vapour discharge, a derivative of the mercury arc lamp. Others also turned to mercury, but the success of the tungsten lamp sealed the commercial fate of these early discharge lamps.

From France came a discharge tube that was commercially successful: the neon tube. Georges Claude perfected a method of liquifying air and separating its components. Neon soon became available at a price suitable for commercial use. Claude also designed and patented a better electrode. Basically the area was made large so that the current density was small, which minimized a previously severe problem, sputtering—the attrition of the electrode material by the constant bombardment of gas ions. Claude found that he had a red light with an efficiency of 10 to 15 lumens per watt and that other colours could be obtained by use of other chemicals. The first neon sign was demonstrated in 1910 at the Paris Motor Show and two years later Cinzano vermouth was the subject of the first neon advertising sign. GE made one of their relatively few mistakes when they declined an offer of an exclusive license for American neon lighting. Instead Claude Neon Lights, Inc., developed the American market itself.

All discharge lamps operate by ionization of a gas and use of the resultant electrons and ions to excite atoms of the gas to a higher than normal energy. The excited atoms quickly lose the extra energy and radiate it away as light. The hot-cathode mercury or sodium discharge lamps developed in the 1930s required much lower electrode voltage drops to achieve this action than the earlier cold-cathode lamps. European countries again took the lead through Philips of Holland, Osram of Germany, and GEC of Britain. High-pressure mercury vapour and sodium lamps were both highly efficient. For example, the sodium lamp gave about 56 lumens per watt in 1932, which has since been pushed beyond 80. Both types of lamp are extensively used for outdoor lighting. The sodium lamp, with the yellow-orange monochromatic light to which the eye is especially sensitive, is widely used for road lighting.

America came back to the fore with the low-voltage fluorescent tube in the late 1930s, although high-voltage cold-cathode tubes had been made in Europe a little earlier. The phenomenon of fluorescence has been known for a few hundred years and was studied scientifically (and named) in 1852 by Sir George G. Stokes. A. E. Becquerel attempted to make a fluorescent lamp from a Geissler tube around 1859 and other attempts followed but were dogged by various problems. In the 20th century the main limitation in the development of the lamp was the need for a gas discharge device to operate at the low voltage of the domestic supply. This problem was solved in Germany and America in the late 1920s.[2]

By the end of the 1920s the technical requirements for the eventual lamp were known and the time was ripe for its development, just as it had been some 50 years earlier for the incandescent lamp. Although there was a technological push, there was as yet no market pull. To most manufacturers the state of artificial lighting was quite satisfactory without a low-voltage fluorescent lamp. Eventually, with the aid of Westinghouse, GE developed the lamp at a cost of over $170 000 and announced it in April 1938. The principle of operation is to start an electric discharge in a tube containing argon and a small quantity of mercury. The mercury is vapourized and visible and ultraviolet light is emitted; the ultraviolet is converted to visible by the fluorescence of the internal phosphor coating of the tube.

Another American company, Sylvania, dissatisfied with its allocated share of the incandescent-lamp market, broke free of GE's licenses by developing its own fluorescent tube, which was announced late in 1938. After World War II an agreement was reached concerning the manufacture of fluorescent tubes in Britain between Sylvania and a small British company, Atlas, founded in 1926 by Jules Thorn. The Thorn company grew in strength and has now absorbed four of the seven major lamp manufacturers of prewar Britain: Mazda (BTH), Metropolitan Vickers, Ediswan, and Siemens. (The four had previously merged to form AEI Lighting.) Thorn Lighting has become the dominant force in the British lighting industry; which includes GEC, Philips, and Crompton (Fig. 6.8).

High-voltage fluorescent tubes found some use in Europe for large-area indoor lighting even in the early 1930s, before the low-voltage tube was introduced. But it is the low-voltage tube, with its high efficiency and diffused lighting, that has now vanquished the incandescent lamp in the battle for the huge indoor lighting market. The almost total domination of fluorescent tubes in offices, shops, and factories has meant that fluorescent lighting now gives us an estimated 80 per cent of our total artificial light.[18] Various colour renderings such as 'daylight' and 'cold-white' can be produced by variations in the phosphor. In 1979 Philips announced a light-bulb shaped fluorescent lamp as a direct replacement for the incandescent lamp. Its high efficiency is its major selling point in these energy-conscious days, but whether it will eclipse the tungsten lamp remains to be seen.

Newer developments in lighting include the tungsten–halogen incandescent lamp. Introduced about 1960, it finally solved that old problem of the blackening of the bulb. A small amount of a halogen in the atmosphere, usually bromine or iodine, causes a regenerative process by which the evaporated tungsten returns to the filament. The operating temperature is high, about $250°C$ for the wall of the bulb, so that an inner container of quartz must be used instead of glass. A longer life, and up to 20 lumens per watt, are achieved. Applications include floodlighting, car headlamps, and projector bulbs. Halogens have also been added to high-pressure discharge lamps to produce the mercury–halogen lamp used for street and floodlighting. Table 6.2 summarizes lamp performance.

Though they are not sources for general illumination, the laser and the light-emitting diode (LED) might be mentioned. The laser grew out of work on the maser, a microwave oscillator, for which two groups of workers shared an award in 1959. One group was led by C. H. Townes of Bell Laboratories and used ruby; the other was led by Nicolaas Bloembergen and used potassium cobalt–chromium cyanide. At the time of the award Townes presented the ruby to his wife as a piece of jewellery. The story goes that on the way home Mrs. Bloembergen asked her husband why he had not done something similar. "But my maser was made of cyanide, dear!" came the reply.[16]

The conditions for laser action were described in 1958 by A. L. Schawlow and Townes working for the Bell Laboratories, and in 1960 T. H. Maiman at Hughes Aircraft won the worldwide race to produce the first laser, again using ruby. It was followed by a helium–neon gas laser the same year and by a

(a)

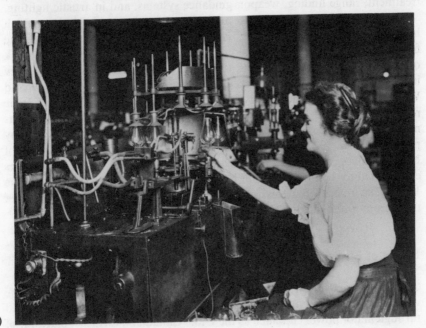

(b)

Figure 6.8 Incandescent lamp production: (a) this modern production line produces 5000 lamps per hour; (b) an automatic exhaust machine for lamp bulbs, 1896

Table 6.2 Lamp Performance

Incandescent		
1881	Bamboo	1.68 lumens/W
1884	Squirted cellulose	3.4
1898	Osmium	5.5
1902	Tantalum	5.0
1904	Nonductile tungsten	7.85
1909	Ductile tungsten	10.0
1917	Gas filled, coiled	10.0–12.5
1936	Gas filled, coiled coil	12.5–16.0
1959	Tungsten–halogen	up to 20

Present-day lamps (typical figures)	
Tungsten incandescent	13 lumens/W
Tungsten–halide incandescent	17
Fluorescent tube	40–90
High-pressure mercury discharge	40–60
High-pressure sodium discharge	90–160

semiconductor laser in 1962. Lasers were slow to find applications but have now been used in diverse ways in communications, surgery, welding, heat treatment, range finding, weapon guidance systems, and in artistic lighting.

The LED, best known for its use in displays, has a fascinating history which has been recounted by Loebner.[17] H. J. Round in England was probably the first to report the basic phenomenon in 1907, when assisting Marconi in work on point-contact crystal detectors. It was rediscovered in 1922 by O. V. Losev in the Soviet Union. Losev lost his life in the siege of Leningrad during World War II after he had ignored advice to flee because he wanted to finish some research. Like Losev, the research was also probably a casualty, since it was never published. Losev obtained four patents on LEDs and published many papers between 1927 and 1942. LEDs re-emerged in 1951 with the work of Kurt Lehovec and his co-workers in the USA and many companies contributed to their development before they reached the marketplace in the 1960s. The arrival of successful liquid-crystal displays, however, prevented them from monopolizing the old and new applications for displays.

Impact of Electric Lighting

One way of viewing the social impact of a product is to try to imagine life without it. Such a mental exercise might be easier for those who have lived through major power cuts such as the U.S.–Canadian Northeast blackout of November 1965, when some 30 million people over an area of 80 000 square miles (207 000 sq km) of Ontario, New York State, and New England were affected. Life without electric lighting would probably be more hazardous and certainly duller than it is today, though who is to say what developments might have taken place with oil and gas lighting?

Electric lighting has exerted strong influence in many areas from building design to signalling, from theatre safety and comfort to crime levels, from low-fire-risk illumination in libraries and factories to medical aids for surgery and dentistry. Through cartels and the manipulation of market forces it influenced international law, trade, and the formation of multinational corporations. As a product that people will buy even in times of economic recession, it has been a reliable and plentiful source of cash to the electrical industry. And to electrical engineers it has provided work in lighting itself and has encouraged the development of other aspects of the profession from power stations to domestic product design.

When the light bulb marched boldly into the home other products, in smaller numbers, crept in slowly behind to help change our domestic way of life. Electricity could be used for heating and motive power as well as to provide light. An electric iron was patented in America in 1882 but the electric fan was probably the first domestic product to be marketed after the light bulb, in 1883. The 1890s saw the earliest models of some electrical goods we now take for granted: the cooker, torch or flashlight, electric kettle, washing machine, and the toaster. The 1900s saw some new outdoor uses for lighting in buses, cars, and even on Christmas trees. Lighting even came to have robot-like control over our actions when traffic lights were introduced in Cleveland, Ohio, in 1914.

One of the most profound indirect effects of electric lighting is not so easily seen, however. It was during studies of problems in early carbon lamps that an effect was noticed that was later exploited in the vacuum diode and triode. The beginnings of vacuum electronics, and the consequent rise of electronics to its powerful role in today's society, are in a way spinoffs from the incandescent light bulb.

Probably the most significant impact of electric lighting is the way it has wrecked the Sun's former influence on man's activities. Night-time working in factories, floodlit football, adult education, and late-night shopping are just simple examples. One wonders how much Edison and Swan foresaw the influence the light bulb would have on people's bedtime. As a student once put it: The fantastic life now begins after sunset.

References and Notes

1. W. T. O'Dea, *Lighting 2*, Science Museum Booklet, H. M. Stationery Office, London, 1967.
2. A. A. Bright Jr., *The Electric Lamp Industry*, Macmillan, New York, 1949. (This careful study has been a major source for this chapter.)
3. *Illustrated London News*, p. 368, 9 December 1848.
4. C. M. Jarvis, *JIEE* 1:145, 280, 566, 1955.

5. B. Lightoller, *Electronics and Power* 15 (No. 1): 3, January 1969.
6. P. Dunsheath, Chapter 5, Ref. 28.
7. G. Siemens, *History of the House of Siemens*, (2 vol.), Karl Alber, Freiburg/Munich, 1957.
8. Refs. 4 and 6.
9. H. C. Passer, *The Electrical Manufacturers, 1875–1900*, Harvard University Press, 1953. (This has been an important source for this chapter.)
10. B. Bowers, *R. E. B. Crompton: Pioneer Electrical Engineer*, Science Museum Booklet, H. M. Stationery Office, London, 1969.
11. Ref. 2, p. 35.
12. R. C. Chirnside, *Electronics and Power* 25 (No. 2): 96, February 1979.
13. P. F. Drucker, *Technology, Management and Society*, Heinemann, London, 1970; Pan, London, 1972.
14. P. Robertson, Chapter 5, Ref. 7.
15. R. Jones and O. Marriott, *Anatomy of a Merger*, Jonathan Cape, London, 1970; Pan, London, 1972.
16. A. L. Schawlow, *IEEE Trans.* ED-23: 773, 1976.
17. E. E. Loebner, *ibid.*, p. 675.
18. C. Phillips, Chapter 1, Ref. 8.

7 ELECTRICAL POWER

No branch of the history of electrical engineering has been studied and written about more widely than electrical power. Several good sources discuss the subject in more detail than can be achieved here.[1-4] Only an outline sketch is attempted in this chapter, with details of just some of the origins.

The outstanding significance of electricity when compared with other power sources is its mobility and flexibility, something to which we are so accustomed that we are apt to forget it. Electrical energy can be moved to any point along a couple of wires and (depending on the user's requirements) converted to light, heat, motion, or other forms. In the previous chapter we related its use for producing light; in this chapter we shall see how its application for the production of motion and heat was developed and how electricity came to be generated and distributed.

Although lighting produced the first demand for the large-scale generation of electrical power, that use was fairly soon matched and surpassed by other applications: electric traction for tramways and railways; electrochemistry for the production of materials, of which aluminium is the obvious example; the use of electric furnaces for electrometallurgy; and—perhaps most important— the provision of mechanical motion at whatever place it is desired, whether on a grand scale to drive a locomotive from one end of the country to the other, or on a small scale to drive a drill or hairdryer. In all cases the basic problems are the same: first, to generate the electrical energy from some other energy source; second, to transmit it to the place of use; and third, to convert it to the form in which it is to be used. The last twenty years of the 19th century brought solutions to these problems that enabled electricity to replace the steam engine as industry's major source of motion, and to oust hydraulics and pneumatics in the bid to provide power distribution systems (Fig. 7.1).

The development of electric current generators from the 1830s to Z. T. Gramme's dynamo of 1870 was reviewed in the previous chapter. The Gramme dynamo and its successors provided a ready source of current in quantity and at a price that no chemical battery could hope to match. They made it possible for large-scale electrical engineering to begin. The first major

"WHAT WILL HE GROW TO?"

*Figure 7.1 King Coal and King Steam ponder the threat of the infant Electricity. 'Punch'
cartoon, 1881*

applications were for arc and incandescent lighting, but they were eventually
surpassed by the use of motors. At first only DC motors (usually generators
operated in reverse) could be used; DC generators and distribution systems,
with back-up batteries, received a boost despite problems associated with what
Lord Kelvin once called the "great evil," the commutator; "a frightful thing."[5]
With the advent of various AC motors, from about 1888, AC could fight back
on equal terms and go on to win what had become known as the Battle of the
Systems, DC versus AC, traces of which could still be found three-quarters of a
century later.

With DC dynamos and AC alternators developed into efficient large-scale
generators of electricity, the concept of a central power station distributing
power to nearby (and even distant) consumers became established. Locating
the generators close to their energy source brought economic savings despite

losses incurred when electricity was transmitted over a distance. The losses of high-voltage AC were lower than those of DC, but both systems existed for many decades. In Britain, S. Z. de Ferranti is remembered as the chief exponent of AC; in America it was George Westinghouse. Ferranti built Britain's first large AC station in 1887–1889 in a bid to supply a major section of London, at a transmission voltage of 10 kV. In 1891 Brown-Boveri, the Swiss firm, built a system in Germany to supply 225 kW over 179 km at 30 kV.[6] In America the famous Niagara Falls station came into operation in the mid 1890s; it operated at 2.2 kV. Less than twenty years later 100 kV systems were operating.

DC Dynamo Design

The Gramme ring armature of 1870 and its successors were the turning point in dynamo design that made the large-scale generation of electric currents a

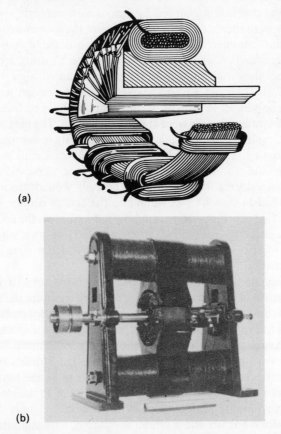

(a)

(b)

Figure 7.2 Gramme ring armature and dynamo (1870): (a) armature; (b) dynamo (source of (a): Ref. 11)

practical reality (Fig. 7.2). For the first ten years practice led theory and design was based largely on trial and error; but by 1890, following the examination of the underlying scientific principles, dynamos were relatively well designed. This progress, together with the discovery that a dynamo could be used in reverse as an electric motor (as well as the improvements made in the design of AC alternators), meant that the technology necessary to enable electrical engineering to grow into a great industry became available in those twenty dynamic years.

Gramme's ring armature was a re-invention of, and improvement upon, Antonio Pacinotti's armature of ten years earlier. The armature consisted of a coil of soft iron wire insulated in some way, such as by bitumen or varnish, around which were positioned insulated copper wire coils connected together as a continuous winding. The junctions were tapped off to a copper commutator. An electromagnet produced the transverse magnetic field in which the armature rotated. The result was a dynamo that produced a satisfactory continuous current and could be operated for long periods without the usual overheating caused by stray (eddy) currents, or Foucault currents as they were then called. Three years later the Siemens company hit back against this commercially successful French-based dynamo with a machine that depended on what was known as Alteneck's drum armature, an improvement on their old shuttle armature and named after their chief designer (Fig. 7.3). The main advantage over Gramme's design was that the coils lay in a plane parallel to the axis of the drum on whose surface they were wound, whereas Gramme's coils were at right angles to it. The part of each coil that lay inside Gramme's ring was inactive: it served only to complete the coil but also increased its resistance, whereas all of Alteneck's coils were on the outside surface and so were fully utilized for the generation process. Originally wooden drums were used with the coils wound around surface pegs. Later insulated iron wire was wound on the drum's surface, or an iron drum was used. Eddy currents were eventually reduced when the solid body of the drum was replaced by one built up from insulated iron wire, which in turn gave way to the mechanically stronger laminated construction (thin iron sheets with paper insulation between).

Both the Gramme and Alteneck designs were manufactured in various places, with or without modifications. As lighting systems were the first major users, dynamos came to be associated with such names as Brush, Edison, and Thomson-Houston in the United States; Emil Bürgin in Switzerland, Jonas Wenström in Sweden; and Crompton and others in Britain. Some firms developed their own particular flair for dynamo design. Brush's arc lighting dynamo was particularly successful and was judged to be the best available when a trial was carried out by the Franklin Institute of Philadelphia late in 1877. Despite invitations sent to various manufacturers, only two other machines, a Wallace-Farmer and a Gramme dynamo, were submitted for the trial. As there was no precedent for the testing procedures, entirely new tests had to be devised.

Figure 7.3 Alteneck's drum armature dynamo, 1872

Thus, by the end of the first of these two dynamic decades, considerable progress had been made. Nevertheless, according to Fleming, "before 1880 or 1882 dynamo construction could hardly be said to have been a scientific art;" it was mostly "an affair of clever guessing in the light of past failures."[7] Even so this clever guessing had produced a variety of useful machines of two basic types: one that generated a constant voltage for incandescent lamps connected in parallel and the other, a constant current to supply arc lamps in series. Either could be driven by a steam engine, either directly or via pulleys or belts. A major problem (according to Fleming) was the low efficiency of the early dynamos, which rarely exceeded 50 or 60 %. The blame was laid squarely at a door labelled 'bad design.'

Most of the problems centred on such factors as the energy lost by the generation of eddy currents in the armature's ironwork, the generation and control of the magnetic field, and frictional losses. However, after 1880 dynamos were designed more scientifically. A major step forward came with the study of the magnetic circuit. If a comparison of the various dynamos is anything to go by, the practical designer was obviously in need of a tutorial on the subject. Long slender field magnets were used by Edison, short stout ones by Siemens and Crompton.[1] In 1885–86 Gisbert Kapp, one of the men who were paving the way, wrote, "Nothing has yet been published on the construction of dynamos, which would be of practical value to the manufacturer. Of theories there are more than enough, but the connecting links between pure science and practical work are still missing."[8,9]

Many designers helped remedy the situation, including Kapp himself, an Austrian-born engineer who later became the first professor of electrical engineering at the University of Birmingham; John Hopkinson, who is particularly remembered for the application of his own theories to the redesign of Edison's dynamo, which he turned into the far more efficient Edison-Hopkinson dynamo; and S. P. Thompson, a famous British engineer who later wrote leading textbooks on the design of generators. All three served as presidents of the Institution of Electrical Engineers in London. Hopkinson became professor of electrical engineering at King's College in London; Thompson, at Finsbury College. Hopkinson, an outstanding mathematician-engineer, and Kapp share priority for the conception and exploration of the vital concept of the magnetic circuit. Hopkinson's paper, which he shared with his younger brother Edward, was published in 1886, just a few months after Kapp's, although he had been at work on the idea for a couple of years.[9] From about 1879 Hopkinson had been highlighting the importance of the graphical presentation and study of dynamo characteristics, especially the curve relating the magnetizing force to the magnetic flux produced. Thanks to the work of these men and others the theory began to lead the design, rather than the other way round.

Two of the little features of electrical engineering still loved by today's students date back to the same period. One is the useful mathematical product we know as ampere-turns (then weber-turns, as current was measured in webers).[1] This idea came from one of Edison's right-hand men, Francis Upton, the man to whom Edison once had to explain basic electrical laws.[2] The second is Fleming's right-hand rule for representing the spatial relationships among current, magnetic field, and motion in a dynamo by the thumb and first two fingers of the right hand.[7] Fleming conceived this famous rule in 1885.

By 1890 efficiencies had been raised from 50–60% levels to 95% and dynamos could be run on load for long periods without overheating or breaking down. The temporary stage of muddling through, the "affair of clever guessing," became a thing of the past.

DC Motors

Although the basic principle of the DC motor had been demonstrated as early as 1821 by Faraday, and many fascinating experimental motors were constructed in the decades that followed, the basic application was quite uneconomic as long as batteries were the major source of current. But once practical and economic dynamos were developed, the electric motor could be seriously considered as a source of motion even though the total efficiency of the dynamo-conductor-motor system was not very high at first.

Some of the later uses for electric motors were presaged in the days of battery power. Thomas Davenport of Vermont constructed a large electric motor in 1837 and used it to drill holes in iron and steel. Two years later a Russian reported fairly successful trials of a boat driven by an electric motor. In Scotland, about 4 mph (6 km/hr) was achieved with an electrically driven railway car, but this record was bettered in America in 1854 when the heady speed of 19 mph (30 km/hr) was achieved. Well before Gramme's dynamo became a reality some basic industrial and traction applications of electric motors had been thus demonstrated as possibilities except—as for early arc lighting—for the cost of the chemical batteries.[1,2]

The traditional birth of the 'real thing' occurred at the Vienna electrical exhibition of 1873, when Gramme and his associate Hippolyte Fontaine demonstrated a Gramme dynamo driving a second Gramme dynamo operated in reverse as a DC motor. That demonstration led fairly quickly to the general recognition that a dynamo could be operated in reverse as a motor. (Some individuals, Lenz and Pacinotti for example, doubtless had known that before.[1])

Passer has pointed to three fundamental facts about the electric motor that were widely known among electricians by 1880.[2] The first has just been mentioned, that the motor was essentially a dynamo operating in reverse, which meant that any significant improvements to dynamo design would also be felt in motor design. The second was that the motor was the key to the transmission of power by electricity; and the third, that it would operate from the same current supply as was used for lighting. With such general realizations the way was open for the motor to spread, at first on the coattails of electric lights.

Taking a cue from the three established types of dynamo construction, designers developed three basic types of DC motor according to the way the field coils were wound and excited: series, shunt, and compound wound. Each had its own advantages and disadvantages. Demand for motors grew as the central power station became more widely established.

The potential markets lay in public transport and in industry. Up to the 1880s public transport was met by steam railways, by a few elevated local railways, and by street tramcars pulled by animals or, in a few cases, by a continuous cable loop driven by a remote steam engine. The electrically driven street tramcar, once the particular electrical engineering problems associated with it were solved, was to become an important area of electrical engineering until midway through the twentieth century. In industry, electric power could be applied to drive heavy or light machinery such as hoists, lathes, drills and so on. Although such applications were found quite early they were not too common. The very early motors were not particularly reliable or efficient, although for some applications their convenience outweighed such factors. Even by 1890 the improved motors, though more common than in the previous decade, were far from universal in either Britain or America. In many respects Britain was especially slow to adopt electrical power in industry. Sales of electricity for power in Britain did not catch up with that for lighting until about 1908, but overtook that for traction a couple of years earlier. From then on power became the major user; by 1912, for example, it took about three times as much electricity as traction.[3, 10]

One interesting historical snippet from this early period tells the tale of what was probably the first electric motor to be a commercial success and sell in large numbers; tens of thousands were sold. It was designed and built by Edison (who else?) and measured about 1 × 1.5 in. It was called the electric pen. The tiny motor ran at 4000 rpm and vibrated a pin that punched holes in waxed paper at a rate of 8000 per minute. The waxed paper could then be used as a stencil for printing onto ordinary paper. Edison's electric pen dates from 1876 and so is unlikely to have any competitors to the claim to be the world's first best-selling electric motor.

AC Generators and Motors

Alternating current generators, or alternators, had been around in one form or another even longer than the DC dynamo, and both types of machine were used almost indiscriminately to provide the supply for the various applications of arc lighting. One form of arc lighting, the Jablochkoff 'candle,' specifically used AC to achieve even burning of the carbons. As incandescent lighting spread both DC and AC generators supplied the new light, but for a time DC was the preferred system.

DC seemed to offer several advantages. Batteries could be used as an emergency back-up in the event of dynamo failures or to supply the current during periods of low demand, and even to top up the supply at peak demand. Dynamos were thought to be cheaper to operate than alternators as they could be directly driven from high-speed engines, whereas alternators were belt or rope driven, which meant some loss of efficiency for the total system. And alternators could not at first be run in parallel—a separate machine had to be used to supply each part of the system. When the load fell each alternator had to be kept running, whereas with dynamos operated in parallel, individual generators could be taken out of service and costs reduced.[1]

Even the recognized advantage of saving copper costs by transmission at a high voltage and use at a low voltage seemed to favour DC at first. Before the problems of AC transformers were sorted out a cumbersome form of DC 'transformer' was available if it was really required. The consumer could have batteries that were charged in series by the high voltage and discharged individually at the low voltage. Later the motor-dynamo combination offered a neater DC transformer: a high-voltage motor driving, on the same shaft, a low-output-voltage dynamo. London had at least two high-voltage DC supplies in the 1890s.[1]

Despite such relatively minor disadvantages AC generation was a going concern and alternator design progressed. Well-known men such as Wilde, Gramme, and Kelvin contributed improvements and a small number of giant alternators were built by J. E. H. Gordon in London to supply a maximum output of 115 kW at 105 V. On the continent, Ganz & Company of Budapest made some large alternators to a design by Charles Zipernowski; and in London, a young man with a rather long name, Sebastian Ziani de Ferranti, formed a company before his nineteenth birthday. Ferranti was manufacturing alternators of his own design but paying royalty to Kelvin (who was still Thomson in the early 1880s) because of patent problems; the end product was known as the Ferranti-Thomson alternator.

The most important of the early problems faced by AC concerned the question of running two alternators in parallel, in synchronism. Even in the late 1860s Henry Wilde had shown that such operation was possible.[8] This realization was ahead of its time; general acceptance came only after John Hopkinson attacked the problem mathematically in 1883–84 and concluded that parallel operation was a practical proposition. He then demonstrated the fact during experiments at the South Foreland lighthouse. Even into the 1890s the practicality of using alternators in parallel for a public supply was still debated. Synchronization could be achieved more easily at a low frequency than a high one, but at that stage there was no thought of standardizing the frequency, probably quite rightly at such an early and almost experimental stage. Even when synchronization was obtained, it was not achieved immediately the machines were connected. "They did not work together until they had jumped for three or four minutes," according to Gordon, "which might take a month's life out of 20 000 lamps."[11] Such a serious consequence helped lead Gordon to the decision to build the giant alternators already mentioned.[1] Nor were these the only difficulties. Two synchronous parallel-operated machines could have large circulating currents that caused problems in the system, usually because of differences between different machines. The typical 'sine' wave output of individual machines could be almost anything from a poor triangular to a poor square wave.[1] Some engineers took the view that a distinct advantage for the whole generation-distribution system might result if alternators were not worked in parallel. Keep them separate and let each supply its own sector of the community. In that way problems would be contained within one sector and would not spread through the whole system. Automatic switches and protection devices did not then exist.

In any AC system today we would expect to find transformers at work. The basic principle dates back to Faraday's ring of 1831; in the decades that followed many induction coils of varying design were constructed. H. D. Rühmkorff, whose name is most closely associated with the induction coil, made his first one early in the 1850s.[8] Others followed and the basic idea was considered in relation to AC lighting systems by Jablochkoff. Patents concerning the application of transformers in distribution systems were taken out by Edison, by J. B. Fuller of Brooklyn, and by Marcel Depréz and

Carpentier, in the period 1878 to 1881.[8] However, credit is usually given to a Frenchman, Lucian Gaulard, and an Englishman, J. D. Gibbs, for launching the transformer on its route to success, even though they themselves really missed its most important application.

Induction coils and transformers were traditionally connected for series operation. Working together, Gaulard and Gibbs followed this trend in 1882–83 when they obtained a British patent for a transformer to act as a current regulator for arc lamp circuits. Their series use (Fig. 7.4) of transformers for high-voltage AC distribution worked up to a point, but proved to be impractical as time went on. Nevertheless, they served to draw widespread attention to high-voltage AC distribution. They achieved considerable publicity. George Westinghouse, the principal American advocate of AC, purchased Gaulard and Gibbs transformers for examination. Some idea of the

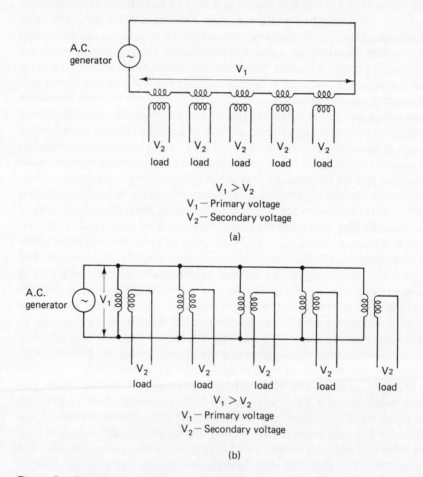

Figure 7.4 Transformer systems: (a) series connection, Gaulard and Gibbs; (b) parallel connection (after Ref. 2)

general state of knowledge of AC, at least in the USA, may be gained from a comment by C. F. Scott, later professor of electrical engineering at Yale University. He had seen his first alternator and the Gaulard and Gibbs transformer in 1887 "and learned that there was such a thing as an alternating current." As he had graduated from college just two years earlier he wondered why he had learned nothing of AC, and decided that it was because the transformer was imported after he had graduated. The earlier AC work, such as Jablochkoff's 'candle' arc lamp, had apparently made little impression on Scott's tutors.[16]

Within two or three years of the Gaulard and Gibbs series operation of their transformer the all-important parallel operation (Fig. 7.4) was tried out by several people. First in line were probably an Hungarian group at Ganz & Co. of Budapest, consisting of Zipernowski, Max Deri, and Otto Blathy. They built an alternator to supply 1000 V and transformers to drop this voltage to 100 V for incandescent lighting. Once the proper operation of the transformer was discovered, high-voltage AC transmission for low-voltage use became a reality. In England, Ferranti turned to the design and use of parallel-operated transformers and redesigned the electrical supply system at the Grosvenor Gallery in London, and in so doing established his growing reputation as a pioneer electrical engineer. In America, William Stanley of incandescent-lamp fame, now working for Westinghouse's Union Switch and Signal Co., had been given the task of developing an AC system even before the arrival of the Gaulard and Gibbs transformer in America.[2] When it did arrive it was judged to be impractical and a new, practical, parallel operating design was evolved in time for the founding of Westinghouse Electric in 1886.

Credit for the idea of the parallel-operated transformer for changing AC voltages cannot really be awarded to a single individual, but the workers at Ganz & Co., led by Zipernowski, and those at Westinghouse, led by Stanley, should share most of the credit. Certainly the modern transformer was a reality by the middle of the 1880s.

As the short lighting-only era of electrical engineering drew to a close AC found itself faced with its biggest disadvantage yet when compared to DC— the lack of a good motor.

When Hopkinson investigated the idea of paralleling two alternators in 1883, he also discovered that one could drive the other as a synchronous motor, just as a dynamo could be driven as a motor. Yet things were not as simple as they might seem. By 1887 there was still no successful AC motor. Two approaches had been tried. The first was the synchronous motor just mentioned, which had the drawback that the motor would run down and stop if synchronization was lost; moreover, it had no starting torque. It only became practical after methods were discovered to give it a starting torque by the adoption of techniques developed for the new induction motor. The second approach was to use a DC series motor and supply it with AC. That was all right in theory, but in practice excessive sparking resulted between the commutator and the brushes and rendered the idea more or less unworkable at that time. Later, with laminated field magnets to reduce eddy currents, it became successful and is known as the commutator motor. One variation, the universal motor, is designed to work off AC or DC and can be used for small (fractional-horsepower) applications.

The breakthrough came with the invention of the induction motor and with the application of polyphase AC. Major credit for these developments is due to Galileo Ferraris in Italy, Nikola Tesla in the USA, and Michael von Dolivo-Dobrowolski in Germany. Their work led to the large variety of AC motors now available, motors which may be roughly classed as single-phase or polyphase motors (usually 3-phase), and induction or synchronous motors.

The basic principle of the induction motor dates right back to the 1820s. D. F. J. Arago, the French scientist, found that when a copper disc was rotated at speed in a horizontal plane then a compass needle magnet, or bar magnet, placed above it would also start to rotate. Faraday's discoveries led to the understanding that eddy currents were induced in the disc, which then reacted to the relative motion of the magnetic poles in such a way as to reduce that relative motion. In other words, the bar magnet was caused to rotate. The experiment also worked the other way round: a rotating magnet would start the disc rotating. Years later, in 1879, Walter Baily showed that he could use four stationary electromagnets to produce a magnetic field rotating in space by electrically changing the polarity of their poles, and that Arago's disc would again rotate. With that the operating principle of the induction motor had been demonstrated. Yet the first induction motors did not appear until 1887–88.

Galileo Ferraris of Turin was the first to announce the construction of such a motor in April 1888,[2] actually three years after he performed his work. He appears not to have fully realized the significance of his discoveries and to have believed that the motor was not really practical.[12] The Ferraris motor was a two-phase machine; he obtained the two phases from a single supply by 'phase splitting'—passing one current through a resistive component and the other through a highly inductive component, and so introducing approximately a 90° phase difference between the two. No winding was placed on the rotor, which was a simple cylinder, so that the induced currents were not guided in any way; they were simply the eddy currents in the cylinder.[8,12] About the same time, in the USA, O. B. Shallenberger of Westinghouse discovered essentially the same phenomenon and developed its use, not as a motor, but as a current meter for use with Westinghouse AC arc lamps.[2] In this way an important gap in the AC commercial system was filled. A year earlier Elihu Thomson, of Thomson-Houston fame, had discovered the basic principle of what would become known as the repulsion-induction motor, that an electromagnet powered by AC repels a conductor. The same phenomenon had been observed previously by Fleming.[7]

Far more important than any of this work was that done by Nikola Tesla (1856–1943), an American born in Croatia (now part of Yugoslavia). Tesla received a technical education in his homeland, briefly attended the University of Prague, and worked in Europe for an Edison company. In 1884 he went to America and for a time worked for Edison in New York City before establishing his own laboratory. Even before going to the States he had worked out the principles involved in producing a rotating magnetic field and had built his first induction motor. In 1887 he began to file for U.S. patents, first for a practical two-phase induction motor and then for a more or less complete three-phase system. The patents were issued between 1888 and 1896 and the first ones were immediately bought up by Westinghouse, who had also purchased the American rights to the Ferraris patents. Tesla in fact moved to Westinghouse with his patents, at least for a time.

As with Ferraris, Tesla's two-phase induction motor of 1888 used phase splitting, but the rotors were specified both with and without windings. Tesla and Westinghouse engineers continued to work on the motor but encountered disappointments. Its potential use as a traction motor for railway use had to be abandoned when it proved to be unsuitable, and then it was found to give poor performance on the relatively high-frequency ($133\frac{1}{3}$ Hz) AC lighting distribution systems operated by Westinghouse. For a short time work slowed down. But in 1891 Westinghouse installed the first American AC power transmission system at Telluride, Colorado, to supply power to a large synchronous motor installed at a mine. As the synchronous motor had no starting torque a small Tesla motor was used to start it.[2] A year later, at the Chicago World Fair, Westinghouse gave a formidable demonstration of the versatility of an AC distribution system. Having won the lighting contract for the Fair, Westinghouse installed twenty-four 500 hp, 60 Hz alternators that supplied a two-phase voltage as the power plant. By the use of an induction motor driving an AC-DC generator, rotary converters to convert AC to DC, and more alternators driven by electric motors, it was shown to be possible for a single AC power plant to provide AC at any voltage and frequency for lighting or power, and to provide DC for traction and other purposes.[2]

Meanwhile the question of the frequency and phasing of AC supplies had been considered. Westinghouse opted for the two-phase system because of its convenience in providing a single-phase supply for lighting, which was then the major market. The frequency was settled at 30 Hz for power though this was later changed to 25 Hz. For lighting the frequency was dropped from $133\frac{1}{3}$ to 60 Hz, a frequency also suitable for running small induction motors. If the frequency was too low it would produce flicker from incandescent lights. The seemingly strange figure of $133\frac{1}{3}$ Hz came about through the original method of stating the frequency as the number of reversals, or alternations, per minute. Based on this system $133\frac{1}{3}$ Hz appears as a more rational $133\frac{1}{3} \times 2 \times 60$ = 16 000 reversals per minute. Later the system was changed to cycles per second, now called after Hertz.

According to B. G. Lammé, an engineer who made important contributions to the AC work at Westinghouse, the problem of commercializing the Tesla motor was solved by a common marketing ploy. The Tesla motor needed a polyphase system at a fairly low frequency. The new 60 Hz frequency was suitable but no polyphase systems were available, so it was suggested that "a 'fad' be made of the polyphase generator, so that when a lot of such machines had been installed throughout the country the sale of the induction motors would automatically follow. This was agreed to . . . ." The ploy worked so well that the demand for motors came sooner than expected and, "in some instances they were actually pulled out of the test room and shipped with only partial tests."[15]

General Electric meanwhile had also turned to the manufacture of AC motors: three-phase with 36 Hz for power and $66\frac{2}{3}$ Hz for lighting and low-power applications. The Stanley Co. settled for two-phase at 132, 125; and $66\frac{2}{3}$ Hz. By 1900 American industry had standardized frequencies down to just two, 25 Hz for heavy power and 60 Hz for general use, although 50 Hz continued to be used in some parts of the country (such as Southern California) until the late 1940s. In the first quarter of the 20th century improvements were made in the design of power machinery that made a single standard of 60 Hz acceptable for virtually all purposes by the early 1920s. As to phase, the advantages of steadier power and lower copper costs of the three-phase system became more important as polyphase systems spread, and the two giants, GE and Westinghouse, both settled for three-phase soon after a patent agreement was reached in 1896.[2]

Independent of the Tesla-Westinghouse work AC induction motors were invented and developed in Europe, especially in Germany by F. A. Haselwander and by Michael von Dolivo-Dobrowolski, then a leading engineer at AEG. It was Dolivo-Dobrowolski who invented the 'squirrel-cage' rotor of solid bars parallel to the axis (in 1889), which quickly became a lasting standard in induction motor design. He also developed large synchronous motors. Dolivo-Dobrowolski's motors were exhibited at the Frankfurt Exhibition in 1891 and were probably the first induction motors available commercially. The Frankfurt Exhibition, held a year before the Chicago Fair, was largely responsible for thrusting the case for AC transmission and use into the consciousness of European (and some visiting American) electrical engineers. Frankfurt itself had been ready to establish a municipal DC electricity supply, but Cologne had just become the first German city to go AC, and Ganz & Co. of Budapest were advocating AC for Frankfurt. Werner Siemens, the grand old gentleman of the German electrical industry, saw the "battle for the city of Frankfurt" as a matter of principle. Siemens & Halske, like Edison in America and Crompton in Britain, favoured DC. The Battle of the Systems was in full swing and AC was about to give a very convincing demonstration of its capabilities.

When the Frankfurt Exhibition opened, lights shone, motors ran, and notices told the public that the power was coming from generators located at a new dam on the River Neckar, 110 miles (176 km) away. The power, up to a maximum load of 240 kW, was transmitted at 15 kV, which could be raised to 25 kV.[12] The generators and transformers were supplied by the Oerlikon Co. of Switzerland and the motors by Dolivo-Dobrowolski of AEG; the whole project was supervised by the consulting engineer Oskar von Miller, formerly a director of AEG, which had itself previously advocated DC transmission. Some idea of the impact on the engineering fraternity may be gained by a review of important AC projects given in Table 7.1.

Table 7.1 Early Examples of High-voltage AC Generation and Transmission

1881	Depréz and von Miller, Munich Exhibition, 1.5 kW, 1.5 to 2 kV, 35 miles (Refs. 12, 1). Standard telegraph line used for transmission
1884	Gaulard and Gibbs, Turin Exhibition, 20 kW, 2 kV, 25 miles (Ref. 12)
1890	Westinghouse, Oregon City to Portland, 3 kV, 14 miles (Ref. 13)
1891	Westinghouse, Telluride, Color., 100 hp, (75 kW), 3 kV, 4 miles (Ref. 2)
1891	Von Miller, Oerlikon, AEG, Post Office; Frankfurt Exhibition, 240 kW, 15 kV to 25 kV, 110 miles (Ref. 12)
1892	Ganz & Co., Rome, 5 kV, 17 miles (Ref. 12)
1892	Westinghouse, Pomona, Calif., 2 by 120 kVA, 10 kV, 14 miles and 28 miles (Ref. 13)
1893	Westinghouse, Chicago Exposition aggregate 9480 kVA, 2 kV, local (Ref. 12)

The Battle of the Systems was also waged in other countries and eventually won by AC because it could be transmitted over long distances at high voltages by relatively thin conductors, especially when the three-phase system was used.

The first large-scale demonstration planned was Ferranti's attempt to supply a large section of London with AC for lighting from what was for its day a giant power station located on the River Thames at Deptford. Outside the city land was cheap, and coal could be easily transported up the river. With high-voltage AC transmission, and the use of transformers, electricity could be generated where energy was cheap and transmitted to the users. The gas industry had already shown the way.

Ferranti, who was born in Liverpool, had been called in by the owners of the Grosvenor Gallery in 1885 to redesign and re-equip their small power station, which supplied current for their own lighting and provided a service to their neighbours. After successfully completing the job he proposed a scheme, based on the economies of scale, to supply current to about one million lamps. The proposed Deptford station would transmit power at 10 kV, about four times any previous transmission voltage. (The Frankfurt Exhibition came four years after this proposal.) The cable would take the route of a private underground railway line to the Grosvenor Gallery, which would effectively become a substation. It would be the biggest power station in existence or planned. Almost everything was to be designed from scratch: five alternators of

7500 kW capacity each, about ten times bigger than any then built; 10 kV transformers; cables of new design with paper insulation; and giant reciprocating engines to drive the alternators. Questions were raised as to the safety of the new cables at such a high voltage and an answer was given in the form of a dramatic "chisel test." A cold chisel was driven right through the live cable. The man holding the chisel is said to have confessed to some fear—as the boy holding the sledge-hammer had never used one before (Fig. 7.5).

Figure 7.5 Ferranti chisel test—on a live 10 kV cable

The Deptford scheme was too ambitious. Planning began in 1887 to supply part of London, yet the whole of London did not reach one million lamps until 1890 and other companies were competing in the supply of electricity. Problems and disasters overtook the project. The Grosvenor Gallery station burned down in one accident. Although the 7500 kW alternators were never finished, smaller ones took their place and the station did begin operations in 1889, only to be knocked out by insulation problems, fires, and other troubles. However, in 1891 the 10 kV supply to London was operational. Much technical ground had been broken but financially the project was unsuccessful. The economies of scale were supposed to provide cheaper power, but did not, and delays led to loss of customers. In 1895 capital was reduced by a third; the first dividend was not paid to shareholders until 1905.[3] Ferranti was dismissed in August 1891. "You are a very clever man, Mr. Ferranti," the chairman is reported to have told him, "but I'm thinking ye're sadly lacking in prevision."[1]

More likely the trouble was that Ferranti had too much prevision. His vision of large-scale power stations was ahead of its time. Even in 1914 only two British power stations were of the size he had proposed for Deptford in 1887. Both used turbines instead of reciprocating engines and neither had generators of 7500 kW capacity.[1,3,10] The major problems involved in building a successful steam turbine had been solved by Charles Parsons in Britain in 1884, and small turbines began to be used as prime movers for electricity generation in the 1890s. Yet not until after about 1904 did turbines really begin to take over from reciprocating engines in Britain.[3]

In America, AC made its bid to oust DC in the Niagara Falls project. Planning began about 1891, after the Deptford scheme and after the dramatic demonstration at Frankfurt. In contrast to the Deptford scheme, where the leading role was played by an electrical engineer, the Niagara scheme was led by financiers who sought advice from scientists and engineers in both America and Europe. At one stage the designs submitted by the two leading American electrical manufacturers, GE and Westinghouse, were both rejected in favour of the Niagara company's own designs, although these were in fact subsequently modified by Westinghouse.[2]

Plans had been made to harness the power of the Niagara Falls from about the middle of the century. The idea that finally came to fruition began in 1886 with a proposal for a canal system feeding water to 238 water wheels of 500 hp each. The water was to be discharged through a $2\frac{1}{2}$ mile (4 km) long tunnel and the power mechanically transmitted to factories along the banks of the canals. In 1890 New York financiers, including J. P. Morgan and W. K. Vanderbilt, took over; the Cataract Construction Company was formed and advice was sought from leading engineers. The aim now was to supply power to Buffalo, about 20 miles (30 km) away, then a city of 250 000 people whose factories used about 50 000 hp a day.[2]

After considering compressed air and DC and AC electricity as methods for transmitting the power to Buffalo, and despite Kelvin's adverse advice (he cabled, "Trust you avoid gigantic mistake of alternating current"), the final decision, made in May 1893, was in favour of two-phase AC. Westinghouse got the major contract in October, presumably because of the company's past experience of AC and because of the success of AC at Frankfurt and Chicago.[2]

Both Westinghouse and GE were to share in the spoils before the project was completed. A local supply began in 1895 and the transmission line to Buffalo opened late the following year. Contrary to expectations the local demand to Niagara exceeded that of Buffalo, since a new electrochemical industry (calcium carbide, sodium peroxide, sodium, chlorine, and so on) developed at Niagara once the power was available there. The new industry provided heavy demand for 24 hours a day and so was an ideal customer.[2] The capacity of the Niagara power station grew from a few thousand horsepower in 1895 to nearly 20 000 hp in 1897; it achieved the design maximum of the discharge tunnel (100 000 hp) in 1904.[2] The 25 Hz frequency used for power applications remained an American power standard in that region for half a century.

The battle between AC and DC was characterized by the great venom with which some companies were accused of pursuing their aims. In the Niagara project, for example, Westinghouse accused GE of industrial espionage, an early example of what is supposed by many to be only a post-James Bond problem.[2] The Edison company, to support its leading position in the use of DC, launched a compaign to attack AC, pioneered by Westinghouse, as a danger to the public. A pamphlet published in 1888 gave details, including a list of people allegedly killed by AC. The decision by New York State to use AC for capital punishment by electrocution has been called "an important victory for the Edison company."[2] Edison is supposed to have suggested calling the process of electrocution 'Westinghousing.' To a public uneducated in electrical matters the New York decision appeared to support the Edison line, held by both the company and the man, that AC was dangerous. Within ten years Edison had changed from a bold innovator to a "cautious and conservative defender of the status quo."[2] In Britain, Ferranti demonstrated the safety of his high-voltage AC transmission cables by the aforementioned test of driving (or rather by getting an assistant to drive) a chisel straight into one.[1] The grounded outer conductor justified the faith of the volunteer.

Though the Battle of the Systems for the transmission and distribution of electricity for general purpose use was won before the turn of the century by AC, DC did not lie down easily. New DC systems were built in Britain in the first decade of the 20th century and experiments were conducted on mixed systems: DC transmitted over a distance and then converted to AC for local distribution. (This scheme came into its own half a century later, when the high price of copper had more than justified the costs of the conversion, at least in long-distance and submarine transmission lines, because DC required only two conductors as against three or four for three-phase AC.) Even after the establishment of the British National Grid in 1933–34, initiated by the Electricity (Supply) Act in 1926, some local DC systems continued unchanged.[10]

The northeast of England gave the lead in the move towards standardization in Britain. By 1914 some 1400 square miles of the area around Newcastle had an interconnected power system operating on three-phase 40 Hz AC. Elsewhere there was a "bewildering variety of voltages, frequencies and distribution systems."[10] Ten years later there were still seventeen different frequencies in use in Britain, though opinion was swinging in favour of a 50 Hz national standard,[10] a frequency that had become common in Europe. The question of the voltage supply for Britain also posed problems. As the National Grid got under way in the mid-1930s, forty-three different supply voltages were in use, ranging from 100 to 480 V. The advantages of a standard voltage were apparent but it was not achieved until 1945, at 240 V. Frequency standardization was completed two years later, at a cost of £ 17.5 ($ 70) million.[17] Such expensive conversions had been largely avoided in America, where most rival companies had agreed to standardize much earlier.

On the railways, the Battle of the Systems continued into the 1920s when DC came out as the victor, mainly because DC motors were better suited for railway use. After World War II the French railways pioneered high-voltage AC transmission, transformed down and rectified for DC traction motors. Thereafter the 'battle' turned in favour of AC.[3] The aforementioned DC bid for a comeback in long-distance high-voltage transmission lines started in the 1950s with a 200 kV experimental line in the Soviet Union and a 100 kV commercial line in Sweden. Since then many more have been constructed and distances up to 1000 miles (1600 km), voltages of 500 kV, and capacities approaching 2000 MW have been achieved or attempted.[14]

Traction

Today electric traction almost invariably means electric locomotives for railway use. Such engines may be on a grand scale for national or continental railways, or on a smaller scale for underground 'mass-transit' (urban) systems and surface light-railway 'rapid-transit' (commuter) systems. In a few places remnants may still be found of the old urban railway streetcar systems, or trams, the municipal mass transit system of the first half of the 20th century. The tram (called streetcar in America) was the first really large-scale application of electric power.

Experiments that were to lead to the successful electric tram, especially as produced by Sprague and the Thomson-Houston company, began about 1879. Previous to that date there had been the usual battery-powered attempts to preempt the electrical revolution ushered in by the dynamo. Such attempts, no matter how ambitious, from the mid-19th century on were doomed to failure: the available batteries were simply not up to the job. From about 1879 the tram went through nearly a decade of birth pangs before emerging as a successful transport system on its way to dominance in urban areas. After World War I the railway tramcar was increasingly challenged by the trolleybus and motor bus and was in rapid decline before World War II (Fig. 7.6.), mainly because of the greater convenience of motor vehicles. (It has been reincarnated in the form of electric mass-transit systems following the great increases in fuel costs of the 1970s.)

The first birth pang of the electric tram is usually said to have occurred in 1879 when the Siemens company demonstrated a small electric railway at the Berlin Exhibition. The small train, driven by a 3 hp motor, has been described as a "good-sized toy."[12] It was big enough though to catch the imagination of both Werner Siemens and the public. Siemens pressed ahead and in 1881, in the Lichterfelde district of Berlin, opened what was probably the world's first commercial tramway (Fig. 7.7). Though it appears to have been largely experimental it was a true tramway and certainly no toy, as it used the conventional cars usually drawn by horses and extended over $1\frac{1}{2}$ miles (2 km).

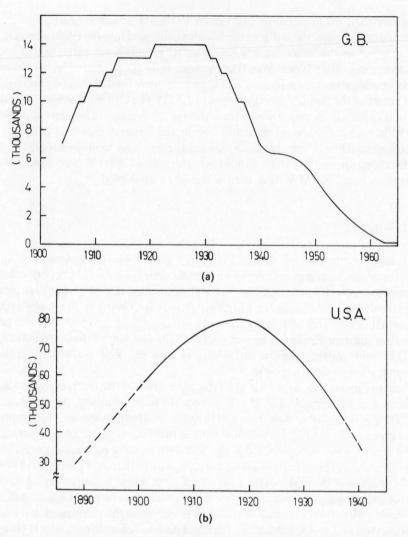

Figure 7.6 The rise and fall of the tram: (a) Britain: tramcars in use; (b) USA: number of passenger cars (includes nonelectric; dashed line based on limited data) (data: Refs. 23, 25)

The line had power fed to the car through one rail at 180 V, with the other rail as the return—not a very safe arrangement. After some years it was converted to an overhead wire supply[12,18]

Meanwhile American interests had turned to the question of electric traction and what were to become the more or less standard systems originated there. Unsuccessful attempts were made by several pioneers before success was achieved, including Edison, E. Julien, Edward Bentley and Walter Knight, and Charles Van Depoele.[2]

Figure 7.7 Electric tram, Lichterfelde, Berlin, 1881

According to Passer[2] three major problems, among the many technical problems of electric traction, had to be solved before electric tramways could become a commercial success. The most fundamental question was how to supply a motor on a moving vehicle with current. Secondary batteries or moving connectors to rails or overhead wires would be required. Batteries were quite unsuitable in an environment subject to continual and sometimes violent vibration. Frequent repairs were needed. One engineer commented, "We patch them up by night, and the following day the patches come off."[3]

The next problem was the design of a motor suitable for driving a vehicle whose speed had to vary and which would have to stop and start frequently. This problem mainly concerned the design of commutator brushes. Copper brushes were widely used in electrical engineering at that time; they caused heavy wear on the commutator and hence excessive sparking, with all its resultant hazards. Even worse, the heavy currents and changing voltages used in starting resulted, in Passer's words, in "electrical fireworks." On stopping, the car would often reverse slightly and the metal brushes would dig into the commutator. Carbon was suggested as a more suitable material for brush manufacture by Charles Van Depoele after he joined the Thomson-Houston Co., around 1888–89. Van Depoele's idea, though it appeared to be an "absurd suggestion" to his colleagues at Thomson-Houston, proved to be a lasting solution.[19] E. W. Rice, another Thomson-Houston engineer, who later became honorary chairman of GE, commented in 1924 that the engineers were "astonished but delighted to find that with . . . [carbon brushes] . . . the motor operated entirely free from sparking." Rice rated the invention of the carbon brush as "the most wonderful and valuable advance ever made in the art of commutating machines," and doubted, even in the 1920s, that engineers appreciated the revolution it brought about in the design and operation of electrical machinery. As a result the commutator, Kelvin's "frightful thing," lost most of its terror.[19]

The third fundamental problem to be solved before electric trams could become successful was that of getting the power from the motor to the axles of the vehicle, which involved the positioning and mounting of the motors themselves, and the attendant problems raised by dirt, rainwater, and (worst of all) the jolting of the tram.

In the 1920s Fleming described a "modern type of electric tramcar." His description tallies exactly with what would still be seen in one of the few remaining trams, some of which are in heavy daily use over 60 years later. The car rests on a pair of bogies each of which holds a DC series motor, chosen for its large starting torque. The driver's control changes the connection of the two motors. For starting from rest they are joined in series with some additional resistance, part of which is then removed to increase speed. They are next switched to a parallel connection and, finally, the resistance is cut out to achieve maximum speed. The DC supply of around 500 V is picked up from overhead wires by a trolley system, a grooved metal block or wheel pressed into contact with the underside of the supply wire by a strong spring. Though the system sounds somewhat archaic it has been in successful service for something like 90 years.

The path from Siemens's Lichterfelde experimental but commercial tramway to the long-established successful system described by Fleming was largely trodden in America in the period from 1880 to the early 1890s. It was taken by two teams. One was a one-man band consisting of Frank J. Sprague; the other, a team of engineers at the Thomson-Houston company, which included Elihu Thomson himself and Charles Van Depoele, whose company had been acquired by Thomson-Houston.

Like many of the other important engineer-entrepreneurs in electrical engineering, Sprague was still a young man, only in his mid-twenties, when he formed the Sprague Electric Railway and Motor Company in 1884.[2] By that time he had graduated from the U.S. Naval Academy; visited Europe, where he had ridden on the smoky London underground railway, then steam driven; worked for Edison; and in his spare time developed an efficient motor suitable for railway work. He left the Edison company and further developed his motor, which soon became commercially successful. For the first few years Sprague's little company was largely restricted to doing research and development while manufacture was left to the Edison Machine Works and marketing to local agents. Within three years motors ranging in power up to 15 hp, very large for their time and a major development in themselves, were enjoying good sales and Sprague opened his own manufacturing plant where special purpose motors could be made. This success with large motors was essential for any future railway work.

After failing to interest the New York elevated railways in the possibility of changing from steam to electricity, Sprague accepted a contract which was to lose him around $70 000 but whose ultimate success proved to be the turning point in the design of electrified street tramcars. For a new street railway at Richmond, Virginia, he built forty tramcars with two motors each, a complete overhead supply for 12 miles (19 km) of track, and a 280 kW generating station. No other electric railway could even approach it in size; the biggest had about ten cars and none was very successful.

The problems were immense.[2] There was the question of the mounting and suspension of the motors, which also tended to burn out when cars tried to climb 10% gradients. Brushes had to be replaced and commutators machined down. Streams of new motors and spares were shipped from the factories to keep operations going. Further problems came from overhead switches, insulation, and lightning strikes. More than fifty different trolley poles were tried before a successful one was developed. Other trials came from the track, which sometimes sank into the mud. Tight corners led to derailments. Yet the problems were solved and in May 1888, a year after the contract was signed, the electrical equipment was formally accepted by the customer who shortly afterwards ordered another forty cars. In solving the problems Sprague achieved a system which, apart from the gearing, became the standard American system for street railways for the next sixty years. The series-parallel motor connection and its controller, operation from either end of the car, a single overhead conductor with return through the track, a single under-running contact to the wire, a 500 V supply, lightning arrestors, and a 'wheelbarrow' suspension for the motors that allowed for movement relative to the car were all used by Sprague and are essentially the same as described by Fleming in 1921.[7,18]

To get the Richmond tramway operational Sprague spent nearly twice what he was paid. Not only had he to find new solutions to problems but he was also working against time. According to his own recollections ten years later, when the contract was signed he had "only a blueprint of a machine and some rough experimental apparatus. . . . Fortunately for the future of electric railways, the difficulties ahead could not be foreseen or the contract would not have been signed."[2]

Nevertheless, Richmond was an outstanding success. Running costs were only 40% of an equivalent animal-powered system and the successful completion brought in new orders. When the Richmond contract was signed there were only eight electric tramways in the USA, which used sixty cars on 35 miles (56 km) of track. Five of them had been designed by Van Depoele. Eighteen months after regular operation began 180 were in operation, or under construction, with nearly 200 cars on 1260 miles (2000 km) of track.[2] Van Depoele had built ten, Sprague had built sixty-seven. Thomson-Houston, Sprague's major rival, had also built sixty-seven.[2]

In 1890, when the various Edison companies were reorganized into Edison General Electric, Sprague's company was absorbed, partly to ensure that Edison's biggest customer did not stray to other suppliers. For a short time

Sprague himself stayed on as a consultant to the new company. After resigning he formed a company to make and install electric elevators.

Sprague's name is now firmly attached to the development of the successful electric street tramcar, yet he was not the only one to achieve this success. Van Depoele appears to have been heading that way, albeit more slowly, when he was bought out by the Thomson-Houston company, which was seeking a way into the traction business. Coffin, the one-time shoe salesman but now head of Thomson-Houston and later of GE, made the decision to enter the railway business by buying one of the existing companies. The firm already had relations with Bentley and Knight but Thomson recommended Van Depoele, and the purchase was made early in 1888. The resulting system was essentially Van Depoele's small-scale one expanded into a large-scale one. Van Depoele, Elihu Thomson, and A. L. Roher were the main engineers concerned.[2] A motor suspension independent of the tramcar body was designed by Thomson to allow for efficient gearing. Two large motors per car were installed and problems with the trolley pole and pick-up were sorted out. With various changes a reliable and efficient street railway resulted. The first installation was made at Lynn, Massachusetts. It became operational one month after Sprague's Richmond line was accepted by its purchaser. By 1889 America had two reliable street railway systems to choose from.

The first large city to adopt electric trams on a large scale was Boston. Since little could be found to choose between the two systems, the decision was made on the basis of which company would undertake to maintain reliable service for five years for a fixed price. If things went wrong the contract could turn out to be expensive for the supplier. Sprague had already lost financially at Richmond and had few reserves, whereas Thomson-Houston had a successful lighting business to fall back on. Sprague refused, Coffin accepted. The equipment proved to be reliable and the fixed-terms maintenance contract brought in extra thousands of dollars for Thomson-Houston before the customer used an escape clause to cancel it after two years.[2]

When General Electric was formed by the merger of the Thomson-Houston and Edison companies the pioneers of electric traction were all under one roof, including Bentley and Knight. The competition now came from, as usual, Westinghouse, who entered the business in 1890. Passer[2] relates interesting tales of would-be customers, with no technical knowledge, organizing tests between rival suppliers. One 'test' apparently was to hold a tug of war between two tramcars equipped by rival manufacturers. Another was to have two trams race up a hill. Often the result depended more on the skill of the driver than the characteristics of the motor, but the tests must have been great fun to watch.

By 1897 88% of American tramways were electrically operated and the building and conversion boom was largely over. Motors had been improved and costs reduced. Costs fell about 70% between 1891 and 1895, partly a result of the intense competition between GE and Westinghouse.[2]

After America came Europe; first Germany with about a five-year lag, then

Britain about ten years behind. Both countries had seen experimental systems but had nothing to match the American systems which, apart from other considerations, enjoyed the cost advantage brought by mass production.[3] In Germany AEG obtained rights to the Sprague patents, and a new company was formed to exploit those of Thomson-Houston. Siemens developed its own system. By 1893 the Germans were constructing 98 miles (157 km) of electric tramway of which 79 miles (126 km) were to use the American systems.[3]

The American invasion of Britain began in the mid-1890s and was spearheaded by the British Thomson-Houston Co. (BTH). Private tram companies in Bristol and Dublin were the first customers and others followed, but the big changeover came when the municipalities decided to act. Leeds was the first in 1895; after two years it had a ten-mile (16 km) system in operation. Other cities and smaller towns followed including Blackpool, which has an extensive electric tram system to this day. The tramway electrification coincided with a house building boom around the turn of the century and there was interaction between the two, although whether one particularly caused the other is open to doubt.[3]

The economic advantage of electric trams was initially seen as a reduction in operating costs when compared to horses (in some cases more than 50%). Accurate figures are difficult to obtain, but it has been stated that traction costs were around 40% and total working costs around 70% of a horse-drawn system.[3] Other advantages may have been even more important, particularly the increase in speed and the increase in carrying capacity. In town centres horse-drawn trams averaged around 5 mph; electric ones achieved 8 mph and did even better outside town. Twenty-five to fifty passengers could be pulled by horse, but sixty to 100 by electricity.[3] Both factors would lead to increased revenues. Another advantage, not always remembered, was that electric trams did not foul the city centres with horse droppings and so reduced by one the hazards pedestrians met while crossing the street.

After leaving the Edison company Sprague, as already noted, turned to the design and installation of electric elevators, but he also continued his earlier interest in the electrification of the overhead city steam railways. It was in this heavier-gauge railway that he was to make his second major improvement. For many years he had failed to persuade elevated railway companies in New York to go electric, but in 1897 he at last got a chance to prove his ideas in Chicago.

Electric railways were not actually new but previously they had only been operated on a small scale: at exhibitions, for hauling trains through tunnels where the smoke and steam were objectionable, and so on. GE obtained the first contract for an overhead electric railway, also in Chicago, and used the standard idea of simply replacing the steam locomotive by an electric one to haul the carriages. Sprague's contribution was in realizing that electric traction could outgrow the locomotive idea, which was essential for steam but not for electricity. He replaced it by what became known as the multiple-unit system of control: each carriage was equipped with electric motors. Carriages could then be operated alone or with others, with a single control, in any sequence or numbers. Each would be powered, lit, heated, and braked independently; and the whole train could be controlled from either end of any carriage. Maximum flexibility could therefore be achieved. There were other advantages, too. On railways tractive power is a function of the weight on the driving wheels. In the multiple unit system the weight of the entire train is used for this purpose. Also there is little wasted power as each carriage provides enough to move itself, whereas in the locomotive approach the same unit may be used to pull two carriages as is used to pull twenty.[2]

As with the trams, Passer[2] relates that Sprague took great financial risks to get his brainchild moving. He personally took all the risks that his idea might fail, signed the contract as an individual, and even agreed to a $100 000 (£20 000) bond for penalties in case he failed to meet the terms. He signed the Chicago contract early in April 1897 and agreed to have six carriages working and ready for test by mid-July, with a further 114 to follow, all to be achieved without interfering with the existing service. For all that he was to receive $300 000 (£60 000). As with the Richmond contract his equipment hardly existed except on paper. He had about 15 weeks in which to do the work, was physically handicapped as a result of a fall made while installing elevators, and was going to London to secure an elevator contract. While he was away in London there was a strike at his factory.

One cannot help but admire his courage and tenacity. Two carriages were operating on the line just one day late, and ten days later he demonstrated a six-carriage train, driven by his ten-year-old son to show how easy it was. By August of the next year all 120 carriages were operating and steam locomotives had been withdrawn from service. It is not surprising that Sprague later referred to 1897 as one of his most difficult years.

The multiple-unit system was a great success and is, with some variations, still in use on the most modern urban mass-transit railways. The net earnings of Sprague's first system were about $3\frac{1}{2}$ times the previous value of the steam railway, since costs declined and revenue increased.[2]

A sequel to Sprague's second success was a suit brought against him by GE claiming infringement of a patent (inherited by GE through mergers) that had originally been granted to Sprague for work on the Richmond tramway. Sprague decided he in turn had a case against GE over the multiple-unit system GE was using. Sprague's case was going so well that GE decided to buy him out for the second time, this time because "in no other way could we get possession of a patent which was absolutely necessary to our business."[2] Sprague received over $1 million, an appropriate reward for his extraordinary boldness, vision, and ingenuity.

Byatt has discussed the development of urban electric railways in Britain, of which the London Underground was the most famous.[3] The first underground railway in the world, the Metropolitan, opened in London in 1863 and showed that steam could be used successfully in this way. Other lines in London followed; by the 1890s many lines were in operation, under construction, or planned. The first decision to go electric was taken in 1888. Mather and Platt, a textile machinery manufacturer who had opened an electrical section under Hopkinson, got the order. Relatively little progress was made until GE and Westinghouse took over around the turn of the century. American subsidiaries in Britain ultimately became major 'British' suppliers of electrical equipment. British Thomson-Houston (BTH) opened its factory at Rugby in 1894; British Westinghouse started in 1899 in Manchester. British Westinghouse became part of Metropolitan-Vickers in 1919. In 1928 another merger took place which brought the former American offshoots together as Associated Electrical Industries (AEI). AEI narrowly avoided the name Empire Electrical, which was a bid to ape Imperial Chemical Industries. The former American companies became part of GEC, the giant British electrical company, in 1967.[20]

Mainline railways were more difficult to electrify though some early efforts were made, usually on a small scale, such as in the tunnel approaches to the mainline New York stations. The major problem appears to have been one of power. A 500 V supply was fine for a tram that took around 50 kW, but quite inadequate for a mainline locomotive that might take ten times that figure and therefore need a much heavier current.[12] Nevertheless progress was made. Some low-voltage DC locomotives were employed in Britain and the USA but the trend was to higher voltages. Siemens-Schuckert used 1000 V on the Cologne-Bonn line in 1905 and 2000 V for an ore carrier the next year. In America DC was used up to 3000 V. Switzerland and Northern Italy, where water power for the generation of electricity was plentiful but coal was in short supply, had particular interests in electric traction and Brown-Boveri achieved some success with high-voltage three-phase AC. Ganz of Budapest also advocated AC. Three-phase supplies had the disadvantage of requiring three conductors, maybe two overhead plus the rails. Siemens, among others, saw the answer in developing a single-phase AC traction motor and eventually evolved a system supplied at 15 kV that was stepped down to a few hundred volts inside the locomotive. The frequency was only a third of the usual 50 Hz and so necessitated the building of separate power stations for the railway. After some successful experimental railways, one of which achieved 80 mph (130 km/hr), disaster struck when the locomotives were found almost literally to shake themselves to pieces at high speed. The problem lay in the machining of certain driving rods. Though many railway stalwarts in Germany nodded wisely at the failure of electric traction, Siemens eventually won through in 1914 when an iron ore train operating between Sweden and Norway ran successfully with an electric locomotive. Loads of 1500 tons were hauled up ice-covered gradients at speeds of nearly 20 mph (30 km/hr), compared with the steam locomotive which, on some gradients, barely managed 6 mph. To Siemens, the ice had been broken in more ways than one.[12]

Electric traction, especially for use in urban tramways, was the first major market for electrical power and second only to lighting in bringing electrical engineering into public use. Its social consequences included the growth of semirural suburbs, transformation of urban transport, and banishment of the horse-drawn tram from the streets of towns and cities. Light-railway companies were often reluctant to change from steam to electricity until the cost and other benefits had been properly demonstrated; then the changeover occurred fairly swiftly and elevated railways as far apart as Chicago, Liverpool, and Berlin went electric. On the mainline railways the change came much more slowly; even today, electric traction still has to compete with effective diesel locomotives. The light railways are still an essential mass-transit system in many cities and in some country areas, in Germany for instance, but the tram has largely gone. Apart from a few systems (Blackpool is perhaps the most famous) the electric tram, which saw the birth of electrical power engineering, has been driven into the museums by the much more versatile, and faster, motor bus.

Industrial Use

Until recently industry has been by far the biggest consumer of electricity. Figure 7.8 shows that in the USA industry has used around 50 % of the electrical power generated since 1910. In Britain, apart from the hectic period of World War II, industrial use has declined from nearly 70 % in 1920 to about 40 % in the late 1970s. The domestic share of the market has shown the biggest relative growth in both countries. Total generation has risen by orders of magnitude (Fig. 7.9) and all consumers have shown very nearly an annual increase in consumption throughout the period; the most notable exception was public

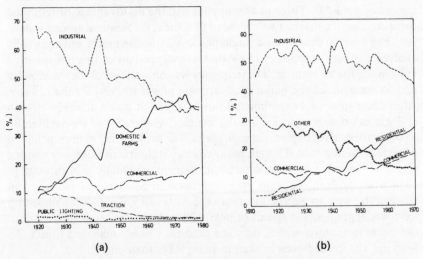

Figure 7.8 Use of electricity: (a) Britain, 1920–1979 (percentage of sales, by consumer); (b) USA, 1920–1970 (percentage of use, by consumer; pre-1920 based on limited data) (data: Refs. 23, 25, 26)

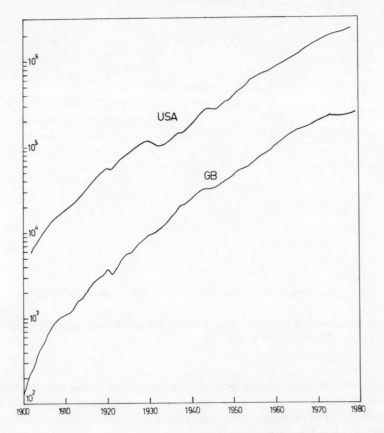

Figure 7.9 Generation of electricity (GWh): Britain: sales; USA: net production (data: Refs. 23, 25, 27)

lighting in Britain during the blackout years of World War II. Consumption by public lighting fell from a high of 367 GWh in 1938 to a low of 17 GWh in 1940.

Though lighting was the first major user of electricity, and traction the second, both were soon eclipsed by industrial use. By 1920 traction took only 11 % of the electricity generated in Britain compared with industry's 68 %. Motors began to be used on a substantial scale in Britain about 1900; by 1907, according to Byatt, factories were using about 50 % of the electricity generated, usually generating it themselves.[3] At that time about 10 % of the total power used by industry was supplied by electricity. Five years later this figure was around 25 % and it had reached 50 % by the mid-1920s, by which time most of the electricity came from central power stations. In the first forty years of the industrial use of electric motors, say from 1890, their increase in use was phenomenal. Figure 7.10 shows the horsepower of electric motors used in American manufacturing industry up to 1962 compared with the

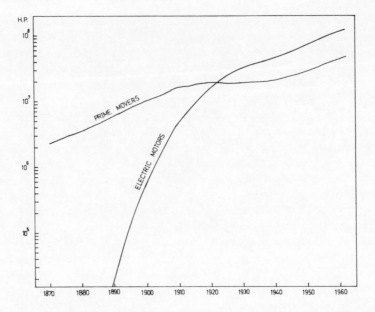

Figure 7.10 Power of equipment used in manufacturing industry: USA (horsepower),
1870–1962 (data: Ref. 25)

horsepower obtained directly from prime movers. The power used by electric motors increased by an order of magnitude per decade to about 1915 and surpassed that of prime movers in the early 1920s. From around 1940 the rate of increase has been about equal.

The first factory motors were usually small DC machines suitable for frequent stopping and starting or for variable-speed type work, such as in cranes or hoists. However, Britain was much slower than the rest of Europe or the USA in adopting the AC induction motor into factory use. In 1894 the British attitude was described by the *Electrical Review* as "a sort of passive resistance."[3] That was partly why the main suppliers were foreign firms, mostly from the Continent. The British shipbuilder Richardson and Co., impressed with the three-phase equipment bought from Brown-Boveri of Switzerland, took out a license to manufacture it. Another British manufacturer, GEC (formed in 1889), took out a license to manufacture polyphase equipment designed by Oerlikon of Switzerland. The German giant AEG moved in with a sales office and Westinghouse and GE also arrived, though initially for traction purposes. By the early 1900s electrical power was well established and prices began to fall. At about the same time the advantages of electricity were becoming more widely known. One example, for a 10 hp motor, quotes the price as more than halving between 1901 and 1905.[3]

The advantages of electrical power over other power sources depended on the type of industry being considered, something Byatt has studied in detail.[3]

Cost advantages could be found in at least four situations: where power was needed at points scattered throughout a factory; where a separate supply would otherwise be needed, as in a travelling crane; where several types of power were needed; and where intermittent power was required. However, lower cost was not the only advantage over steam engines, with their rambling belts and shafts, or even over compressed air. One major advantage was that the layout of a factory need no longer be dictated by the positioning of the power plant and its array of shafts and belts. Electrical cable was flexible enough so that the shop floor could be laid out to permit optimum handling of materials and ease of access to machinery.[3]

Some industries electrified more quickly than others. The engineering industry in general turned to electrical power fairly quickly; textiles and mining, rather slowly. Figure 7.11 gives a very rough indication of the progress made by several industries in the USA and Britain. The production of heat was a major use of electricity in the chemical, iron and steel industries. Resistance heating has been known for almost as long as man has been able to generate an electric current, and if one allows a lightning strike to qualify as resistance heating it has been known far longer. Induction effects can also be used for heating; so can the electric arc, as Davy so convincingly showed.

One of the first products to be made on a large scale by electric heating was carborundum, or silicon carbide, which was first made in 1891 by E. G. Acheson when he was attempting to make artificial diamonds. Plants near the Niagara Falls station produced it in large quantities. Another large-scale product of electrical heating was calcium carbide, which reacts with water to

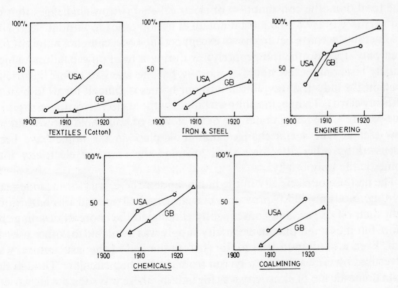

Figure 7.11 Approximate percentage of electrification by industrial group, to 1924 (data: Table 15 of Ref. 3)

produce acetylene gas. It has also been used to produce a fertilizer, although fertilizers (and explosives) are more usually produced with nitric acid, itself once an electrothermal product.[7] The iron and steel industry has enjoyed the benefit of electric furnaces, of both the arc and induction varieties. Aluminium is probably the most widely known example of a material that was once scarce and expensive but became plentiful and cheap thanks to electricity, this time by the use of both electrolysis and heating. According to Fleming the annual production of aluminium before the electrical process was discovered was about three tons per annum, hardly the metal to make cooking pots from.[7] In 1906 the electrolytic process produced 12 000 tons and the cost had fallen to about 3 % of its former value. Aluminium pots and pans became commonplace. Other electrochemical processes have been widely used in many industries. Electroplating is probably the oldest and has been used for depositing a wide variety of metals to produce a large number of products, almost anything from a silver-plated teapot to tinplated steel cans.

The industrial uses of electricity have been widespread, far reaching, and not confined to producing motion. The first use was battery-powered electroplating, though only on a small scale. Lighting was next; then came the numerous applications of motors, electrochemistry, and so on. Electricity as a source of power has had an enormous effect on industry in general and has engendered some industries that would otherwise never have existed.

Domestic Use

The total domestic consumption of electrical energy now challenges that of industry (see Fig. 7.8). To most if not all of us domestic life without electricity is something we have never known except perhaps on a camping holiday (and then only if we do the job properly) or during a total power failure, which usually lasts only a few hours at most. What life was like before electricity brought the industrial revolution into our homes is difficult for us to picture with any clarity. To imagine our own homes without electricity does not give a true picture as we remain unaware of the dozens of household chores that we now escape. The social changes that took place in the home have been discussed by others[21,17]; we shall examine the entry of electricity into domestic life only briefly.

The first use of electricity in the home, as elsewhere, was for incandescent lighting. Not only did it provide a cleaner, smoke-free, and less hazardous light than oil or gas, and consequently eliminated the chore of cleaning the lamp, but it ensured that an electricity supply was available for other uses as well. Even a small industrial motor, before the end of the last century, was advertised by GE as meant to be run from the lighting circuit.[22] Though the main domestic use of electricity was for lighting there was soon a wide variety of domestic appliances available for those who desired them. A catalogue published by Crompton and Co. in Britain in 1894 included ovens, hotplates, hot cupboards, irons, kettles, heaters, radiators, and coffee urns.[11] Even so almost all homes with electricity were using it for lighting only.

Many of the small appliances that followed the electric light into the home possibly had a bigger impact than the light, especially on the housewife. The electric iron was the first to be widely adopted and soon replaced the old flatiron, which had to be heated indirectly. A lightweight electric iron was advertised by GE in America in 1905.[22] It drew its current from the light outlet and salesmen would leave it for a full month for a free trial, by which time it was presumed no housewife or domestic servant would be without it. Irons were more widely advertised after World War I. In a 1929 survey of 100 Ford workers in the USA, 98 were found to have an electric iron at home.[21]

Another domestic electrical appliance to gain early acceptance was the vacuum cleaner. The first vacuum cleaner is believed to have been made by H. C. Booth in London in 1901. He took it around in a horse-drawn carriage from which long hoses could be dragged into people's homes to suck up the dirt from furniture and carpets.[11] Small versions soon followed and in 1907 an American leather manufacturer, W. H. Hoover, developed one that was to

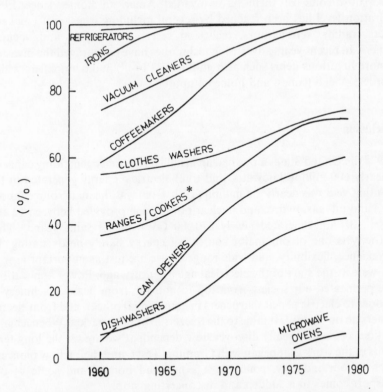

Figure 7.12 Percentage of homes wired for electricity with certain appliances, USA (*exclusive of hotplates and buffet ranges) (data: Ref. 27)

make his name synonymous with vacuum cleaners. (It was invented by J. M. Spangler, whose name is almost forgotten.) Another household name, Hotpoint, was invented for a new iron whose point was heated to aid the ironing of inaccessible cuffs and collars.

Cowan tells us that advertisements for coal or wood stoves in the American magazine *Ladies' Home Journal* disappeared after 1918.[21] Thereafter the choice was between gas, oil, and electricity. Though electric cookers had been available for many years they were not very widely used. Fleming in his book about the first fifty years of electricity, published in 1921, went to great lengths to explain the niceties of domestic electrical appliances such as heaters, cookers, kettles, and so on, and described cooking by electricity as a 'thoroughly practical matter.'[7] However, he warned against the dangers of a 220 V supply, then widely used in Britain, and recommended transforming it down to 25 or 50 V. "It is nothing short of a crime," he concluded, "to place in the hands of ordinary domestic cooks electric cooking apparatus worked at 220 volts off one side of a 440 volt three wire system of supply."

A great many domestic appliances have made the progression from being expensive luxuries to becoming relatively inexpensive necessities. Figure 7.12 shows how some of them have invaded American homes since 1960. Saturation level has been reached by several including irons, coffeemakers, mixers, radios, refrigerators, television receivers, toasters, and vacuum cleaners. In recent years advertising techniques have encouraged the invasion by more luxurious items such as high-fidelity ('hi-fi') audio amplifiers, video recorders, video games, and home computers.

Conclusion

In 1978 the United States alone had an installed electricity generating capacity of nearly 600 million kilowatts, and total electrical energy generated in the USA that year was nearly 2.3 million GWh. Coal was the main source (about 44%) and oil, gas, hydro, and nuclear power each provided between 12 and 17%.[27] Britain generated nearly 270 000 GWh in the same year.[26] That electricity is one of our major sources of energy barely needs stating. Its convenience, flexibility, and wide range of use are just as important now as they were at the start of the electrical age a century ago. Power applications have permeated our society from top to bottom; from heavy machinery in factories to electric pencil sharpeners in commercial offices, and from electric toasters on our breakfast table to the freezers in a supermarket. When a major blackout occurs we really discover how dependent we are on the long-term results of the work that began with Gramme, Tesla, and their fellow pioneers. Much of our industry, commerce, farms, and homes, and some of our transport comes to a sudden and disconcerting stop.

References

1. P. Dunsheath, Chapter 5, Ref. 28.
2. H. C. Passer, Chapter 6, Ref. 9.
3. I. C. R. Byatt, *The British Electrical Industry, 1875–1914*, Oxford University Press, 1979.
4. C. M. Jarvis, *JIEE* 1: 280, 566, 1955; 3: 310, 1957; 4: 298, 1958.
5. G. W. O. Howe, *Report of the British Association for the Advancement of Science*, Toronto, 178, 1924.
6. D. S. Landes, *The Unbound Prometheus: Technological Change and Industrial Development in Western Europe from 1750 to the Present*, Cambridge University Press, London, 1969.
7. J. A. Fleming, Chapter 5, Ref. 16.
8. C. C. Knox (unpublished manuscript).
9. D. G. Tucker, *Gisbert Kapp, 1852–1922*, University of Birmingham, 1973. (The original comment appeared in G. Kapp, *Proc. ICE* 83: 123–154, 1886.)
10. L. Hannah, *Electricity Before Nationalisation*, Macmillan, London, 1979.
11. C. Singer, Ed., *A History of Technology*, Oxford University Press, Oxford, 1958, vol. 5.
12. G. Siemens, Chapter 5, Ref. 9.
13. H. W. Cope, *Electric J.* 33 (No. 1): 3–16, January 1936.
14. F. J. Ellert and N. G. Hingorani, *IEEE Spectrum* 13 (No. 8): 37–42, August 1976.
15. B. G. Lammé, *Electrical World* 84: 601–603, 1924.
16. C. F. Scott, *Electrical World* 84: 585–587, 1924.
17. C. Singer, Ed., *A History of Technology*, Oxford University Press, Oxford, 1978, vol. 6, pt. 1.
18. F. J. Sprague, *Electrical World* 84: 576–578, 1924.
19. E. W. Rice Jr., *Electrical World* 84: 581–584, 1924.
20. R. Jones and O. Marriott, Chapter 6, Ref. 15.
21. R. S. Cowan, *Technology and Culture* 17: 1–23, 1976.
22. B. Gorowitz, Ed., *The General Electric Story: The Steinmetz Era, 1892–1923*, Elfun Society, Schenectady, 1977, vol. 2.
23. B. R. Mitchell and H. G. Jones, *Second Abstract of British Historical Statistics*, Cambridge University Press, London, 1971.
24. B. R. Mitchell, *Abstract of British Historical Statistics*, Cambridge University Press, London, 1971.
25. U. S. Bureau of the Census, *Historical Statistics of the United States, Colonial Times to 1970*, Bicentennial Edition, Pt. 1, Washington, D.C., 1975.
26. Central Statistical Office, *Annual Abstract of Statistics*, H. M. Stationery Office, London, 1981.
27. U. S. Bureau of the Census, *Statistical Abstract of the United States*, 100th edition, Washington, D.C., 1979.

8 RADIO

The quest to apply electromagnetic waves to communications is a little less than 100 years old; yet it has brought incredible changes to our way of life. Radio and its uses have continued to hit the headlines from its earliest days. Whether the story described experiments by Hertz, the arrest of Dr. Crippen, the rescue operation for the *Titanic*, the bombing of Pearl Harbor, the arrival of television, or men walking on the moon, it was followed avidly by the people of the day and brought to them by our mastery of electromagnetic radiation. In this chapter we shall attempt to review the highlights of the development and use of radio from the first radiotelegraphs, through wireless and the beginnings of electronics, to radar, television, and other present-day applications of the radio art. As in other chapters, some things will have to be left out if we are to cover the era from the early struggles to communicate across a few feet to the use of radio in an optimistic attempt to search for extraterrestrial life. Yet it is hoped that the chapter will present an informative summary of the development of radio in its most important forms.

Prehistory

The history of radio is usually said to have begun with the experiments of Hertz, although that is not the whole truth. Hertz climaxed what may be thought of as the prehistory of radio, a period that included both the theory of electromagnetism and a number of experimental observations. The theory came from the work of Faraday, Maxwell, and others and has been fully reviewed in Chapter 4, where mention was also briefly made of some of the experimental observations.

These observations occurred in a region described by Süsskind[1] as being where scientific discovery and practical invention overlap. It is near certain that they were genuine radio wave observations though not understood at the time, and were mostly made by men who had probably never heard of Maxwell's theory, let alone understood it. Besides these observations there is a long history of proposals for 'wireless' communications based on induction or conduction; for instance to communicate with lighthouses, ships, and so on by conduction through earth and water. Although these attempts have an inherent historical interest in themselves they are not part of the technical history of radio.

Ever since men have been able to generate reasonably large sparks there have probably been observations of the effects of the propagation of electromagnetic waves. Sparks are electrical discharges and are oscillatory in nature, a feature essential for the generation of electromagnetic waves. Their oscillatory nature was suspected in the 1820s, proved qualitatively in the 1840s, and mathematically analyzed and experimentally verified in the 1850s. It was this background that supported Fessenden's later suggestion of discharging a capacitor to produce an alternating current in a loop of wire so as to generate waves, and which led to the first commercial spark transmitters for radiotelegraphy.

Some baffling observations were made long before Hertz.[1,2] Some time between 1780 and 1791 Galvani is thought to have witnessed the effects of electromagnetic radiation as he struggled to understand electrical conduction. Over half a century later Joseph Henry was able to magnetize steel needles 10 m away from a 1-in. spark, which led him to think of an 'electrical plenum' and compare the transmission of this unusual effect with that of light. In 1875, after another third of a century, Edison thought he had discovered a new force of nature which he termed the 'etheric force.' He found that a telegraph key when used to interrupt an electrical circuit produced sparks that could then be picked up between any metal object (a local 'ground') and the free end of a piece of wire attached to a metal plate (an antenna). Edison's theory of his discovery was disproved the next year by Elihu Thomson, then still teaching in Philadelphia. An account of Thomson's work was written up by his older colleague E. J. Houston.

Thomson used a Rühmkorff coil to produce his sparks. One side of the spark gap was grounded and the other was connected to a large tin still insulated from ground, which (intentionally or not) acted as an antenna. His receiver was a small spark gap between two graphite pencil points, one of which was connected to a large brass knob. Sparks were obtained with this receiver up to at least 30 m from the transmitter. Thomson now had a base from which radiotelegraphy could have been developed and both he and Edison came to regret missing their respective opportunities. Edison is said to have puzzled as to why he never thought of using the results he had obtained.

Edison and E. Thomson were not the only ones to approach, and then turn away from, the threshold of radiotelegraphy. News of Edison's etheric force reached England and led a postgraduate student, S. P. Thompson, to investigate the phenomenon. He showed that the received sparks were oscillatory but wrongly concluded that the phenomenon was due to electrostatic induction. Induction was also the basis of explanations, from leading scientists in England, of demonstrations given by D. E. Hughes. In 1879

Hughes noticed that his microphone reacted whenever current was inter-
rupted in a nearby coil. The results were reproducible and, though it was not
realized at the time, the effects of standing waves were demonstrated. Visiting
scientists were impressed but disagreed with Hughes's theory of the phenomenon; his notebooks indicate that he thought conduction was taking place
through the air. Hughes became discouraged and did little more in this field.

In America, A. E. Dolbear went as far as applying for a patent in 1882 for a
new type of telephone system that operated without wires. In doing so he not
only stole a march on radiotelegraphy but on radiotelephony as well. The
transmitter and receiver were each connected between a free wire and ground,
but not to each other. Dolbear believed that conduction took place through
the earth, but it is more likely that it was via radio waves. Reception was
improved if the free wires were elevated, which would have made them into
better antennas; a gilt kite was used at the transmitter and an insulated tin roof
at the receiver. It was also found that reception was possible even when the
receiver was insulated from the earth, a demonstration that literally as well as
metaphorically took the ground out from under Dolbear's theory. The
receiver was based on an electrostatic microphone of his own design. In a
demonstration given at a meeting in London in 1882 Dolbear showed that the
microphone was sensitive enough to operate without being attached to the
receiving circuit. When the receiver was held to the chairman's ear he reported
that he heard the transmissions, which up to that point had included 'God
Save the Queen' and 'Yankee Doodle,' as loud "as the cry of a new-born
kitten."

However, as far as the development of radio is concerned, Dolbear's kitten
was stillborn. It was only after the scientific demonstration of 'Hertzian'
waves that radiotelegraphy was given life.

Radiotelegraphy

Hertz's transmitter was a simple spark gap across the secondary terminals of a
Rühmkorff induction coil and it set the pattern for the coming generation of
spark transmitters. Soon many people were to speculate whether these newly
found waves could be used for telegraphy. One of the first was William
Crookes in England.[3]

Writing in 1892 on 'Some possibilities of electricity' Crookes listed the
three main requirements as (in modern terminology) reliable transmitters,
sensitive tuned receivers, and directional aerials. Thirty-one years later Oliver
Lodge described Crookes's article as "an anticipation of genius."[2] Of these
requirements the poor sensitivity of the first receivers was the biggest
drawback, but a temporary solution was soon found in the coherer detector.

Coherers became the basis of radio reception for the first ten years or so of
radiotelegraphy. They were based on the phenomenon that light particles,
dust for example, stick together or 'cohere' in the presence of an electric field.

Observations have been dated back to 1850 and the effect is used in some modern electrostatic dust precipitators. The first to use the principle to detect radio waves was the Frenchman Édouard Branly, who achieved such distinction that when he died in 1940 he was given a state funeral. In 1911 he just beat Marie Curie, who discovered radium, in an election to the Academy of Sciences. He may also have been the first to use the word 'radio' in its present sense when he suggested the name radioconductor for the device which Lodge named the coherer. The term 'Radio-Telegraphy' was suggested by J. Munro in *The Electrician* in 1898.[4]

Branly discovered that the resistance of a glass tube of metal filings fell from a few megohms to a few hundred ohms when a discharge occurred nearby. Mechanical shock restored the coherer to its original state and electro-mechanical tappers were developed for this purpose. Spark transmitters produced highly damped broadband pulses which the coherer, though crude, could detect effectively. The coherer was popularized by Lodge, especially as a result of a lecture and demonstration he gave in 1894 in honour of Hertz, who had just died. Lodge himself had come close to anticipating Hertz, except that Lodge had experimented with waves transmitted along wires rather than through space. The lecture received wide publicity.

At a second lecture given in 1894 Lodge demonstrated not only coherers, but the effects of tuning as well; another of Crookes's requirements. In what he called the 'syntonic-jar' experiment he used Leyden jars with pairs of wires as the transmitter and receiver. The length of one pair of wires could be adjusted so as to bring the two circuits into resonance. He described it as a system of 'syntonic telegraphy,' emphasizing harmony or tuning. The transmitter and receiver had to be accurately tuned to achieve any response. Claims have been made that Morse code was transmitted during the demonstration but that is uncertain and it can only be said that 'signalling' took place. Süsskind[3] states categorically that there was "no attempt to transmit intelligence." Lodge himself, in 1923, merely claimed that something "akin" to signalling took place.[2] He also commented on his own lack of foresight in perceiving the importance of radiotelegraphy, something of whose potential he must have realized. At any rate, he did secure the fundamental tuning patent in 1897.

Lodge and Crookes are prime examples of serious scientists who braved criticism by examining areas outside the accepted realm of science; in their case it was psychical research. Meanwhile inside the realm of science, radio was soon to become a technology. Crookes had defined the requirements for radiotelegraphy and Lodge had brought it to the brink of achievement. Lodge's work inspired others to seek extensions and applications; Jackson in England, Popov in Russia, Slaby in Germany, and Righi in Italy. Through Righi that inspiration reached a young man who was to make radiotelegraphy his own, Guglielmo Marconi.

Marconi

Though others performed important work, Marconi is the man to whom we owe the greatest debt for the early application of electromagnetic waves to communications. He was born in 1874 in Bologna of a well-to-do Italian father and a Scots-Irish mother. Much of his education came from private tuition and he is said to have been a rather solitary child. His father was not impressed when he failed the Italian Naval Academy's entrance examination, still less so when he capped that by failing the matriculation examination of the University of Bologna. Marconi's scientific interests were not thwarted, however. Augusto Righi, at the university, was known to the family and allowed Marconi access to his lectures and laboratory, a relationship that could hardly have been bettered from the point of view of influencing Marconi in the direction of radio. Perhaps Marconi did better by going to university through the back door than if he had gone in through the front. Righi has been described as a master scientist, one of the few people who really understood what Hertz had accomplished. Righi's obituary of Hertz is traditionally said to have convinced Marconi that electromagnetic waves could be used for telegraphy. The year was 1894. As Hertz's work died, Marconi's was born.[5]

Marconi learned from Righi how to generate, radiate, and detect electromagnetic waves. From the very beginning it would seem that his desire was to communicate over ever greater distances, possibly to impress his father, who was paying for his tinkering.[5] He improved the coherer (Fig. 8.1) by reducing the distance between the contacts, sifting the filings to a uniform size, and partially evacuating the glass tube; and he improved on Lodge's and Righi's transmitters. A telegraph key placed in the primary of the induction coil enabled him to generate long or short trains of sparks. Soon he rediscovered the principle previously known to Dolbear and others of grounding one side of his transmitter and connecting the other to an elevated metal object which acted as an antenna (Fig. 8.2). The "grounded-vertical" antenna greatly

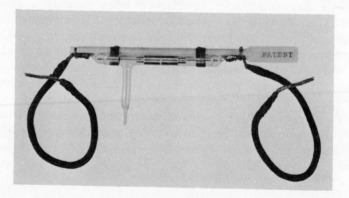

Figure 8.1 The Marconi coherer

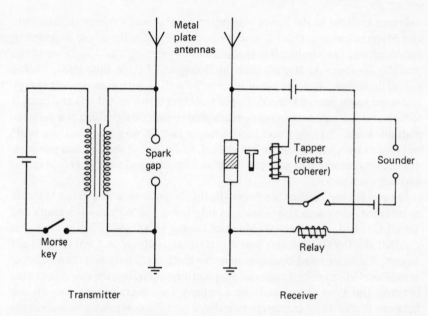

Figure 8.2 Marconi untuned transmitter and receiver with ground and antenna (1896)

increased his distances and was to continue to serve him well in the future. Its importance, though he did not then know it, lay in the fact that it generated long waves whose ground wave followed the earth's surface. That bigger antennas meant longer distances was quickly established as a rule of thumb. It has frequently been said that Marconi excelled not so much at creative invention as at improving the work of others. Equally important was his vision of radiotelegraphy in the service of mankind.

When he could operate reliably over a range of 1–2 km Marconi approached the Italian Ministry of Posts and Telegraphs. He got the same reply the British Admiralty had given Ronalds when he approached them with his telegraph system earlier in the 19th century: no, thank you.

As ships cannot trail telegraph wires behind them marine communications seemed to offer the next best opportunity and so Marconi turned to Britain, the world's greatest maritime power. There too lay his mother's extensive family connections, with the families that made Jameson Irish whiskey, and the Haig and Ballantyne Scotch whisky.

He arrived in England in February 1896, and though his equipment was damaged by a customs inspection his entrepreneurial drive was not. A meeting was arranged with W. H. Preece, the chief engineer of the Post Office and a man who had entertained a long interest in wireless communication by induction, a system that had proved itself impossibly cumbersome. After several demonstrations, including one over several miles on Salisbury Plain and one across the Bristol Channel, Preece gave the young Italian his considerable support.

Preece lectured to the Royal Institution on Marconi's system in June 1897 and Marconi came to enjoy considerable publicity. On the whole engineering comment was favourable, but Lodge's initial reaction was not. "One of the students in Prof. A. Righi's class at Bologna," Lodge thundered, "being gifted, doubtless, with a sense of humour as well as with considerable energy and some spare time, proceeded to put a coherer into a sealed box and bring it to England as a new and secret plan adapted to electric signalling at a distance without wires."[3] Lodge was hardly being fair. It was true that the basic techniques were widely known but as Prof. Slaby, one of the German pioneers, put it, everyone else had got "just as far as 50 metres and no farther." Marconi had achieved miles.

Despite Preece's efforts the British Post Office was slow in making Marconi an offer for his system and instead, in July 1897, The Wireless Telegraph and Signal Co. Ltd. was formed. Marconi took £15 000 ($75 000) in cash plus £60 000 ($300 000) in shares and a three-year contract at £500 ($2500) per annum; the money had been put up by his mother's family and friends. Some at the Post Office felt betrayed and a certain amount of mutual mistrust existed between the Post Office and the company for several years, though not between Preece and Marconi personally. From the company's viewpoint the Post Office was also culpable: it had been responsible for Slaby's unwanted presence at some of the early trials. Slaby's eventual patents were used by Marconi's German rival, Telefunken.

At the beginning the sales of Marconi's (as the company came to be known) were nonexistent. Some equipment was sold to the British Army in 1898 for use in the Boer War but for a time the obvious markets, such as ship-to-shore and competition with submarine cables, did not materialize. But in 1900 a contract was signed with the Admiralty to provide equipment and to train operators. Perhaps that was the key. After all, not even Marconi could hope to sell equipment to people who could not use it. A change of policy followed. Instead of equipment, a radio communications *service* was to be offered. Equipment would be leased together with Marconi personnel who would operate and maintain it. This arrangement also overcame the question of possible infringement of the Post Office's monopoly, which was meant to cover wired services but which might be construed to cover ship-to-shore as well. As yet there was nothing in the law to prevent one branch of a private company communicating with another branch of the same company, even if one was ashore and the other on a ship, even someone else's ship. A new company was formed, the Marconi International Marine Communications Company, and the original firm changed its name to Marconi's Wireless Telegraph Company.

Shipowners now had access to a communications system that used standard equipment and techniques, but one that refused contact with any rival except in emergencies. The new policy was a success and Lloyds of London, the world's major marine insurance firm, gave its stamp of approval with a contract signed in 1901. In practice this meant that any shipping line that used

Lloyds' vast network of shipping agents also had to use Marconi equipment. This monopoly, which was ultimately broken by an international convention on communications at sea, lasted seven years and effectively sealed the market against competitors; the Marconi Co. achieved dominance in marine radio.[5]

Marconi himself felt there was another market to be reached; competition with submarine cables. The first permanent installation was on the Isle of Wight, whence messages could be sent a few miles to the mainland at about 12 words per minute. When Kelvin visited the station in 1898 he sent telegrams to Preece and Stokes; he insisted on paying for them, which made them the first commercial radiotelegrams. The next year a radio link was established across the English Channel and Branly received a message acknowledging him as the inventor of the coherer. The next step was the attempt to span the Atlantic. If that could be achieved then full coverage of the North Atlantic shipping lanes could also be achieved, and that was probably a more important market than competition with the cable companies.

After more than a year of preparation, during which huge transmitting and receiving antenna arrays were built only to be destroyed by gales, the test began. A simpler transmitter antenna was constructed at Poldhu in Cornwall and a kite was used to hoist a receiving wire at St. John's, Newfoundland. The transmission consisted of the single Morse letter 's' sent at intervals during the day. The kite, bobbing up and down in the wind in December 1901, varied the capacitance and hence the resonant frequency of the receiving circuit. Reception was irregular and, as recording equipment had been replaced by a telephone receiver to increase the sensitivity, sceptics suggested that Marconi and his men had deceived themselves into believing that the clicks of atmospheric effects were the signal.

A month later all doubts were dismissed when automatic equipment recorded reception on board ship as Marconi sailed back across the Atlantic. The difference in transmission range between day and night was also noted: 700 and 1653 miles (1120 and 2650 km), respectively. At the age of 27, like Edison before him, Marconi had become a living legend.

Among scientific and mathematical comments on Marconi's achievement, which beforehand had generally been dismissed as impossible because of the earth's curvature, were those made independently by Heaviside and Kennelly suggesting the existence of a conducting layer in the upper atmosphere. Kennelly even calculated its height at about 50 miles, basing his work on the assumption that the air would conduct because of the low pressure. (That the air had to be ionized was realized later.) In 1924 E. V. Appleton in England measured the Kennelly-Heaviside layer at 60 miles (96 km) high and two years later discovered other layers at about 150 miles (240 km). They are now named after him. Robert Watson-Watt was probably the first to suggest the name ionosphere for this region of the atmosphere.[6]

In April 1900 Marconi was granted his most important patent, number 7777, his master tuning patent. Four tuned circuits were used: the transmitting and receiving antennas, the exciter circuit in the transmitter, and the detector

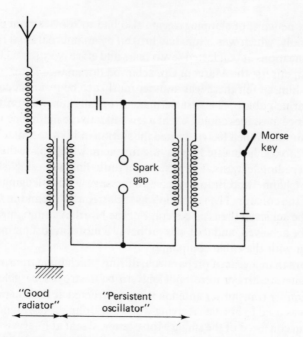

*Figure 8.3 Marconi's 'four-sevens' tuned transmitter based on two tuned circuits (1900)
(after Ref. 32)*

circuit in the receiver (Fig. 8.3). Until it expired in 1914 the 'four-sevens'
patent proved to be very valuable. However, Lodge's previous tuning patent
was Marconi's Achilles heel, as it has been described.[5] Though the company
paid no royalties to Lodge they did purchase his patent in 1911, after its life
had been extended. Ownership of both patents strengthened the Marconi
Co.'s position so that it gained almost total control of commercial radio in
America and was able to reach an agreement with Telefunken in Germany.[5]

Other Inventors

The only important early competition to Marconi came from two groups in
Germany.[3] One was led by Adolf Slaby, supported by AEG and favoured by
the Navy. The other was led by Karl Ferdinand Braun, supported in part by
Siemens & Halske and favoured by the Army. The two groups were rivals and
at one time Braun's group considered joining with Marconi against Slaby. But
in 1903, following government pressure, a new joint company was formed,
Telefunken, and all work on radiotelegraphy was transferred to it. Braun, who
is also known for his discovery of the rectifier effect in semiconductors and for
his invention of the cathode-ray oscilloscope, shared the 1909 Nobel Prize for
physics with Marconi for their work on radio.

In America John Stone Stone was probably the man whose thinking most closely approached or even surpassed that of Marconi. After mathematical analysis, Stone became convinced that a pure sine wave of uniform periodicity was needed if true tuning was ever to be achieved. Since spark-gap transmitters did not produce this waveform Stone saw the need for tuned circuits to act as filters. Later this idea became commonplace, but according to Aitken the first to appreciate it was Stone.[5] A patent was issued in 1902.

In Russia, an instructor at the Naval School at Kronstadt, Aleksandr Stepanovich Popov, was another of the men inspired by Lodge's early lecture. In 1895 he recorded atmospheric electrical disturbances with a coherer and an ink recorder; ever since then, there have been Russian claims that he demonstrated radiotelegraphy the following year and so 'invented radio.' After he visited Germany and France in 1899 he returned home confident that his work was not much behind that of others. Meanwhile, in England in 1896, Captain Henry Jackson carried out the Royal Navy's first radio signalling between two ships; his range of $3\frac{1}{2}$ miles ($5\frac{1}{2}$ km) was similar to Marconi's at the time. Later he collaborated with Marconi in conducting tests for the Royal Navy.

Arcs and Alternators

As spark radio was refined, and the Marconi company's position strengthened, the technological successors were already on the horizon. Not that spark radio was yet in any technological crisis, but that would come as stations proliferated and interference became a problem. Continuous waves (CW) were a necessity if tuning was to become really effective and in their wake would come amplitude modulation, radiotelephony, and broadcasting. The question was how to generate the CW at the high frequencies required. Three types of systems were developed, based on arcs, alternators, and electronic vacuum tubes.

In 1900 W. D. Duddell in Britain, who was associated with the development of the oscillograph, suggested a method of producing continuous waves that depended on a resonant circuit coupled to a DC arc. The negative-resistance characteristic of the arc made it possible for oscillations to be generated. A couple of years later a Danish engineer, Valdemar Poulsen, patented an arc transmitter that operated at frequencies up to 1 MHz, a performance he achieved by striking the arc in an atmosphere of hydrogen or hydrocarbon gas rather than air. A telegraph key could be used to short together a few turns of the antenna inductor; in this way messages could be sent by periodic shifting of the signal off frequency. The transmission speed was also increased; one claim was for 300 words per minute with use of a punched-tape input.[7]

Arc transmitters were used more by Americans than Europeans, perhaps because spark systems had been more extensively developed in Europe. Lee de Forest, for example, secured the American rights to the Poulsen arc and attempted a transatlantic link. Later his company was sued for fraud, although De Forest himself was innocent, and the rights passed to the Federal Telegraph Co., which installed an arc transmitter at the U.S. Navy's Arlington transmitter. Arc transmitters eventually provided powers up to around 500 kW and some stations used arcs for several decades.

The other major system developed was based on a high-frequency alternator. From the late 1870s onwards low-frequency alternators were developed for power applications and in the 1890s Nikola Tesla pioneered high-frequency alternators with a view to transmitting electrical power without wires. It was Reginald Fessenden in America, however, who patented an HF alternator for radio use. It was developed at GE, initially by Charles Steinmetz, but mainly by Ernst Alexanderson.

Fessenden did some work on a wireless system for the U.S. Weather Bureau but broke with the Bureau and persuaded two Pittsburgh capitalists to form a company to develop his work on radiotelegraphy and radiotelephony. As a result the National Electric Signaling Company (NESCO) was founded. Fessenden had transmitted speech in 1900 using a spark transmitter but the results were not satisfactory and it was clear to him that CW was needed.[5] Steinmetz built the first machine to Fessenden's design but the 10 kHz frequency was on the low side. By 1906 50 kHz had been achieved. This alternator, at Fessenden's insistence, used a wooden armature. Alexanderson favoured an iron armature and GE obtained a separate patent on that design.[7]

In January 1906 two-way radiotelegraphy was achieved across the Atlantic, though the results were no better than Marconi had achieved with sparks. A gale destroyed one of the antennas the next year and it was not rebuilt.[8]

However, radiotelephony progressed more rapidly. A demonstration was given over 11 miles (17.6 km) in December 1906, and on Christmas Eve a broadcast of speech and music was picked up by some doubtless rather surprised ship's wireless operators in the North Atlantic. NESCO's backers wanted to sell the company and its patents but there were no takers. Fessenden wanted to compete with Marconi and his relationship with his backers deteriorated, as it had with the Weather Bureau before; indeed Fessenden is said to have had a 'choleric' personality. He broke away to form a Canadian company. NESCO dismissed him. He then sued for breach of contract and won, and NESCO went into receivership to conserve its assets and continue its work. Eventually the company and its patents were sold, to Westinghouse after World War I, but neither of the backers lived to see it.

Coherers were useless for detecting CW and Fessenden developed an electrolytic detector which is said to have set the standard of sensitivity for years. A similar device was developed in Germany. High-frequency alternators were also developed in Germany and France, and some German models were in service in America in 1912. Fessenden's greatest contribution to radio, however, was his work on heterodyne reception.[8] While seeking to develop a better detector he hit on the idea of mixing the incoming high frequency with a locally produced signal of almost the same frequency so as to produce an audible beat note. A 100 kHz transmitted signal, for example, mixed with a local frequency of 101 kHz at the receiver, would produce an audible 1 kHz signal. Tests by the U.S. Navy in 1910 and 1913 proved the system to be worthwhile, but the noisy arc generator used as the local oscillator did not show the principle to its best advantage. Not until the advent of electronic oscillators did the heterodyne come into its own.

Fessenden's second order to GE for a high-frequency alternator was entrusted to Alexanderson, a 26-year-old graduate of the technical university in Stockholm and a man who had also studied under Slaby. His work at GE continued despite NESCO's near bankruptcy. Ideas were developed, patents were taken out, and a new customer was sought. In the course of time a magnetic amplifier was built to modulate the carrier with the human voice. Electronic amplifiers were also built when the triode became available and by 1915 GE had a complete system of CW transmission and reception, but still no customer.

GE's desire was for design and manufacture, not for competition with Marconi or AT&T in the communications business. The Marconi Co. was an obvious potential customer and Marconi himself examined the alternator in 1915. Although the Marconi Co. was still very strong, it had no access to this alternative system, and so an agreement was reached whereby GE would have sole rights to manufacture the Alexanderson alternator and Marconi would have sole rights to its use. The agreement was postponed because of World War I.

After the war the U.S. Navy became increasingly worried that its radio communications would depend entirely on foreign interests (cables linking America and Germany had been cut by the British in 1914), and pushed for an American rival to the Marconi Co. The result was the Radio Corporation of America (RCA), in which four companies were involved: GE, American Marconi, AT&T, and Westinghouse. All four had a big stake in radio. Three were blocked in international communications by the British Marconi Co. and none could get the best designs because of patents held by the others. In the end RCA held rights to some 2000 patents, including virtually all the patents important to the radio science of the day and including agreements with the major companies in Britain, Germany, and France. An American radio giant had at last been born.

As CW systems were developed Marconi sought to use his spark expertise to achieve a semicontinuous timed spark that approximated to a continuous wave. In a sense this was the ultimate Marconi spark transmitter and was used at the international transmitter at Caernarvon for a few years. It was noisy and a Poulsen arc was held in standby. Eventually it was replaced by an Alexanderson alternator.

However, the new star on the horizon was not the alternator, but the radio valve or tube; and it was to glow undisputed for well over thirty years. When triodes made electronic radio a possibility the basic principles of radio—

selective tuning, the modulation of continuous waves, and heterodyning—
had already been laid down by courtesy of the older technologies.

Electronic Radio

The thermionic valve, or vacuum tube, was the base on which electronic radio,
and with it the electronics industry, was built. Its genesis can be traced back to
the incandescent-lamp industry of the 1880s when problems with broken
filaments and blackened bulbs were investigated, the cause of which, it was
generally held, was the ejection of charged particles of carbon from the
filament. When Edison probed inside the light bulb with a wire (Fig. 8.4) he
found that a current would flow if the wire was connected to the positive end of
the filament. This flow of current through a vacuum, the Edison effect, was
investigated by several people after its discovery in 1880. In 1889 J. A. Fleming
in London had some special bulbs made by Ediswan and resumed his own
study of the effect. He established that one-way conductivity was a basic
property of an evacuated space containing a hot negative and a cold positive
electrode, thus confirming the previous results of Edison, Preece, and several
German workers.[9] Meanwhile Edison had used the effect in a voltage
indicator and in 1884 obtained what may be viewed as the first patent in
electronics.[9,10]

Figure 8.4 The Edison effect; from Edison's notebook, 1880

It was not until October 1904, when Fleming was acting as a consultant to
the Marconi Co., that he had "a sudden very happy thought." Telephones and
meters were too slow to register the positive-negative cycling of a high-
frequency radio signal and therefore only indicated the zero average value.
Yet his Edison-effect lamps could rectify the signal, and so make it possible for

the resulting intermittent DC to be registered. Fleming's lamps, or oscillation valves as he now called them, could be used as radio detectors. The next month he wrote to Marconi, "I have been receiving signals on an aerial with nothing but a mirror galvanometer and my device."[9] The vacuum diode, the first valve or tube, had arrived (Fig. 8.5). However, Fleming got relatively little joy from his valve. It was no panacea for all detection problems and played only a small role in the early years of radio. A couple of years later H. H. Dunwoody, of the De Forest Wireless Company, produced an important rival, the crystal detector. Meanwhile the Marconi Co. manufactured and used some diodes, and they held the patent. Worst of all for Fleming and Marconi's, much of the limelight was stolen by Lee de Forest's invention of the triode.

In America, De Forest had been working on the problems of radio wave reception for several years. He was determined to challenge Marconi's in wireless and had fame and riches as his goal, with Tesla, Edison, and Marconi as his examples. To do it he needed a detector that did not infringe the patents of the others. In his student days he is said to have written, "I always seem lost in the financial woods," and the affairs of his companies would seem to support that statement.[8] He had little enthusiasm for working for others and instead launched many companies, several of which foundered. Though he was not an entrepreneur on the scale of Edison or Marconi he did at least have a touch of the showman in him. In 1908 he transmitted from the Eiffel Tower and two years later staged the world's first opera broadcast, from New York's Metropolitan Opera House, with Caruso as the star. Reception was poor and interference came from "an operator somewhere who was carrying on a ribald conversation with some other operator, greatly to the detriment of science and an evening's entertainment."[7] In 1912 De Forest and his associates of the De Forest Radio Telephone Company were charged with fraud. It was stated that the company's only assets were "De Forest's patents chiefly directed to a strange device like an incandescent lamp which he called an Audion and which device had proven worthless."[8] De Forest was exonerated but two of his associates were jailed.

After hearing of Fleming's diode De Forest took out patents on modifications to its circuitry and made diodes himself. The path to the invention of the triode has been clouded by the De Forest 'legend,' and it has been claimed that there was often only a superficial relation between De Forest's recollections and the actual sequence of events.[11] He is also said to have made broad claims about his diode which have since been described as "technically unjustifiable,"[11] perhaps to build up a defence against a possible suit for infringement of Fleming's patent. Such a suit came nine years later and was won by the Marconi Co. Considerable personal animosity also developed between Fleming and De Forest. Patent litigation made matters worse, with American courts at one time ruling that neither side could make triodes without infringing the other's patent!

De Forest experimented with many electrode configurations and applied for patents for three electrode tubes in 1905 and 1906. The best results, he decided, were obtained when the third electrode was made from a fine wire grid and placed between the other two. A patent applied for in 1907 related to this grid-triode, which became known as the audion, a term De Forest used loosely for all his diodes and triodes (Fig. 8.6).

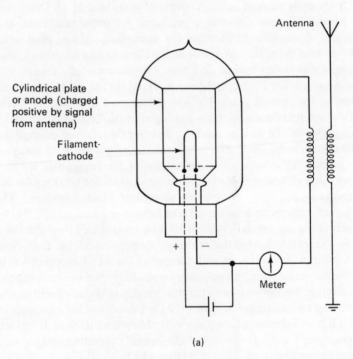

(a)

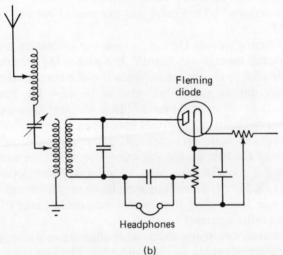

(b)

Like the early diode, the early triode was not a great commercial success and probably was not much better than the diode as a detector. Crystal and electrolytic detectors were more popular. The inventor had little understanding of how it worked (a not uncommon plight among inventors) and made none of the important changes that transformed it into the basis of electronics. Nevertheless De Forest had considerable inventive talent. It was he who discovered how to use the triode in a circuit as an amplifier and as an oscillator.

In 1912 he built a cascade amplifier using first two and then three of his audions. Then with the aid of his friend John Stone Stone he demonstrated an electronic amplifier to the telephone company. Though the performance was erratic and the output distorted, the potential of the equipment was evident. At AT&T, H. D. Arnold was given the task of investigating it. He quickly noticed the glow discharge caused by the ionization of the residual gas in the bulb. Correctly he rejected the gas as unnecessary, though De Forest had

(c)

Figure 8.5 (above and opposite) Flemings diode: (a) 1905 patent (redrawn); (b) typical circuit diagram (after Ref. 29); (c) experimental diode

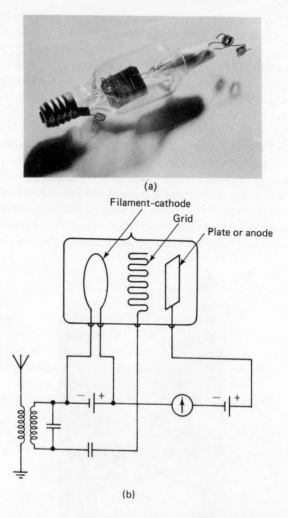

Figure 8.6 (a) De Forest's audion (triode); (b) from patent application, 1907 (after Ref. 11)

thought it important, and with the aid of a high-vacuum pump the triode was improved. It was thanks to the work of large industrial laboratories, especially those at AT&T and GE, that the crude triode was transformed into a reliable and efficient device. Less than a year after the De Forest-Stone demonstration vacuum-tube repeaters had been built and tested on commercial telephone lines, and De Forest had received $50 000 (£10 000) for the telephone-repeater rights of the triode.

Meanwhile De Forest discovered how to use the triode as an oscillator, a third use for the device and one that opened up the possibility of competition with arcs and alternators. With the heterodyne principle already established the triode now had a chance to be used as an amplifier and oscillator at both the transmitter and receiver, and also as a detector at the receiver. AT&T bought the radio receiver rights to the triode in 1914 for $90 000 (£18 000). The age of electronic circuitry had begun.

The first important circuit was that using positive feedback, or regeneration, which was investigated for its effects on both amplification and oscillation; 1913 saw several claimants to its invention in America and Europe. Edwin Armstrong described his 'regenerative' circuit to the new Institute of Radio Engineers (founded in 1912 and a forerunner of the present IEEE) and shortly afterwards De Forest applied for a patent for the same principle. Irving Langmuir in America and Alexander Meissner in Germany were also inventors. The question of who was first became important because of the commercial applications. In America the ensuing patent litigation dragged on for twenty years and led one electronics magazine to lament the millions of dollars that had been spent on lawyers instead of on research. The Supreme Court eventually ruled in favour of De Forest, who had sold his patent to Bell in 1917. The engineering community instead honoured Armstrong, whose patent had been purchased by Westinghouse in 1920.[8]

Progress in electronics was rapid from the very beginning. The triode was extensively studied and improved in the first decade of its use and World War I encouraged the development; over one million were used in that war. Harold Arnold at AT&T used an oxide-coated cathode, invented by A. Wehnelt in Germany, which provided a longer life because it operated at a lower temperature. By the middle of 1913 a filament life of 1000 hours had been achieved. Circuitry was developing also. Examples from AT&T include the push-pull amplifier from E. H. Colpitts in 1912, feedback circuit modulators, an inductive oscillator by R. V. L. Hartley, and its capacitive equivalent by Colpitts.[12] Meanwhile the GE engineers were not idle. Langmuir also developed the high-vacuum triode and he too understood that electron emission was impeded by the residual gas. It is interesting to note that O. W. Richardson's scientific theory of thermionic emission (1903) predates all this work but was not then widely accepted. GE acquired rights to the German tungsten lamp filament and then developed ductile tungsten, which made tungsten available for cathode filaments. Work was also in progress on other devices such as x-ray tubes.

With similar work progressing at AT&T and GE, patent collisions were inevitable. Between 1912 and 1926 there were no fewer than twenty patent interferences between the two companies over tube improvements and circuit techniques such as modulation, feedback, carrier-wave suppression, current limiters, and sideband transmission.[8]

Nor were developments and patents limited to the United States. In Vienna Robert von Lieben applied for a patent in 1906 for a 'cathode ray relay,' a device similar to Fleming's diode.[13] Four years later he patented the idea of a third electrode. Though Lieben tubes were far from perfect, and included mercury vapour in a mistaken belief that it helped the operation, AEG and Siemens & Halske were both interested. As a compromise Telefunken, their joint venture, got the patents. The German Post Office received its first valves from AEG about the same time that AT&T acquired the audion. In Britain the Marconi Co. held Fleming's original diode patent.

The circuit that was to become the standard radio circuit, the super-heterodyne (Fig. 8.7), was not subject to the claims and counterclaims of some other circuit techniques. Its undisputed inventor was Edwin Armstrong, the American who did so much for radio. While serving in France in World War I Armstrong pondered the gunnery problem of locating enemy aircraft. The solution seemed to lie in detecting the high-frequency radiation emitted by the aircraft's ignition system. It could then be heterodyned twice, first to an inaudible intermediate frequency that could be amplified, and then to an audible frequency. The superhet was patented in 1920 but was not used for aircraft detection as originally conceived. This important patent was acquired by Westinghouse and became available to RCA. In 1920 Westinghouse handed over \$335 000 (£84 000) for Armstrong's regenerative and superhet circuits and other patents.[8]

Meanwhile receiving sets were on sale based on the 'neutrodyne' circuit. During the war the U.S. Navy asked a professor of electrical engineering, L. A. Hazeltine, to develop its radio equipment. After the war Hazeltine was approached by a group of small radio manufacturers who were seeking a way of moving from crystal sets to tube sets without infringing RCA's feedback patents. For a time the neutrodyne solved their problems and sold well after its commercial introduction in 1923, a year ahead of the superhet. In the end it was judged to be dependent on an earlier RCA patent.[8]

Meanwhile the design and manufacture of triodes also improved. Better vacuum was obtained by outgassing of the electrodes during evacuation and by use of magnesium as a getter. Thorium added to the tungsten improved the cathodes. Later, platinum coated with alkaline-earth metal oxides was used until the vapour process was developed in 1924 to produce oxide-coated cathodes without expensive platinum.[14] Indirectly heated cathodes made it possible for AC to be used for the heater and so eliminate the need for substantial batteries. By 1925 several years' service could be expected from some valves. The advantages of electrostatic screening were also realized from the early days. At first this screening was achieved by placement of the valve inside a metal can; later the outside of the glass bulb was coated with zinc or copper.

The neutrodyne had evaded the capacitance defects of the triode by circuit techniques, but this artifice was rendered obsolete by the screen-grid tetrode, introduced about 1927, only to be superseded by the pentode in 1929. Pentodes (three grids between cathode and anode) yielded improvements in low-frequency amplification and stability.

Completely new types of vacuum tubes became available and the variety was astonishing. The 1930 *RCA Tube Handbook* listed fifty-nine types of

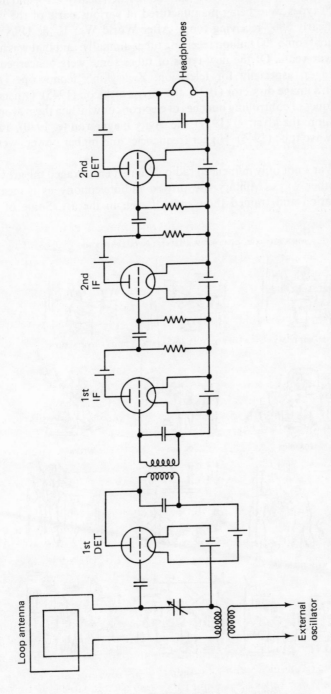

Figure 8.7 Early circuit of Armstrong's superheterodyne receiver (after Ref. 29), with two detector and two intermediate-frequency stages

tubes.[15] Twenty years later another handbook listed nearly 1200, and in 1949 over 10 000 types were being manufactured in various parts of the world, including nearly 5000 receiving tubes. After World War II the USA alone manufactured some 200 million receiving tubes annually, and that was a drop from the war years. Of the new types of tubes some were concerned with optical images, especially for television: Zworykin's iconoscope (1933), Farnsworth's image dissector (1934), the image orthicon (1945). Others used new techniques of controlling bunches of electrons or whirling them around in helical fashion: the klystron (1939), the cavity magnetron (c. 1940), and the travelling-wave tube (1946). Higher frequencies and higher powers were the goals.

World War I not only influenced radio design, it encouraged the growth of radio in other ways. Military use proved radiotelephony as a means of communication and trained thousands of men in the art. Some of them

Figure 8.8 'Punch' cartoon, radio, 1924

continued their interest in radio as amateurs on returning to civilian life (to the amusement of readers of *Punch*, Figure 8.8), and some became broadcasters on a limited scale.

Even earlier, amateur enthusiasts had grouped together to discuss and experiment.[16] The first wireless club in Britain was formed at Derby in 1911; as a method of exchanging technical information, members collected books and cuttings from magazines such as *The Electrician* and *The Marconigraph* (later renamed *Wireless World*). British clock manufacturers became interested in radio because of the time signals transmitted from the Eiffel Tower, an alternative to that from Greenwich which came via the railway telegraph. Amateur radio enthusiasts formed a national society, The Radio Society of Great Britain, in 1922. The professional society now called the Institution of Electronic and Radio Engineers was formed in 1925.

Broadcasting by amateurs eventually led others to perceive the market for radio broadcasting. For example, in Pittsburgh Frank Conrad, a Westinghouse engineer, started to broadcast a two-hour program twice a week. The response was such that a local department store started to sell crystal sets and the net outcome was a regular broadcasting station, Westinghouse station KDKA, which opened in November 1920. In Europe regular 'Dutch Concerts' were broadcast from the Hague in Holland earlier the same year, and in June Dame Nellie Melba broadcast from the Marconi station at Chelmsford. In Derby one member of the local club squashed the town's Salvation Army band into his home and broadcast a Sunday afternoon concert.

Broadcasting mushroomed and, especially in America, the situation became somewhat chaotic.[8] In the USA at the end of 1922 there were thirty licensed stations; two years later there were over 500 and the issue of licenses had to be halted. RCA (under the leadership of the visionary David Sarnoff) and AT&T both sought dominance. AT&T had by that time sold its RCA stock. Any station competing with an AT&T station was not allowed to use the telephone connections of the Bell System; rivals therefore turned to Western Union for help with their outside broadcasts. In 1926 peace was achieved. Mutually beneficial agreements were made and Bell dropped out of entertainment broadcasting.

RCA, GE, and Westinghouse then came together to form the National Broadcasting Company (NBC). Other networks followed, Columbia (CBS) in 1927 and Mutual in 1934, for example. In Britain, the British Broadcasting Company came into existence as station 2LO in November 1922, the same year that a Soviet station opened in Moscow. The BBC was then privately owned. In 1927 it became a government-owned monopoly under Royal Charter and was changed from a company to a corporation. It held its radio broadcasting monopoly until 1973, when commercial radio made its debut in Britain under the Independent Broadcasting Authority. Earlier, in 1954, it had lost its television monopoly after 18 years of continuous service.

In 1937 the wireless stations of the world paid a unique honour to Marconi.

After news of his death, in July, radio stations throughout the world closed down for two minutes, an impressive gesture to the memory of radio's foremost pioneer.

FM Radio

One problem that plagued radio from its earliest days was that of 'static.' Since static is amplitude modulation, any AM radio system is prone to static interference and it was natural for radio engineers to try to reduce this interference. Many tried and failed. Pupin is reported to have summed it up with the words, "God gave men radio and the Devil made static."[17]

As early as 1914 E. H. Armstrong, a man who became a millionaire through his radio inventions, had studied ways of beating static. In 1922 he called it a terrific problem, "the only one I ever encountered that, approached from any direction, always seems to be a stone wall."[18] In the end it was Armstrong himself who climbed the stone wall when he obtained four patents in 1933 dealing with frequency modulation (FM). Yet FM was not a new principle. It was first tried out in 1902 and again in the early 1920s, but in general it was treated as if it were like AM and required a narrow band of frequencies. Traditional thinking was to keep the bandwidth as narrow as possible, to let the signal through while keeping interference to a minimum. Used in that way FM had little to offer and was discarded.

Amstrong's answer was to step completely outside the existing state of the art by going the other way. In his own words, "The invention of the FM system gave a reduction of interfering noises of hundreds or thousands of times. It did so by proceeding in exactly the opposite direction that mathematical theory had demonstrated one ought to go to reduce interference. It widened instead of narrowed the band." He also commented that learned mathematical treatments had discarded FM as "totally useless or greatly inferior to amplitude modulation."[18] There is a lesson here that every engineer should remember.

A transmitter and receiver were built with Armstrong himself paying the equipment and salary costs. The transmitter used a stable crystal-controlled oscillator whose output was modulated by phase shift. At the receiver the signal was heterodyned to an intermediate frequency and clipped to remove any amplitude modulation, static interference included. A discriminator then converted the frequency variation into an amplitude variation for detection and reproduction in the usual way. A demonstration was given in 1935 at a high frequency, 110 MHz. Such high frequencies had been held to be of little use but in fact helped Armstrong to achieve success. The signal-to-noise ratio of around 100:1 was a lot better than the 30:1 of the best AM stations.

In 1934 Armstrong had brought his new invention to the attention of RCA. Trials were held but RCA had other ideas. To them the future lay with television, and television and FM radio would be competing for the same frequency band. Also, to make FM broadcasts a reality, new transmitters and new and more expensive receivers would be needed. It was alleged that FM would not operate beyond the horizon and that a similar reduction in static could be achieved with AM at higher frequencies. For two years Armstrong had given RCA exclusive information about his system, but now came the break. RCA announced its decision to develop electronic television and Armstrong withdrew to go it alone, selling a block of his RCA shares as he went.[8,17]

Again he used his own money, this time to build his own station. Single-handedly he battled on until he got an FM station onto the air in July 1939 and proved that FM really did work. He then found an ally in a New England network that was suffering badly from static. FM broadcasting began to take off. By January 1940 some 150 applications had been made to operate FM stations in the USA and twenty were operating or near completion. Westinghouse, GE, and Zenith were among companies that applied to manufacture receivers and pay royalties. RCA refused to pay royalties but offered a single payment of $1 million ($250 000). Armstrong refused; he had already spent $700 000 or more to get that far.

World War II delayed legal proceedings but they began in 1948. When the patents expired in 1950 $4.5 million had been paid in royalties. Armstrong felt he owed it to the firms that had paid royalties and to himself to prosecute RCA and other firms that had not. The cases, over twenty of them, dragged on; the last crucial one, against Motorola, was not settled until 1967. But the strain proved too much for the man who gave the industry feedback, the superhet, and frequency modulation. In January 1954 he ended it all by committing suicide. Shortly afterwards, across the Atlantic, the BBC announced its decision to build a nationwide FM network to provide high-quality radio broadcasts. Armstrong's widow (who had been Sarnoff's secretary at RCA) continued the fight but had to finance it by accepting RCA's $1 million as a once-and-for-all payment—the same deal her husband had rejected fourteen years earlier. It must have been a bitter moment for her. Eventually Armstrong was vindicated and a $10 million settlement was received from various companies on just three crucial years of the patents' lives, 1948–1950.[17]

The whole sorry story of FM radio appears as a sad comment on the radio business's treatment of its greatest inventor, E. H. Armstrong; the little but richly inventive man who made frequency modulation work. With his outstanding achievement one cannot help feeling that, as far as FM is concerned, he really did deserve better treatment than he received.

Television

Writing in 1934 one author expressed the view that "The future of television seems now to be more hopeful to a degree which, only a short time ago, would have been decidedly optimistic."[19] His view was soundly based. High-definition television was becoming a reality and the official opening of the BBC's service, the world's first regular high-definition television service, was just two years away.

Still pictures had been transmitted by telegraphy as early as the 1860s, and several proposals for 'seeing by electricity' followed the discovery of the photoconductive properties of selenium in 1873. Mechanical picture scanners depending on a series of apertures arranged in a spiral were suggested. Two of them, Paul Nipkow's revolving disc (1884) and Lazare Weiller's mirror drum (1889), were eventually used in electromechanical television systems in the 1920s and 1930s. But the technology of the 1890s could not match the ideas. The response time of selenium was much too slow, and Kerr cells, polarizers, electromechanical shutters, and so forth could not be developed into television receivers.[20] Though cartoonists in *Punch* magazine could sketch wide-screen television (Fig. 8.9), the real thing still lay far in the future.

EDISON'S TELEPHONOSCOPE (TRANSMITS LIGHT AS WELL AS SOUND).

(*Every evening, before going to bed, Pater- and Materfamilias set up an electric camera-obscura over their bedroom mantel-piece, and gladden their eyes with the sight of their children at the Antipodes, and converse gaily with them through the wire.*) Paterfamilias (in Wilton Place). "BEATRICE, COME CLOSER, I WANT TO WHISPER." Beatrice (from Ceylon). "YES, PAPA DEAR."

Paterfamilias."WHO IS THAT CHARMING YOUNG LADY PLAYING ON CHARLE'S SIDE?"
Beatrice. "SHE'S JUST COME OVER FROM ENGLAND, PAPA. I'LL INTRODUCE YOU TO HER AS SOON AS THE GAME'S OVER?"

Figure 8.9 'Punch' cartoon, television, 1879

Low-definition electromechanical television became a reality in the 1920s after J. L. Baird in Britain and C. F. Jenkins in America demonstrated television 'shadowgraphs' in 1923, both using Nipkow discs. Baird collected an impressive string of TV achievements, including telephone-line transmission, colour TV, experimental broadcasts with the BBC (1929), and an on-location broadcast of the Epsom Derby in 1931. American companies were not behind. H. E. Ives (Bell) hit the headlines in 1927 and Alexanderson of alternator fame (GE) followed the next year. Jenkins formed a television company in 1929 and programs were scheduled for 1930. When it became obvious that profits could not be expected the company quickly closed.[8]

All these systems used mechanical scanning—and they were doomed. Baird's 30-line, 12.5-frames-per-second picture of 1929 was pushed to 120, 180, and even 240 lines, and probably represented the peak achievement of electromechanical television. But in 1936 when it competed against a Marconi-EMI electronic system for adoption as the BBC's standard it was soundly beaten.[21] The future lay with electronics (Fig. 8.10).

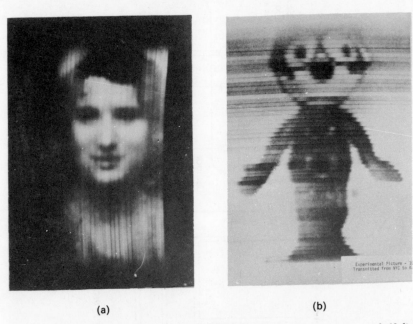

(a)   (b)

Figure 8.10 (a) Baird's 30-line 'Televisor' picture, 1929; (b) experimental 60-line transmission by RCA-NBC, 1928

Electronic television required a camera, improved cathode-ray tubes (CRTs), and new circuitry for synchronization, time bases, and wide-bandwidth amplification.

The long haul to electronic television was begun by Boris Rosing in Russia. Ferdinand Braun in Germany had developed the cathode-ray oscilloscope in 1897 from the earlier Crookes and Hittorf tubes by adding deflection electromagnets to a CRT. In 1907 Rosing used a CRT to display images from a mechanical transmitter and thereby laid the foundations of scanning by electrons (Fig. 8.11). The next year A. A. Campbell-Swinton in Britain suggested using a CRT at both transmitter and receiver. "It is an idea only," he wrote later, "the apparatus has never been constructed."[20] Nor was it constructed for many years. The solution of some of the problems that had to be overcome became feasible as electronic circuitry was developed, but the

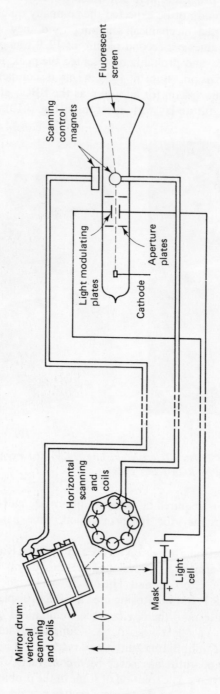

Figure 8.11 Principle of Rosing's television, 1907 (after Ref. 8)

major stumbling block was how to transform an image into a video signal at the camera end.

The first success came from Vladimir Zworykin, who, together with Westinghouse and later RCA, was largely responsible for making electronic TV a reality. Zworykin was a former student of Rosing and it was from Rosing he caught the television bug. After the Bolshevik Revolution he went to America where in 1928, while working at Westinghouse, he at last produced a working photoelectric television camera tube, the iconoscope (Fig. 8.12). The idea had taken about nine years to develop but the iconoscope was to revolutionize television. Its principle was ingenious. The optical image was

Figure 8.12 Zworykin and his iconoscope

focussed onto a photoemissive mosaic formed on a sheet of mica, each photocell insulated from the rest. The resulting positive charge on each cell was neutralized by the scanning electron beam and the output signal was picked up by capacitive effects on a metal backing plate.

Meanwhile Philo Farnsworth, an independent inventor, was developing the 'image dissector' at a private laboratory in California. As expenses rose his original backers sought help and Philco provided funds for two years, 1930–1932. Eventually a company was formed in 1938 and the wartime radio boom put the company onto its feet. Over the years a strong patent position was built up and RCA felt the need to take out Farnsworth licenses.[8]

In 1930 the radio research of GE and Westinghouse was transferred to RCA where David Sarnoff, RCA's general manager, gave Zworykin enthusiastic backing. By 1932 RCA could demonstrate 120-line all-electronic television. Experimental broadcasts were made in 1936 and limited commercialization was authorized in 1939, by which time RCA had spent more than $9 million on television work. In Britain meanwhile the Marconi-EMI electronic television had been adopted by the BBC in 1937 in preference to Baird's mechanical TV; the crucial factor was an improvement of the iconoscope, which had been acquired via a link with RCA. The British work was led by another Russian emigré Isaac Schoenberg, and Alan Blumlein (one of the best electronics engineers Britain has ever produced, according to Bernard Lovell) was a member of the team.[30,31] Blumlein also made pioneer contributions to stereo sound recording and radar. The EMI team pioneered many of the television engineering techniques that became standard throughout the world, including the interlaced picture and the composition of the video signal. This work was performed independently of RCA and there was no full exchange of technical information until 1937. EMI's 'emitron' camera set the British standard at 405 lines with a picture repetition rate of 25 (half the power-line frequency). The American standard was recommended in 1941 by the new National Television Systems Committee (NTSC): 525 lines at 30 frames per second.

The war years spurred electronic research in many areas; the CRT for example was improved for radar use, but in general television went into hibernation. At the end of the war RCA announced the image orthicon, an improvement over the iconoscope which suffered from some problems especially in low lighting conditions. Again it was RCA that developed the next camera tube, the vidicon, announced in May 1950, the first camera tube to use photoconductivity rather than photoemission. The vidicon's major advantage was its small size, a byproduct of its simple construction. It was ideal for portable use. In the early 1950s the English Electric Valve Co. developed a 4.5-in. (11.4 cm) image orthicon for the BBC. It was so successful that America was soon importing a British development of an American invention.[21]

After the war television services grew and within twenty years many came to consider a TV receiver an essential item of household equipment. Colour slowly became the norm. Colour had of course been demonstrated in several of the electromechanical systems in the late 1920s and 1930s, and such schemes reached a peak about 1940. Meanwhile various alternative approaches to the design of a colour CRT (or kinescope in America) were being explored. By 1942 Baird could demonstrate all-electronic two-colour pictures by forming two images on the screen of a single CRT, and then combining them optically to present a single picture. RCA took a similar approach but used three separate CRTs to obtain primary colour images which were then superimposed. Many other ingenious schemes were also devised for producing a single three-colour image on a CRT screen by use of either one or three electron beams. Some used velocity or current modulation of the beam; others used electronically controlled deflection plates at the screen.[22] But it was the invention of the shadowmask tube by RCA, demonstrated in 1950, and the development of techniques to ensure compatibility between colour and monochrome transmissions, that made commercial colour TV a reality.

The shadowmask tube uses a technique sensitive to the direction of the three electron beams used. An array of dots of phosphors of the three primary colours is deposited on the screen. Each of the three phosphors can be excited by only one electron beam and is shielded from the other two by the shadowmask. Probably the earliest 'direction-sensitive' method was proposed in Germany by W. Flechsig in 1938 and used a grid to produce the shadowing; one or three electron beams could be used. Baird demonstrated a two-colour direction-sensitive system in 1944, using widely separated electron beams, and he had proposals for extending his scheme to three colours. RCA's system was proposed by A. N. Goldsmith and A. C. Schroeder and required a team effort to make it a reality. Special printing techniques were developed to lay down the thousands of phosphor dots required.

In 1953 America adopted as a standard the NTSC dot sequential system, using the shadowmask tube and two image signals (one for brightness, colour or black and white, and one for colour.) In Europe two variants of the NTSC system were developed: the German PAL system from Telefunken, and the French SECAM. Germany and Britain began broadcasting using PAL in 1967; France and Russia opted for SECAM. At the same time Britain began to adopt the European 625 lines and phase out its obsolete 405-line standard. NTSC, PAL, and SECAM have remained as rival standards.

Television, like all areas of electronics, continues to be developed. Charge-coupled devices have been suggested as a potential successor to the CRT, laser-based holography has been examined as a possible source of 3-D television, and high definition television has been built using more than 1000 lines. Meanwhile home videotape recorders, with competing standards, have been marketed for several years, far cries from the first studio machine of 1956 (Ampex) which obviated the need for film for telerecording. Home televisions have been turned into 'electronic magazines' either by the slotting of data into unused broadcast lines (Teletext, Britain, 1975) or by hooking up to a computer via the telephone line (Viewdata, Britain, 1979).[23] But not every technical advance is accepted by the public. Bell's Picturephone (a scheme for transmitting a speaker's image along an ordinary telephone line) flopped when introduced in 1970. The public will only buy what it wants, though sometimes it can be persuaded to want things it did not know it wanted.

Radar

The acronym 'radar,' for radio detection and ranging, has been credited to the U.S. Navy, which used it officially towards the end of 1940, but the concept of radar is somewhat older.[24] Hertz showed that metals would reflect electromagnetic waves and Tesla is said to have suggested using this phenomenon in a radar-like manner in 1899. A few years later a German, Christian Hülsmeyer, received patents for a ship's anticollision device. Also many radio engineers and experimenters observed that passing aircraft or ships, and in one case a steam roller,[25] interfered with their experiments.

Although these features are all suggestive of radar none was actually radar unless the term is very loosely defined. In the 1930s, however, several of the major powers became aware of the military possibilities of radar and work started in the USA, Britain, France, Germany, Italy, Japan, and the Soviet Union. By the end of World War II military radar, and military radio navigation aids too, were well developed.

R. M. Page and his colleagues at the U.S. Naval Research Laboratory take credit for what was probably the first true radar equipment, which gave some mediocre results in 1934.[24] Improvements led to "spectacular success" in April 1936.[26] The project had originally been suggested in 1931, but Page's work began in 1934 following the suggestion to use pulse techniques to overcome the need for a wide separation between transmitter and receiver, a problem more acute to the Navy than, say, the Army. A move to higher frequencies (from 28.6 to 200 MHz), together with the use of duplexing to enable the transmitter and receiver to use the same antenna, led to successful shipborne radar in time for America's entry into the war.

Meanwhile in Britain and Germany aircraft detection radar had come into service by 1939.

A British committee in the mid-1930s examined methods for detecting approaching aircraft at a distance, and previously observed accidental radio reflections from aircraft were analyzed. The BBC's short-wave transmitter at Daventry was used for demonstrations in 1935 and the committee then authorized work to begin along lines laid down by Robert A. Watson-Watt, an engineer who boasted James Watt as one of his ancestors. Through his interest in meteorology and the location of thunderstorm atmospherics by radio, Watson-Watt had a near ideal background from which to develop radar. With a team of six assistants he began work at Orfordness on the Suffolk coast and could soon detect aircraft 50 miles away. "Distance was already in the bag; vertical angle was quite another thing and certainly no 'piece of cake,'" he wrote.[27] Even so a chain of five stations was authorized to cover the bombing approaches to London, and by mid-1940 the radar net covered the British coast from the Isle of Wight to the Orkneys. It achieved near immortal fame as a result of the Battle of Britain in 1940. In Germany, meanwhile, a 600 MHz radar was in use in 1940 to help direct anti-aircraft fire; the Germans may well have had the most accurate radar equipment at that time.

The big breakthrough came with the invention of the multicavity magnetron. Better radar needed higher powers and much narrower beams, and narrower beams could only be obtained by use of frequencies well above the few hundred megahertz used early in the war.

In 1939 the British Admiralty asked the University of Birmingham to develop a high-power microwave transmitter. Some work centred on the klystron, but J. T. Randall and H. A. H. Boot instead applied the resonator principle, as used in the klystron, to the magnetron.[28] The result was the multicavity magnetron, a device that revolutionized radar by offering very high power at centimetric wavelengths. An idea of its effect can be gained from the story of its introduction to America, where it was sent in 1940 following an agreement between the two countries to exchange scientific and technical information, even though America was not yet in the war. Bell Laboratory workers had pushed radar frequencies up to 700 MHz and had obtained 2 kW of pulse power from research-type vacuum triodes, mainly by overheating the cathodes and operating the anodes at ten times their rated value.[24] Triode technology was being pushed to its limits. Enter the primitive multicavity magnetron, still wide open for development, yet already producing around 10 kW of pulse power at 3 GHz. Centimetric radar quickly became a reality and a closely guarded Allied secret. The magnetron fell into German hands, almost literally, in 1943 when first a British plane carrying a 10 cm radar set and then an American plane carrying 3 cm radar were shot down. Subsequently German workers developed centimetric radar, too. Though not the most usual method of transferring technology, it was certainly an effective one.

Postwar radar has been developed for an enormous range of uses from the motorist's bane, police radar speed traps, to the strategist's delight, the ballistic missile early warning systems. At sea it is used on ships of all sizes from the supertankers down to pleasure craft, and in the air it guards military and civilian aircraft against collisions. It is even used to keep track of the orbital junkyard created by innumerable space shots.

Radar found an unexpected use in astronomy and space navigation. Radar signals were bounced off the moon in 1946 and reflections were obtained from Venus and the sun in the late 1950s. Subsequently radar maps were made of the moon and Venus—not that such long ranges are essential for radar maps to prove themselves useful. For example, satellite-borne radar aimed at the earth has led to the discovery of previously unknown remnants of a Mayan canal drainage system in Central America.

One of the many wartime uses of radio and radar that continued to be developed and widely used after the war is its use as a navigation aid, especially for aircraft. Starting with beacon and transponder systems such as the British Oboe, these aids developed to produce hyperbolic-map systems based on radio nets such as LORAN, still widely used forty years later. Omega is a more modern system that offers greater range and uses a lower frequency.

Radar astronomy's older cousin is radio astronomy, whose basic phenomenon was discovered by Karl Jansky in America in 1932. Jansky was investigating the sources of interference noises that affected Bell's then recently introduced transoceanic radiotelephone circuits. He identified three categories of noise, one of which was described as "a very steady hiss static, the origin of which is not yet known." In 1933 he identified its origin as the centre of the Milky Way. This was an epochal revelation for astronomy, man's oldest science; it opened up the wide realms of the electromagnetic spectrum to the astronomer's gaze. Radio astronomy began in earnest after World War II and has yielded such fruits as the discovery of quasars and the identification of molecules in space. To some it carries the hope of some day discovering evidence of extraterrestrial intelligent life. What would the early radio pioneers have thought of that, I wonder?

References and Notes

1. C. Süsskind, *Isis* 55: 32, 1964. Also see *IEEE Spectrum* 5 (No. 8): 90, August 1968.
2. O. Lodge, *Nature* 111: 328, 1923.
3. C. Süsskind, *IEEE Spectrum* 5 (No. 12): 57, December 1968; 6 (No. 4): 69, April 1969; 7 (No. 4): 78, April 1970; 7 (No. 9): 76, September 1970.
4. J. Munro, *The Electrician* 40: 428, 21 January 1898.
5. H. G. J. Aitken, *Syntony and Spark: The Origins of Radio*, Wiley, New York, 1976.
6. I. Azimov, Chapter 3, Ref. 9.
7. R. F. Pocock, "History of transmitters: Some aspects of early radio," IEE London Group lecture, 7 April 1976.
8. W. R. Maclaurin, *Invention and Innovation in the Radio Industry*, Macmillan, New York, 1949.
9. G. Shiers, *Sci. Am.* 220: 104, March 1969.
10. G. W. O. Howe, *The Engineer* 745: 26 November 1954.
11. R. A. Chipman, *Sci. Am.* 212: 92, March 1965.
12. M. D. Fagen, Chapter 5, Ref. 19.
13. G. Siemens, Chapter 6, Ref. 7, Vol. 2.
14. S. P. Mullard, *JIEE* 76: 10, 1935.
15. D. G. Fink, *Electronics* 23: 66, April 1950.
16. F. C. Ward, "The first decade of amateur radio in the Midlands, 1911–

1921," 6th IEE Weekend Meeting on the History of Electrical Engineering, University of Nottingham, 1978.

17. A. Hope, *New Scientist* 81: 306, 1 February 1979.
18. A Santoni, *Electronic Design* 18: 58, 1 September 1977.
19. W. T. O'Dea, *Handbook of the Collections Illustrating Electrical Engineering: II. Radio Communication, Part I,* Science Museum, London, H. M. Stationery Office, 1934.
20. S. Runyon, *Electronic Design* 18: 70, 1 September 1977.
21. K. Geddes, *Broadcasting in Britain: 1922–1972,* Science Museum Booklet, H. M. Stationery Office, 1972.
22. E. W. Herold, *Proc. IRE* 39: 1177, 1951.
23. R. N. Jackson, *IEEE Spectrum* 17 (No. 3): 26, March 1980.
24. M. D. Fagen, Chapter 5, Ref. 19, vol. 2.
25. G. A. Isted, *Electronics and Power* 20: 315, 2 May 1974. (It was an experiment conducted by Marconi.)
26. R. M. Page, *Proc. IRE* 50: 1232, 1962.
27. P. W. Kingsford, *Electrical Engineering: A History of the Men and the Ideas,* St. Martin's Press, New York, 1970.
28. J. Jewkes, D. Sawyers, and R. Stillerman, *The Sources of Invention,* Macmillan, London, 1958.
29. D. McNicol, *Radio's Conquest of Space,* Murray Hill, New York, 1946; Arno Reprint edition 1974, New York.
30. B. Longman and B. Rhodes, *BST Journal* 64: 386, 1982.
31. B. Fox, *New Scientist* 94: 641, 1982.
32. E. Eastwood, *Electronics and Power* 20: 308, May 1974.

9 SOME THEORIES AND DISCOVERIES

This chapter ties up a few loose ends. It aims to show how some of the theories and discoveries that helped shape modern electrical engineering were made. Not all the important theories or discoveries are included but many of them are. They include the discovery of the electron and the question of its nature, which leads us into quantum theory; modern magnetism, a subject that was almost removed from the sphere of interest of most electrical engineers until rejuvenated by the interest in magnetic bubble memories; communication theories and the path towards Shannon's information theory, which includes advances in circuit and network theories and the invention of filters and the negative-feedback amplifier; and the development of our system of units. Justice cannot be done to any of these topics in such a short space but, it is hoped, the basic story can be put across.

The Electron

Knowledge of the electron as a fundamental particle carrying negative charge dates from the very end of the last century when studies of cathode rays identified the negative charge carrier as a 'corpuscle' with a mass of about 1/1800 that of the simplest atom, hydrogen. Well before the electron was identified, however, belief in a natural unit of electricity was fairly widespread, though probably not universal. This belief originally came from studies of electrolysis and, in some minds, was reinforced by the discovery of cathode rays. In 1891 the Irish physicist G. Johnstone Stoney gave the name electron to the postulated "single definite quantity of electricity."

The laws of electrolysis were discovered by Faraday in 1833. He found that the rate of decomposition of an electrolytic solution was proportional to the electric current that flowed through it or, to put it another way, that the mass of the chemical deposited was proportional to the total electric charge. From this result he was led to state that "there is a certain absolute quantity of the electric power associated with each atom of matter."[1] Weber subsequently constructed a theory of electrodynamics (1846) based on the concept of charged particles, but this important idea found little favour and such concepts received little further attention for the next thirty years. In Faraday's words, "Though it is very easy to talk of atoms, it is very difficult to form a clear idea of their nature, especially when compound bodies are under consideration."[2] When attention was again given to the subject of electrolysis both G. J. Stoney and Hermann Helmholtz concluded that electricity came in discrete amounts.[2] In 1881 Helmholtz stated that "electricity, both positive and negative, is divided into discrete elementary particles that behave like atoms of electricity."[3] Even earlier, in 1874, Stoney had postulated his "single definite quantity of electricity" which at first he termed the "electromagnetic electrine."[4] This postulate only appeared in print in 1881 and Stoney renamed his 'electrine' the 'electron' ten years later. The modern concept of a particle of electricity, therefore, and the first modern use of the word electron, were associated with the natural unit of electricity as found in electrolysis, ions as we would call them; but early suggestions vaguely along these lines can be traced right back to the early fluid theories of electricity (see Chapter 2). Benjamin Franklin had concluded in 1747 that electricity consisted of "particles extremely subtle."[1, 3] As a three-word definition of electrons that would still hold true today, at least as a starting point for discussion.

The path that led to the isolation of electrons and to measurements of their charge and mass came by quite a different route: through the study of electric discharges in gases.[2] Such studies began early in the 18th century but little real progress was made until big technological improvements were achieved in the manufacture of vacuum pumps in the mid-19th century. Table 9.1 lists some of the events which led to the 'discovery' of the electron by the British physicist J. J. Thomson (1856–1940) in 1897.

In Germany Heinrich Geissler developed the mercury air pump in 1855 and used it to make glass tubes that enclosed what was for those days a good vacuum.[2] Shortly afterwards, at the University of Bonn, Julius Plücker and J. W. Hittorf used the new vacuum tubes to study electrical discharges. Geissler tubes, or Crookes tubes, with a gas pressure of 1 or 2 Torr, are still used to demonstrate gas discharge phenomena. Plücker discovered that an electrical glow discharge could be deflected by a magnetic field, just as Davy had discovered for an electric arc in air about 50 years earlier. He also found that platinum from the cathode was deposited on the walls of the tube, a problem that was later to plague the new incandescent lamp industry and indirectly led to the invention of the electronic vacuum diode. Hittorf, once Plücker's pupil, then discovered (in 1869) that when an object was placed in front of the point-source cathode it cast a shadow in the glow discharge. This finding suggested that whatever was leaving the cathode was being propagated as a straight-line ray. Hittorf named these rays 'glow rays.' They were renamed as cathode rays seven years later by Eugen Goldstein, when he demonstrated that they were emitted from the whole cathode and not just from a point source. Hittorf is usually credited with the discovery of cathode rays, although, somewhat like

Table 9.1 Events Leading to the Discovery of the Electron (sources: Refs. 2, 3)

1838	M. Faraday	— Faraday dark space
1858–1859	J. Plücker	— deflection of a gaseous discharge by a magnet
1869	J. W. Hittorf	— shadow cast in the glow by an object, point cathode used
		— rectilinear propagation of 'glow rays'
1871	C. F. Varley	— rays deflected by an electrostatic field
1876	E. Goldstein	— Hittorf's shadow is cast even when cathode is not a point source. Cathode rays
1876 onwards		debate on the nature of the rays; disturbance in the ether or particles
1879–1883	W. Crookes	— extensive investigations of electrical discharges in vacuum tubes. Crookes dark space
1892	H. Hertz	— rays pass through metal foil
1894	J. J. Thomson	— velocity of cathode rays is much less than that of light
1895	J. Perrin	— metal collector of cathode rays becomes negatively charged
1896	P. Zeeman	— broadening of spectral lines in strong magnetic field
1897	J. J. Thomson	— measurement of charge to mass ratio of electrons

Ohm, he received little recognition for his work until later in life. In 1878 in England, William Crookes began to report on his own extensive work on cathode rays and electrical discharges in vacuum and received wide acclaim, so much so that evacuated glass tubes are often still known as Crookes tubes.

By the late 1870s the existence of cathode rays was clearly established. It was known that they travelled in straight lines and that a magnetic field would deflect them into a curved trajectory. C. F. Varley had shown that they were also deflected by an electric field though others, including Hertz, had been unable to verify this result because of poor vacuum in their discharge tubes.[2] According to Süsskind,[2] Varley concluded that the rays consisted of negatively charged corpuscles or, "attenuated particles of metal, projected from their negative pole by electricity." Not everyone agreed; for the next twenty years or so the nature of the rays was the focus of much discussion. Many British and French physicists leaned towards the corpuscular theory of streams of charged particles, whereas many German physicists took a more electromagnetic view and saw the rays as a vibration of the ether. Hertz discovered that the rays could, at least in part, pass through thin metal foil. This finding was confirmed by Philipp Lenard and appeared as a strong objection to the corpuscular ideas. However, the end of the ether theory and the vindication of the particle theory were drawing near. In 1894 J. J. Thomson, using the established rotating mirror technique, measured the velocity of the rays and found it to be well below that of light.[3] A year later Jean Perrin collected the rays in a metallic cylinder, a Faraday cage, which was then found to be negatively charged.[3] That showed that cathode rays consisted of a stream of negatively charged particles. Very little was yet known about the putative particles. They would create a glow discharge in a partial vacuum and pass through thin metal foil, but not through anything thicker; they could be deflected by magnetic and electric fields; and they carried negative charge. The size of the particles was not known but the general assumption was that they were charged atoms, or, to us, ions.

A year later, in a seemingly unrelated experiment, Pieter Zeeman, a Dutch physicist, found that the spectral lines of sodium are broadened when a sodium flame is placed in a strong magnetic field. From this result Lorentz was able to obtain support for his own theory of electromagnetism which, unlike Maxwell's, was based on the assumption of hypothetical particles having both mass and charge and moving through a stationary ether (see Chapter 4). As yet these particles were not identified with those assumed to make up cathode rays.

A convincing proof that cathode rays were made up of charged particles came from the experiments conducted by J. J. Thomson at the Cavendish Laboratory in Cambridge in 1897. The rays were known to be deflected by a magnetic field and believed to be deflected by an electric field. Thomson first verified electrostatic deflection. He then deflected the rays with a magnetic field B and neutralized the effect with an electric field E, so that the forces on the assumed particles were equal and opposite. This configuration enabled Thomson to calculate the velocity V from the equation relating the forces ($eE = Bev$, where e is the charge on the particle). He then calculated the deflecting force produced by the magnetic field alone, using measurements of the radius r of the circular path taken by the cathode rays when only a magnetic field was present. The resulting equation ($mv^2/r = Bev$, where m is the mass of the particle) enabled him to calculate the ratio of the charge to the mass of the particle ($e/m = v/Br = E/B^2 r$).[2] The result was astonishing. The ratio was far bigger than the value obtained for the monovalent ions liberated in electrolysis. If it was assumed that the charge was the same as that on an ionized hydrogen atom, which seemed to be a reasonable assumption, then the mass of the newly discovered particle must be about three orders of magnitude smaller than that of the hydrogen atom, the lightest particle then known. The consequences were startling: the atom was not, after all, the smallest subdivision of matter. "I had for a long time been convinced that these rays were charged particles," Thomson said later, "but it was some time before I had any suspicion that they were anything but charged atoms."[2] His new suspicion was now just about proved. Shortly afterwards he verified the assumption that the charge on the electron was about the same as that on a positive ion by measuring the charge. The tiny mass of the electron was thereby verified. Though Thomson cannot be said to have split the atom, in the words of one writer, 'chipped' might be a better word.[5]

Thomson performed his work with a poorly evacuated discharge tube and a cold cathode. He and others subsequently made more accurate measurements of the e/m ratio for the electron, and R. A. Millikan in 1913 obtained accurate measurements of the charge. At present the accepted value for its mass at rest is 9.1096×10^{-31} kg (or 1/1836 that of the hydrogen ion), and its charge is 1.6022×10^{-19} Coulombs. Within just a few years of its discovery it was shown to be involved in a number of physical phenomena including thermionic emission, the photoelectric effect, and beta radiation from radium.[3]

By 1900 the existence of fundamental particles of electricity was established beyond reasonable doubt and the name electron had been ascribed to them. That same year Max Planck gave a satisfactory explanation of the spectrum of radiation emitted from a black body, a problem that was puzzling many physicists. However, the explanation was satisfactory only if one accepted Planck's idea that energy emitted as electromagnetic radiation could only be emitted in fixed quantities, or quanta; i.e., in integral multiples of hf, where h is a constant and f is the frequency of the radiation. This was the beginning of quantum theory.[7,8] Five years later Albert Einstein used the same idea of fixed amounts of energy to solve another physics puzzle, the photoelectric effect— the emission of electrons from a substance caused by the arrival of light, usually said to have been discovered by Hertz in 1887 and by Wilhelm Hallwachs the next year. Einstein, in effect, was saying that light can be absorbed or emitted only in finite amounts. For nearly a century light had been viewed as a wave phenomenon; the old corpuscular theory had been debunked. Now it was seen to behave also as if it were made up of particles. But could light really consist of particles, or photons as they were named in 1926, when it was known to be an electromagnetic wave? Experiments conducted by Millikan between 1912 and 1917 on the energy of light quanta and photoelectrically generated electrons supported Einstein. The idea that cathode rays were some form of wave-like disturbance of the ether had been disproved by the discovery of the electron, the particle of electricity. Now light, which definitely had a wave nature, was also shown to act like a particle.

About the same time the Danish physicist Niels Bohr proposed his model of the atom, which enabled another physics puzzle to be explained with the aid of quantum theory: line spectra. It was also to earn him a Nobel prize. The discovery of lines in the spectra of incandescent gases is linked with the names of W. H. Wolaston, and Joseph von Fraunhofer; and in 1885 J. J. Balmer observed that a simple mathematical series could be used to relate the lines in the spectrum of hydrogen. This series was generalized in 1889 but classical physics was unable to explain it, a situation that persisted until 1913. It had been shown, on the basis of the known properties of electromagnetism, that for positive and negative point charges to remain in a stable state the electrons would have to orbit the nucleus, but according to Maxwell's equations orbiting electrons radiate energy. If electrons did lose energy in this way, they should spiral inwards and eventually fall into the positive nucleus, which clearly did not happen. In five papers published from 1913 to 1915 Bohr proposed a quantum condition that allowed electrons to exist in certain orbits without radiating energy. Line spectra could then be explained by the emission or absorption of quanta of energy when electrons changed orbits.

Another major step in the advance of quantum theory, and one that was to have a bearing on electrons, came in 1923 when Louis de Broglie turned to the idea of symmetry in physics, so well used earlier by Faraday, and suggested extending the idea of the dual nature of light, its wave-like and particle-like behaviour, to all matter. Ordinary particles of matter, he suggested, would have a wave nature and a specific wavelength given by h/p, where h was the same constant Planck had used and p was the momentum of the particle. As h is a very small quantity (about 6.6×10^{-34} Joule seconds) the wavelength of even a tiny particle would be extremely short indeed. This seemingly absurd but extremely bold idea of the particle-wave duality of matter was verified in 1927 at the Bell Laboratories by C. J. Davisson and L. H. Germer who obtained diffraction effects, very much a wave phenomenon, with a beam of electrons.[6] Davisson had spent several years exploring the way in which ions and electrons were scattered after they impinged on various metals. In 1925 while bombarding polycrystalline nickel he and Germer discovered a startling difference in the scattering distribution produced by a sample which included, by accident, small single crystals of nickel. Figure 9.1 shows what they found. This discovery led to a careful study in which single crystals were used and, two years later, to the conclusion that electrons do behave as waves as well as particles. That same year J. J. Thomson's son, G. P. Thomson, at the University of Aberdeen, arrived at the same result. He and Davisson shared the 1937 Nobel prize in physics for this discovery. De Broglie's wave mechanics received much attention in the 1920s from men such as Werner Heisenberg, Erwin Schrödinger, and Paul Dirac. In 1926 it reached new heights in the formulation of Schrödinger's wave equation for the electron, an equation that describes the wave-like behaviour of the electron in mathematical terms. Dirac's relativistic quantum theory followed in 1928 and introduced the positron, so named by its discoverer Carl Anderson in 1933, the antimatter equivalent of the electron. Other antiparticles followed later.

Studies of electrons and their interaction with fields and other matter have of course continued. Electrons as particles ushered in the age of vacuum electronics with the diode, triode, pentode, klystron, magnetron and all the

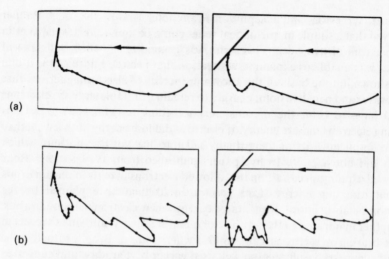

Figure 9.1 Davisson and Germer's electron-scattering curves from nickel (1925): (a) polycrystalline material, (b) with small single crystals. The difference between (a) and (b) was the first of the evidence that eventually led to the discovery of the wave nature of the electron

other vacuum devices. The study of electrons in solids led to semiconductor physics, semiconductor devices, and the silicon chip. The cathode-ray tube employs electrons as particles to produce the picture on oscilloscope and television screens, and electron diffraction cameras and microscopes use electrons as waves to provide information about small quantities of matter. The word electron has been adapted to describe the branch of electrical engineering specializing in the use and control of electrons. Süsskind[4] tells us that E. F. W. Alexanderson of GE is believed to have referred to an amplifier built with De Forest's triode as 'an electronic amplifier.' As this was probably soon after 1912 he may have been the first to use the word 'electronic' in this sense, though it had been used earlier in a scientific sense by Fleming in 1902 in 'the electronic theory of electricity.' AT&T (Bell) engineers used the word in 1919 in classifying elements of telephone repeaters as electrodynamic, electronic, and gaseous; and it seems to have passed into fairly general technical use by the early 1920s.[4] Indirect evidence suggests that 'electronics' with an 's' was used in Britain before 1930; however, the American magazine of that name (published by McGraw-Hill) is usually credited with the first use of the word, as the title of its first issue in April 1930.

Magnetism[3,9]

For centuries the origin of magnetism remained a mystery. The key to unlocking this mystery was provided by Oersted when he discovered the link with electricity, after which Ampère suggested the existence of molecular currents as the source of permanent magnetism. Though this idea was later developed by Weber, a deep insight into the origins of magnetism had to await the discovery of the electron, an understanding of the structure of the atom, and the revelations of quantum theory, especially the discovery of what are called 'exchange interactions.' From Oersted to the discovery of exchange took a little over 100 years. From an experimental study Faraday found that all materials can be magnetized to some extent and he then provided three of the five known classes of magnetism. A bar of heavy glass was found to lie across a line joining the poles of a magnet instead of, as was more usual, parallel to the line. This finding led to the naming of diamagnetism, for across, and paramagnetism, for parallel. The word ferromagnetism was reserved for materials that behaved like iron and steel. The two remaining classes, antiferromagnetism and ferrimagnetism, came after 1930 with the deeper insights given by the mature quantum theory.

Ampère's idea of circulating currents causing what became known as ferromagnetism was adapted and developed in 1871 by Weber. He assumed that the atoms or molecules of iron or steel were themselves like tiny magnets that could rotate around fixed centres and so orient themselves in various directions. In unmagnetized iron or steel the molecules were assumed to be randomly oriented and their magnetic fields neutralized one another. When an external force was applied the molecules turned around so as to line up in such a way that their net effect was no longer zero and the material was now magnetized, a theory recognized by most school children.[9] In the early 1890s J. A. Ewing took dozens of small bar magnets, pivoted each of them, and used them to imitate on a large scale Weber's molecular magnets and so simulate the properties of ferromagnetic materials. In this way he demonstrated many of the properties of ferromagnetism, including hysteresis. Shortly afterwards Pierre Curie studied the thermal properties of magnetic materials. Among his discoveries was the critical temperature, now named after him, at which ferromagnetics degenerate and begin to act like paramagnetics.

After the discovery of the electron, the simple molecular magnet theory of Weber, and the molecular current ideas of Ampère and Fresnel, were placed on a surer foundation when Paul Langevin suggested that Ampère's molecular currents were actually electrons in orbit around the centre spot of an atom. It followed that if the effects of the orbits of many electrons in an atom were balanced, the atoms would have no net magnetic moment, but they would have a weak reaction to an applied magnetic field as suggested by Lenz's law. Such materials are diamagnetics. If the effect of the electron orbits were not balanced, the diamagnetic property would be masked by a much stronger effect. The unbalance produced a net magnetic moment which, Langevin explained, was the property of paramagnetism. By statistical means Langevin was able to explain the basic temperature variation of paramagnetism discovered by Curie. Two years later, in 1907, another Frenchman, Pierre Weiss, extended the theory to explain ferromagnetics. He suggested that each molecule was acted upon by a 'molecular field' caused by the interaction between the other molecules. This internal field caused the atomic or molecular magnets to align themselves. Weiss was able to explain the Curie point, a temperature at which ferromagnetics cease to act as such. Another product of his theory was the suggestion of magnetic domains, now of vital concern in magnetic bubble memories. The Weiss theory has been described as the first modern theory of magnetism, though it was not widely acclaimed at the beginning. However, scepticism was vanquished in 1919 when Heinrich Barkhausen amplified the audio clicks of the domains as they were magnetized with an external field. In 1931 Francis Bitter developed a technique of covering the surface of a ferromagnetic with a colloidal suspension of magnetic material. The boundaries of the domains were then revealed under the microscope. With transparent ferromagnetics Faraday rotation of light can also be used to observe the domains. In modern studies an electron microscope would usually be employed.

In 1967 Andrew Bobeck [10,11] and his colleagues at Bell Laboratories realized that the magnetic domains of certain ferromagnetic materials could be used to store data as ones and zeros. In some single-crystal materials, such as some orthoferrites and garnets, the domains can be made to shrink down to small circular cylinders when an external magnetic field is applied. These cylinders, or magnetic bubbles, extend through the width of a thin platelet of the material and can be easily ordered and moved about by magnetic means. The presence or absence of a bubble at a given place can be used to represent a 1 or a 0. At a time when semiconductor memories were storing a few hundred bits on a chip, bubbles were touted as capable of storing a million or more. However, advances in silicon chips and problems with bubble memories meant that silicon reigned supreme through the 1970s. Yet by the end of the decade bubble memories had become commercially available with storage several times greater than the latest silicon random-access memory elements.

The cause of Weiss's strong internal molecular field, which resulted in the magnetic domains, was never explained by classical physics; nor were the more fundamental permanent atomic magnetic moments. Instead they became further trophies for the rapidly developing quantum theory.

A satisfactory model of the atom was needed before further progress could be made. Bohr provided such a model in 1913 and suggested that the angular momentum of the electrons, which orbited the nucleus in Lord Rutherford's earlier model of the atom, should be quantized. The fundamental unit of the atomic moment was then named the Bohr magneton. The American physicist A. H. Compton suggested in 1921 that the electron has a spin that yields another magnetic moment. Both suggestions were verified experimentally: the first by Otto Stern and Walther Gerlach in 1922, the second by S. A. Goudsmit and G. E. Uhlenbeck in 1925. In the mid-1920s Heisenberg and Dirac independently revealed the existence of an unexpected term called the exchange energy, which is associated with the ability of two electrons in an atom to keep exchanging their places and roles, which would be impossible under the laws of classical physics. In 1928 Heisenberg showed that Weiss's molecular field was a consequence of this exchange energy, whereupon it at last became possible to explain ferromagnetism. The two remaining classes of magnetic materials were subsequently identified. Following the discovery of exchange interactions it became possible to contemplate negative molecular fields as well as Weiss's positive field found in ferromagnetism. This concept led to the discovery of antiferromagnetism by Louis Néel in 1932 and by L. D. Landau in 1933. Néel was also responsible for naming ferrimagnetism, of which ferrites, including the lodestone, are the obvious examples.

On the more practical side[12] the naturally occurring lodestone was for centuries the only permanent magnet available. In the 18th century Gowin Knight in England invented a method of making bonded powder magnets by stirring iron filings into water to form a sludge of iron oxide (Fe_3O_4) which was then mixed with linseed oil, shaped, and baked. More recently Kato's oxide, a mixture of iron and cobalt oxides held together by an adhesive, was developed by Y. Kato and T. Takei (1933) to produce the first modern ferrite ceramic magnet. Barium ferrites were intensively developed by a Philips team at Eindhoven during World War II, which led to their importance after about 1950. Magnets made from Kato's oxide had an energy product about four times greater than lodestone, whereas anisotropic barium ferrites are about 28 times greater. The Alnico alloys (Al-Ni-Co-Fe) were also discovered in the early 1930s. In the 1970s the rare earth-cobalt magnets received much attention; they display much higher energy products than any other permanent magnet, some over 200 times more than lodestone.[12]

Towards Information Theory[6,13]

In the mid to late 20th century electrical communications have reached a high state of sophistication. We need only think of the colour television pictures received from the moon and displayed live in millions of homes, or of the stimulating photographs of other planets sent back by unmanned spacecraft. This degree of sophistication has been achieved with the aid of a strong background of mathematically based theory. Such mathematical theory, even if the practising engineer has not always been able to take immediate advantage of it or even been aware of it, has been relevant to electrical communications right from the start (Ohm preceded Cooke and Morse, for example) and the 19th century theoreticians placed electrical communications on a sound theoretical basis by the turn of the century. That work included Kelvin's theory of simple electrical circuits (1853) and his theory of telegraph transmission by cable (1855), Maxwell's electromagnetic theory (1864), and its subsequent development by Hertz and Heaviside. In the early decades of the 20th century the setting for much of the progress shifted from Europe to America where researchers at the Bell Laboratories and elsewhere came to grips with some fundamental problems of communications.

A typical example of cartoons exploiting the mysteries of mathematical theories

The mathematical treatment of circuits has passed through several stages of development.[14] Originally mathematical statements were made describing the properties of individual components and simple groups of these components, which was sufficient for much of the 19th century work on telegraphs and telephone networks. Later researchers analyzed more complex circuit structures by breaking them down into smaller simpler sections for which mathematical statements could be made; these sections were then used to analyze the whole complex structure. This stage began around 1900 with the analysis of telephone networks and continued through the hectic decades after World War I. Out of this work emerged such disciplines as line theory, circuit theory, and network analysis. Perhaps the final stage, as exemplified by network synthesis, is the ability to synthesize a circuit by mathematically stating the function it is to have and then evolving the design that will achieve it. From a communications point of view the mathematical disciplines of line, circuit, and network theories have been tied in with the engineering problems of the time as new technologies came to be considered; telegraphy, telephony, radio, video, pulse techniques, and so on. One culmination of much of the mathematical work was Claude Shannon's information theory, which was first published in 1948. It is a general mathematical theory of communications.

The 19th century mathematical work had a bearing on telegraphy and telephony but was often understood by only a small circle of people. Kelvin analyzed the discharge of a capacitor through a circuit containing resistance and inductance and demonstrated that the discharge could be oscillatory, as had previously been suggested. In 1868 Maxwell performed a somewhat similar analysis on a circuit to which a sinusoidal voltage was applied and calculated the conditions necessary to achieve resonance. In so doing he used descriptive mechanical analogies as in the early days of his electromagnetic theory. This time he illustrated the inductance of a coil by analogy with the inertia of a large boat, capacitance with the spring-loaded buffer on a railway carriage, and resistance with the effect of a viscous fluid on a body moving through it. In this way he was able to show analogies between electrical science and Newtonian mechanics.[6] Later, in the 1880s, Hopkinson used Maxwell's ideas to explain the resonance effects of a capacitor when used in series with an alternator. Little of this mathematical knowledge of AC properties percolated down to the ordinary engineer; the most notable examples of the early use of electrical resonance fell to Hertz, himself a mathematical physicist, and to Lodge. Both generated and detected electromagnetic waves; Hertz transmitted through the air, Lodge along a wire. Apart from resonance considerations, Ohm's and Kirchhoff's laws were generalized to take account of reactive components and frequency, and in 1883 L. Thévenin in France formulated his famous network theorem to facilitate analysis of a complex circuit by the derivation of a simpler equivalent circuit and voltage source. Its current-source companion theorem by E. L. Norton of Bell Laboratories was not formulated until 1926. However, both Norton and Thévenin had been anticipated by Helmholtz in 1853.[6]

After the invention of the telephone, communications engineers faced new problems in attempting to achieve the long-distance transmission of speech. Lord Rayleigh, in 1884, calculated the distance that speech could be transmitted by cable and in doing so was probably the first to consider the exponential decay of a sinusoidal voltage applied to a cable. His results indicated a serious practical limit for speech on the Atlantic telegraph cable, a limit which "would not be likely to exceed fifty miles."[6] A year later it was shown by Blakesley of the Royal Naval College at Greenwich in England that the various harmonics of a fundamental frequency would be attenuated differently; therefore severe distortion might be expected in the transmission of a human voice through a long capacitive cable. "The ear has not the synthetic power of reconstructing a composite tone from the wreck of variously degraded components," wrote Blakesley.[6] These two constraints posed a grave problem. It was the application of Heaviside's fundamental work on the role of the magnetic field in the propagation of a wave along a wire which rescued the situation, as did the development of the electronic amplifier and of carrier transmission. Heaviside's major work was performed between 1873 and 1901 and opened the door for the inductive loading of cables, which was to compensate for the capacitance effects. Telephone transmission theory was to be pushed beyond the limits that had proved satisfactory to the telegraph engineers, and the mathematical analysis would have to move on from simple to complex circuit structures.

Several researchers were able to understand and apply Heaviside's theory; prominent among them were M. I. Pupin of Columbia University and G. A. Campbell of AT&T (Bell). Other contributions came from A. Vaschy in

France, S. P. Thompson in England, and C. E. Krarup in Denmark. Though Pupin won the legal proceedings between his patents and Campbell's, by virtue of two weeks' priority of disclosure, it was Campbell who had the greater influence on the Bell telephone network in America.[6] Elsewhere, in Britain for example, the loading coils became known as Pupin coils and the invention is said to have earned Pupin $1 million.[15] Rayleigh's 50-mile (80-km) limit for speech transmission on submarine cables was shattered by inductive loading. In 1921 the longest such cable stretched 115 miles (184 km), whereas open wire on land, with loading coils, had achieved 670 miles (1072 km) as early as 1900.[6]

A small team of outstanding scientist-engineers was slowly forming at Bell who were to contribute much to the theory and practice of communications. There and elsewhere, a mathematical attack began on the theory of electrical communications that was to bring many advances. The band (rather than single-frequency) nature of modulated signals was recognized by a handful of people, especially by Campbell and E. H. Colpitts, of oscillator fame. Fourier analysis was employed so that the study of telephone networks could make use of the complex algebra already used by pioneers, such as C. P. Steinmetz of GE, to analyze AC power systems. The concepts of reactance and impedance evolved, with much help from Heaviside, and the exponential function e^{jx} was brought into service, even though the mathematician in Campbell led him to protest against the use of j as the imaginary coefficient ($\sqrt{-1}$) instead of the mathematicians' i.

One product of Campbell's work was his invention of the electric filter in 1909. The German engineer K. W. Wagner has also been credited with the invention of the filter in World War I,[16,17] but it now seems to be established that priority should go to Campbell, even though his patents were not issued until 1917.[18] However, in 1882 a capacitor and an inductor had been effectively used as a filter by François van Rysselberghe of Belgium to separate a telephone and a telegraph channel on the same wire, and the next year he had used what 40 years later would be called a low-pass 'tee' filter.[14,19] However, van Rysselberghe did not analyze his circuits mathematically as Campbell did; proper realization of the capabilities of filters came from Campbell. Campbell's patents were actually written by another Bell mathematician, J. R. Carson, whose major claim to fame was the recognition (also realized by H. D. Arnold) that only one sideband was required for transmission of a high-frequency carrier modulated by low-frequency speech. (This scheme results in three signals: the carrier, and two 'sidebands' each of which contains the entire intelligence.) Carson also proposed to suppress the carrier, and he takes priority as the inventor of single-sideband transmission (1915).[6] In March 1910 Campbell was able to present drawings and characteristic curves of low-pass, bandpass, and high-pass filters to J. J. Carty, the chief engineer at Bell. Two years later a low-pass filter was used for the first time at the input of a telephone repeater to reject high frequencies. The invention of the vacuum-tube amplifier and its subsequent use as a telephone repeater made the understanding of frequency-dependent networks even more important, especially as telephone networks moved towards carrier transmission. Shortly afterwards Colpitts and others began to study the use of bandpass filters with a view to selecting individual channels in carrier telephony,[18] and the next decade or so saw much of the major work on passive filter design. The older use of simple AC or DC blocks was giving way to a more mathematical approach based on filters.

Until the 1920s filters were of the type the textbooks call ladder networks, from the way loading coils split up a telephone line into regular sections. Each section of line could be viewed as an equivalent circuit containing resistance, capacitance, inductance, and leakage conductance; a line could be viewed as a continuous ladder of such circuits, rather like a daisy chain. In 1923 O. J. Zobel, also working at Bell, considerably advanced the progress of network theory with the invention of what is known as the 'm-derived' type of filter, a method of designing selective filters with an unlimited number of reactances.[17] He also sorted out the remaining problems associated with the connections between the various sections of the ladder networks, especially the matching of the 'image' impedances, and he established the conditions for ideal transmission. Zobel's work, which combined mathematical skills with engineering ingenuity, made filter design an established art.

From World War I to the 1930s was a very active period in circuit design, far too active for any full assessment to be made here. (See Refs. 6, 16, 17.) New circuits were developed and network analysis gradually moved towards the even more mathematically dependent network synthesis. Contributions came from a number of people whose names are still remembered in present-day textbooks in connection with the basic concepts they invented. G. A. Campbell, J. J. Carty, E. H. Colpitts, K. W. Wagner, and O. J. Zobel have already been mentioned; other well-remembered names include W. H. Eccles and F. W. Jordan, R. V. L. Hartley, Wilhelm Cauer, R. M. Foster, S. Butterworth, and many others. The two-port black-box concept dates back to 1921 at least, and ABCD matrices were used in continental Europe in the early twenties.[16] The decibel unit was introduced in 1924.[16] Advances in network theory came as demands were made for amplifiers with flat frequency responses and steep cutoffs. Broadcasting gave a further boost to the mathematical design of networks. The invention of the negative-feedback amplifier by Harold Black in 1927 led to new and severe demands on the phase and loss characteristics of networks and was to demand a new understanding of the stability of dynamic systems. H. W. Bode and Harry Nyquist were to be prominent in meeting the new demands. As for filters, the work of Zobel and its development by others such as Foster, H. W. Bode, and O. Brune was not seriously challenged until the 1940s when, independently of one another, Sidney Darlington in 1939 and Cauer in 1940 published theories that were fundamental to exact mathematical synthesis.[17] Another decade was to pass though before the Cauer-Darlington theory began to be used widely, and then it was with the help of the new electronic computers.

Of the circuit techniques invented in the 1920s probably the most important, in terms of its impact on later electronics, was the invention by H. S. Black in 1927 of the negative-feedback amplifier. A special fiftieth-anniversary issue of *Electronics* in 1980 included Black's invention in a list of twelve circuits regarded as 'classics.'[20] Feedback was not new, as it had been used in thousands of mechanical contrivances for centuries; a fin was used in windmills to keep them turned into the wind, for example. In electronics positive feedback, in which the output signal is fed back to the input in phase with the input signal, had been used for more than a decade in regenerative radio receivers. Even negative feedback, where the signal is fed back in antiphase, had been used in electronics for control purposes before Black studied it. H. A. Wheeler at the Hazeltine Corp. used it in 1926 in his invention of an automatic gain control for radio receivers.[20] But it was Black who discovered how to use feedback to minimize the distortion in an amplifier over a wide frequency band, a problem he had been pondering for years, while stabilizing the amplifier's gain. This discovery, according to one employee at Bell, "had all the initial impact of a blow with a wet noodle."[20] Negative feedback was also found to widen the bandwidth of amplifiers, which meant that more channels could be used when frequency-division multiplex was employed. The small price to be paid for these vast improvements was a loss of gain by the amplifier.

For many years, probably since the first oscillators were built, it had been wrongly believed that instability must occur if the loop gain $|\mu\beta|$ was greater than unity (Fig. 9.2). This error was corrected in 1932 by Harry Nyquist, also at Bell, when he worked out a general rule, now known as the Nyquist criterion, for avoiding instability in a feedback amplifier. This rule was experimentally verified two years later, in the same year that Black published the details of his feedback discoveries. Final clarification of these ideas was provided by Bode in 1940. By that time the 'wet noodle' had been somewhat stiffened.

Black's idea for reducing distortion in an amplifier was radical in the

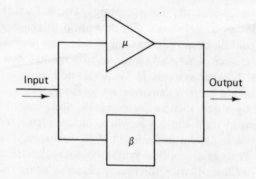

Figure 9.2 Amplifier with feedback

extreme and Black himself has described how he had searched for an answer to this problem from 1921 onwards.[21] A lecture by C. P. Steinmetz in 1923 impressed on him the need to get down to the fundamentals and the following day he invented the feedforward amplifier. For four years he struggled to get what in theory was an excellent solution to work with simplicity and perfection in a practical amplifier. Then, while travelling to work on a ferry across the Hudson River, the solution came to him "in a flash." The first diagrams and equations for the negative feedback amplifier were sketched while he was still on the ferry, across a page of the *New York Times* on 2 August 1927. By the end of December the distortion in a practical amplifier had been reduced by a factor of 100 000 (50 dB), more than enough to crack the problem of the vast distortion previously encountered in a telephone line when many amplifiers were used. A trial was conducted in 1930 at which a simulation was made of a 7000-mile circuit with nine channels instead of the usual three. The results were excellent, despite the fact that an attenuation of some 12 000 dB was being compensated for by an equal amount of amplification. Black has noted that each feedback amplifier used occupied a 19 in. rack and had a feedback loop more than a yard long.[21]

One example that illustrates just how radical the feedback amplifier was is given by its passage through the U.S. Patent Office. It took nine years to obtain the U.S. patent, and several years to get a British patent. Initially it was claimed that the invention would not work and the old erroneous belief that the loop gain must always be less than unity was quoted. In Britain, according to Black, the patent application was at first treated in the same way as those for perpetual-motion machines, and a working model was demanded.[21]

Black's invention, the work on filters, and carrier transmission were all important in developing the art of electrical communications. However, at about the same time as such circuit and other technical developments were taking place, both theoretical and practical, work was also progressing towards a mathematical understanding of communications in general, electrical or otherwise. This work occupied the labours of men such as Nyquist, K. Küpfmüller, R. V. L. Hartley, and Dennis Gabor and reached what may be viewed as a peak of perfection when, in 1948, C. E. Shannon published in the *Bell System Technical Journal* a two-part paper called 'The mathematical theory of communication.'[22] Since then it has simply been called information theory and, in the words of one writer, it "cast about as much light on the problem of the communication engineer as can be shed."[13]

The route that led to information theory involved both mathematics and engineering practice.[6,13,23] In the 1920s the role played by the bandwidth of signals was increasingly studied mathematically. Bandwidth considerations had become important when carrier transmissions began to be used, especially during World War I, and the idea of allocating different signals to different frequency bands (frequency-division multiplexing) became the mainstay of electrical communications and remained so, virtually unchallenged, until World War II. The desire to keep the bandwidth at a minimum became established, though J. R. Carson in 1922 showed that modulating a signal would apparently always preserve or widen the bandwidth. In the early television experiments of the mid and late 1920s it was soon realized that a very wide bandwidth was needed in order to send, in the very short time periods allowed, the masses of information required to give the impression of a moving picture. This requirement led to problems with wide-bandwidth amplifiers and the control of noise and transients; and further problems as phase distortion suddenly became important because of the eye's intolerance to it.

In 1924 Nyquist in America and Küpfmüller in Germany independently arrived at a basic law of communications that to transmit telegraph signals at a given rate a definite bandwidth was required. Four years later Hartley stated the law in more general terms and attempted to formulate a mathematical theory of the transmission of information. In his theory he rejected 'meaning' as such as a mere subjective factor. Instead, information was defined as a successive selection of symbols or words. A message of N symbols, chosen from a code or alphabet of S symbols, was shown to have S^N possibilities and the 'quantity of information' was defined as $N \log S$. To transmit a given quantity of information required a definite factor, bandwidth times time.[23] A unit of information was also suggested. This unit has been called the Hartley and was subsequently defined as equal to 3.219 bits.[3] Though Hartley's work has been seen as the genesis of the modern theory of the communication of information[23] and as providing the guiding rules for transmission engineers for the next 20 years,[6] some regard Nyquist's work as having been more fruitful.[13]

Bandwidth considerations again caused trouble in the 1930s when Armstrong demonstrated his wide-deviation frequency modulation (FM) system in which he showed, not only that it actually worked, but that the signal-to-noise ratio was greatly improved, something not anticipated by the previous FM theories of Carson (1922) and Balthazaar van der Pol (1930). Contrary to the expectations of others Armstrong had reduced noise and interference by widening, rather than by narrowing, the bandwidth.

In 1946 Dennis Gabor in Britain turned to the mathematics of quantum theory for help in understanding the theory of communication. Other workers had just shown, by the examination of speech spectrograms which displayed patterns of speech as plots of frequency against time, that the resolution of the frequency could be increased but only at the expense of time resolution.[13] Similarly, the time resolution could be increased but only at the expense of the frequency resolution. Both resolutions could not be increased simultaneously. One may think of attempting to measure the frequency of a signal; the longer the time element allowed (δt), the smaller the error in the measurement of the frequency (δf). By analogy with Heisenberg's uncertainty principle Gabor was able to show that $\delta f \times \delta t \approx 1$. If therefore the frequency band was widened, as had been done by Armstrong, one could expect the time resolution to be improved. From these ideas Gabor was also led to the concept of a unit of information, a unit he called the logon.[23] Shannon's information theory was now just two years away.

Noise

One consideration that had been ignored in the previous theories of Nyquist and Hartley was the contribution made by random noise. Their work was basically concerned with a noise-free medium. Shannon was to contribute a statistical theory that took into account the previous understanding of both bandwidth and noise considerations.

For many years background noise had been recognized as imposing limits to electrical measurements and communications. Brownian motion had been studied by Einstein in 1905 and by 1912 it was known to set a limit to the sensitivity of a galvanometer. Noise in electronic amplifiers also set limits on the sensitivity of radio receivers. In 1918 Walter Schottky at Siemens & Halske published a classic paper on noise in amplifiers and suggested that there were two fundamental sources of noise in an amplifier. Possibly little could be done about either of them. The first became known as thermal noise and resulted from the random movement of electric charge in conductors and resistors caused by the thermal motion of their molecules. The second, known as shot noise, was associated with the fact that electric charge is not continuous and therefore there is always an electrical noise associated with the random fluctuations in the number of electrons emitted from a cathode. Schottky felt that the latter gave the greater effects and so directed his laboratory studies towards it. However, in the mid-1920s J. B. Johnson in America discovered that in amplifiers in which very 'quiet' vacuum tubes were used the noise was proportional to the amplification. This finding led to the practical discovery of thermal (or Johnson) noise and its association with resistors in the input circuit. Nyquist then mathematically analyzed the noise, using thermodynamics, and calculated the noise to be $4kT$ watts, where k is Boltzmann's constant and T the absolute temperature.[6] These discoveries were published in 1928. Since then many researchers have extended the study of noise in vacuum and semiconductor electronics. Aldert van der Ziel is probably the best known recent contributor.

Shannon's synthesis of previous theories into a comprehensive theory of communications was published in 1948 and has been described as "something of a delayed-action bomb."[13] Communications theory has not been the same since that bomb exploded. During and after World War II novel communications techniques were tried based on pulse techniques, including pulse-length and pulse-position modulation. Pulse-code modulation (PCM), invented in France by A. H. Reeves in 1938–39, proved to be the most successful. The contribution of Shannon's theory towards such schemes, in describing PCM and correcting the theory of FM, for example, has been profound as it has provided the theoretical basis for understanding them.

Electrical Units

Our present system of units is based on the metric system of units and measures which was a product of late 18th century France, and it was there that Coulomb established the inverse-square law for electrical and magnetic attractions by careful measurements with a torsion balance. When Volta's invention of the primary cell was made known in 1800 it became possible to try to establish some electrical units based on the somewhat fluctuating voltages of those early batteries. Better cells were researched throughout the 19th century so that standard voltages could be achieved, and in 1872 Latimer Clark announced his zinc-mercury cell which became a standard for voltage measurements. Meanwhile, Gauss in 1833 and Weber in 1840 had made the first moves towards establishing absolute units for electrical science by discovering that measurements of magnetic phenomena and electric current could be reduced to measures in terms of mass, length, and time. In 1851 Weber added resistance and voltage to the list of quantities whose measurement could be expressed absolutely in the dimensions of mass, length, and time.[24]

Meanwhile practical telegraph engineers were feeling the need for some practical units of their own, and in 1861 a scientific committee was formed to deal with the question of electrical and magnetic units for scientific use. It was the first of many. This committee of the British Association for the Advancement of Science (BA) presented its first report in 1863. Kennelly tells us that it recommended the metre-gramme-second (MGS) system of absolute units and formally established the ohm and farad as the units of resistance and capitance.[24] The volt was proposed as the unit of emf and the weber was suggested as the unit of electrical charge (now the coulomb). Current would have been measured in webers per second. But these names were already favoured by telegraph engineers (except that the weber was used for current as well as for charge), whereas the scientists rarely used the names at all, preferring to use simply 'BA units' of resistance, current, etc. Practical engineers also used the emf of the Daniell cell as a unit of voltage, and had used fixed lengths of standard telegraph wire (a foot, a mile, a kilometre) as units of resistance. The wire lengths were replaced in the 1860s by the siemens, based on the resistance of a fixed column of mercury and proposed by Werner Siemens. This unit in turn gave way to the BA ohm in the early 1870s, according to Kennelly. The word ohm was a shortened form of the originally proposed 'ohmad.'[15]

A second BA committee in 1873 scrapped the MGS absolute system of units in favour of the CGS (centimetre-gramme-second) system, which survived until quite recently. It also recommended the new names dyne and erg for the CGS units of force and work, and also suggested the adoption of the metric prefixes mega, kilo, milli, micro, etc. (We must be thankful that one of its suggestions has died out. Cumbersome expressions such as metre-nine and ninth-metre, and the like, were to have been used for 10^9 metres and 10^{-9}

metres, etc.) The CGS system and the practical units of the ohm, volt, and farad steadily gained acceptance and were formally adopted internationally in 1881, at the first International Congress of Electricians in Paris. The same Congress formally accepted our present names of ampere and coulomb for current and charge. Although these practical units were seen as subordinate to their fundamental CGS equivalents, as Kennelly tells us, it was the practical units that survived.

Although the names were settled confusion still existed over the definitions. By 1885, for example, three 'ohms' were recognized: an 'absolute ohm' of 10^9 CGS units, a 'BA ohm' of 104.8 cm of a specified column of mercury, and a 'legal ohm' of 106.0 cm of mercury.[24] In general, however, practical measurements of capitance and resistance had been satisfactory for about twenty years, and current and voltage for roughly ten years. The terms voltmeter and ammeter were widely accepted.

The point of no return was reached in 1893 at the Fourth International Electrical Congress in Chicago.[15,24] A new and better specification was accepted for the ohm and (together with the volt, ampere, coulomb, farad, joule, watt, and henry) was recommended to governments for adoption as the legal unit. Once these units had reached the statute books of the major countries it would require a phenomenal effort to change them. The joule and the watt had been previously adopted at the Second Congress in 1889.

The development of dynamos and motors emphasized the need, at least to practical engineers, for the adoption of named units for magnetic quantities. The Third Congress in 1891 suggested gauss and weber, and the Fifth Congress in 1900 adopted the gauss and the maxwell, for field intensity and magnetic flux. A Seventh Congress in 1911 agreed that impedance should be represented algebraically as $R + jX$ and not as $R - jX$. Ohm's law, it was also agreed, would henceforth be written as $I = E/R$ and not, as English-speaking countries had it, $C = E/R$.[24]

With time being devoted to such matters it was clear that the main task had been achieved. The practical units were here to stay but, as scientific units, they were not 'rational,' a fact pointed out by Heaviside in 1882. Dimensionally they were correct, but the constant 4π plagued the system, which was itself split into two subsystems: electromagnetic units and electrostatic units. The capacitance of a parallel-plate capacitor with an air dielectric was $1/4\pi d$, for example, whereas in a more rational system it would simply be $1/d$. Many suggestions were put forward to 'rationalize' the system and the one eventually adopted (in 1950) was based on that suggested by Giovanni Giorgi in 1901.[15] The CGS base for the absolute units was replaced by the MKS base (metre-kilogram-second) together with the ohm. By careful choice of the constants for the permeability and the permittivity of free space, a rationalized system was achieved which could replace the two former systems of electromagnetic and electrostatic units and incorporate the practical units of electricity and magnetism as the fundamental units of the system.

Weights and measures have been important to societies for thousands of years and legal definitions of mass, length, and volume have long been essential to all societies for purposes of trade. It was one measure of the progress of electrical science and engineering that definitions of electrical units were on the statute books of many countries soon after the recommendation by the International Congress of 1893. As the buying and selling of electrical equipment made the transition from an exchange of curiosity items among a handful of specialists to an important international trade affecting the masses, the understanding of exactly what was meant by, say, 5 A at 10 V became financially as well as technically important.

References

1. *Encyclopaedia Britannica*, "Electron," 15th edition, 1974, vol. 6, p. 665.
2. C. Süsskind, *IEEE Spectrum* 7 (No. 9): 76, September 1970.
3. R. Taton, Ed., *Science in the Twentieth Century*, Thames and Hudson, London, 1966.
4. C. Süsskind, *IEEE Spectrum* 3 (No. 5): 72, May 1966.
5. *Encyclopaedia Britannica*, "Thomson, Sir Joseph John," 15th edition, 1974, vol. 18, p. 348.
6. M. D. Fagen, Chapter 5, Ref. 19.
7. F. Hund, *The History of Quantum Theory*, Harrap, London, 1974.
8. W. H. Cropper, *The Quantum Physicists and an Introduction to Their Physics*, Oxford University Press, New York, 1970.
9. D. C. Mattis, *The Theory of Magnetism*, Harper and Row, New York, 1965.
10. A. H. Bobeck, *Bell Syst. Tech. J.* 46: 1901, 1967.
11. A. H. Bobeck and H. E. D. Scovil, *Scientific American* 224: 78, June 1971.
12. D. Hadfield, *Powder Metallurgy* 22: 132, 1979.
13. J. R. Pierce, *Trans. IEEE* IT-19: 3, 1973.
14. T. I. Williams, Ed., *A History of Technology*, Oxford University Press, London, vol. 7, 1978.
15. P. Dunsheath, Chapter 5, Ref. 28.
16. V. Belevitch, *Proc. IRE* 50: 848, 1962.
17. A. I. Zverov, *IEEE Spectrum* 3 (No. 3): 129, March 1966.
18. L. Espenschied, *IEEE Spectrum* 3 (No. 8): 162, August 1966.
19. D. G. Tucker, *Technology and Culture* 19: 650, 1978.
20. *Electronics* 53 (No. 9), 17 April 1980.
21. H. S. Black, *IEEE Spectrum* 14 (No. 12): 55, December 1977.
22. C. E. Shannon, *Bell Syst. Tech. J.* 27: 379, 623, 1948.
23. E. C. Cherry, *Proc. IEE* 98 (Part 3): 383, 1951.
24. A. E. Kennelly, *Proc. 36th Ann. Meeting Soc. Promotion Eng. Ed.* 36: 229, 1928; also *J. Eng. Ed.* 18: 1, 1927–28.

10 MINIATURIZATION OF ELECTRONICS

Miniaturization has made it possible for electronics to penetrate society more widely and deeply than ever before. Pocket calculators, electronic watches, miniature colour television receivers and the like are only some of the examples of the miniaturization of electronics of which the general public first became aware. Even before they came along, miniaturized electronic systems had made a significant impact in military, industrial, and commercial areas. Miniaturization helped in the exploration of space, in communications, in the control of machinery and processes, and in the handling and processing of data. The miniaturization of electronics is sometimes regarded as a somewhat late development that derives from the integrated circuit; yet miniaturization on the grounds of size, weight, and power requirements was under way long before the integrated circuit was invented and even before the transistor became commercially available. Valve (vacuum-tube) manufacturers were remarkably successful in producing miniature and subminiature valves, some of them smaller than a present-day power transistor; and the screen printing of resistive and other passive components, and the concept of electronic modules, helped to bring about smaller electronic systems. Yet the big acceleration towards microelectronics did indeed begin with the invention of the integrated circuit, when at first small and later large circuits were formed on a single chip of silicon. The net result was systems far larger and far more complex than could even have been dreamed of before.

The purpose of this chapter is to review the miniaturization of electronics, mainly as achieved by the use of semiconductors. We begin by noting the discovery and early use of semiconductors and follow the subsequent improvement of materials, and our deepening understanding of them through scientific study, which led to the invention of the transistor and onwards to microminiaturization.

Three fundamental properties of semiconductors differentiate them from metals and insulators; all three can be discovered by simple experiments. They are the negative temperature coefficient of resistivity, the photoelectric effect, and the use of semiconductors to achieve rectification. All three were discovered in the nineteenth century but were not fully understood until the 1950s.

As temperature increases, the number of charge carriers (that is, the number of electrons and 'holes' or positive carriers available for conduction) also increases. The result is a drop in resistivity, the opposite of what happens under similar circumstances in a metal. This negative temperature coefficient was observed in silver sulfide as early as 1833 by Faraday. A. E. Becquerel (the father of A. H. Becquerel who discovered radioactivity) reported the observation of a photovoltage in an electrolyte in 1839; a photocurrent was observed in selenium by Willoughby Smith in 1873 in England. In 1874 rectification, the ability to conduct better in one direction than the other (and so to rectify or 'correct' an alternating current) was discovered when it was found that the resistance of contacts between some materials depended on the applied voltage. K. F. Braun observed this phenomenon for metallic contacts to pyrites and galena, and Arthur Schuster for contacts between untarnished and tarnished copper wire. All three important effects—negative temperature coefficient, photoelectricity, and rectification—had therefore been discovered by the last quarter of the 19th century. By the end of the 19th century photoelements had been fabricated and C. E. Fritts had made a large-area selenium rectifier. A further effect particularly noticeable in semiconductors, the Hall effect (a change in conductivity under the influence of a magnetic field), also dates from the 19th century (1879).

In the early 20th century radio technology developed rapidly. Detectors were among the weakest parts of early radio receivers and new detectors were needed to replace the old coherers. Point-contact rectifiers were found to be good detectors and became known as crystal detectors or 'cat's whiskers;' the 'whisker' was a springy metal wire placed in contact with a "crystal" of galena, silicon carbide, or silicon (Fig. 10.1). Silicon was found by experience to be the

Figure 10.1 Cat's whisker, 1920s

most stable. The advent of the vacuum tube or valve rectifier eventually provided an even better device and scientific interest in semiconductor rectifiers lagged, although crystal sets were used extensively by enthusiastic amateurs and became the first commercial application of solid-state electronics.

Improvement of Understanding and of Materials[1,2,3]

The 1920s saw further growth in solid-state electronics, with the commercial development of copper oxide and selenium rectifiers and photocells; and their use as rectifiers, battery chargers, photographic exposure meters, and so on. The greater use stimulated scientific investigation, which in turn benefited from the development of quantum mechanics that took place in the late 1920s and early 1930s. Quantum theory was introduced by Planck in 1901 and advanced by many scientists later, but the quantum mechanics used in semiconductor theory was largely developed after 1925. A brief summary is given in Table 10.1.

Table 10.1 Summary of the Development of the Quantum Mechanical Theory of Semiconductors (1926–1931)

1926	Schrödinger's equation to describe electron behaviour
1926	Born's interpretation of Schrödinger's equation in probability terms
1926–1928	Heisenberg's theory of the molecular field (Principle of Indeterminacy or Uncertainty Principle, 1927)
1927	Davisson and Germer's electron diffraction experiment, electron wave behaviour (also G. P. Thomson, 1928)
1928	Sommerfeld's application of wave mechanics to conduction in metals
1928	Bloch; start of the band theory of solids
1929	Peierls; concept of holes
1930	Dirac; synthesis of quantum mechanics
1931	Wilson's semiconductor theory based on quantum mechanics and band theory

By the early 1930s some important theoretical points had been pinned down with the aid of Hall effect measurements. The conductivity σ was known to depend on the density n of charge carriers and on their mobility μ; the expression $\sigma = ne\mu$, where e is the electronic charge, was known to be valid for semiconductors. The wide variation of n with temperature could be used to explain the negative temperature coefficient, the first of the 19th century discoveries, and pinpointed the difference between metals and semiconductors. From high-temperature experiments an 'activation energy' had been discovered that was constant for a given semiconductor regardless of its physical structure and served as a measure of the energy 'band gap,' possibly the most important characteristic of any semiconductor. Hall-effect results also bewildered many a scientist since they showed that the current was carried by positive charge carriers as well as by negative electrons. These positive carriers, now called holes, were conceived by Rudolf Peierls in 1929 and were named defect electrons.[4,6]

A. H. Wilson investigated the quantum-mechanical theory of semiconductors and used the so-called band theory to study the processes by which conduction occurs. His theory is still fundamental to present-day understanding of semiconductors, even though it took about 15 years for its implications to be properly appreciated. When this understanding dawned, during and after World War II, it became one of the factors that led towards the invention of the transistor, which was not accidental but the outcome of goal-oriented research. Wilson's theory introduced the concepts of acceptor and donor states associated with the deliberate 'doping' (the admixture of carefully measured quantities) of otherwise pure semiconductors with two types of impurities. The resulting two types of semiconductor were known as defect and excess types until renamed p-type and n-type in 1941 by J. H. Scaff of Bell Laboratories.

The photoeffect was explained in 1935 by I. I. Frenkel of Russia in terms of light energy creating pairs of free electrons and holes, each of which then diffused away at different rates. With this suggestion the second of the three 19th century discoveries was on its way to being understood. The third, rectification, had to wait a little longer.

It had still not been decided whether rectification took place at the junction of the metal and semiconductor (that is, whether it was an interface property) or whether it occurred inside the bulk of the semiconductor. Gradually it was recognized as a surface or junction effect. The next big challenge was to explain that. One puzzle was why rectifiers did not rectify in both directions if all that was needed was a metal-to-semiconductor contact. It was recognized that rectifying contacts needed to be explained; what was not recognized until later was that nonrectifying (ohmic) contacts also needed explanation.

Quantum mechanics predicted that electron waves would have a probability of penetrating an electrical potential barrier. This prediction was examined as a possible explanation of rectification. A potential barrier exists at any junction between a metal and a semiconductor. In the late 1930s it was realized that this barrier would be relatively wide in the semiconductor. Such a wide barrier would be a real barrier to electron waves, whereas the similar (but much narrower) barriers in metal-to-metal junctions would be easily penetrated by electron waves. Various theories, especially by N. F. Mott, Schottky, and Hans Bethe, now purported to explain rectification in metal-to-excess (n type) semiconductor junctions and in metal-to-defect (p type) semiconductor junctions. However, the theories did not always agree with experiment. As R. W. Pohl once remarked, "Theories come and go; experimental facts are there to stay."[5] Even though the theory was now on the right track, it still needed considerable development. In particular, the existence of energy states at or near a semiconductor's surface had yet to be discovered and difficulties over the measurement of a metal's surface properties (notably the so-called 'work function') had yet to be recognized.

The Russian B. Davidov made two vital suggestions in 1938, although his work attracted little attention at the time.[3] He suggested that in metal-to-semiconductor rectifiers there was a change from excess type to defect type in the semiconductor and that it was here, at this interface, that rectification took place. Davidov was in fact writing about what we now call a p-n junction. In developing his theory he also recognized the importance of the role played by minority carriers in both rectification and in producing a photovoltage. Minority carriers (holes or electrons, depending on their relative numbers) are not only important in p-n junction theory but are vital for an understanding of transistors.

This brief description of the development of semiconductor theory is of course a simplification and has neglected the roles played by many other workers. However, it does show what success scientists had in coming to terms with some baffling phenomena within a little over a decade. It was not a storybook tale of success after success—scientific research rarely is—but as World War II started a realistic picture was emerging. Yet for the development of actual semiconductor devices, improvements were needed in materials as well as theories.

When Hertz generated and received radio waves in 1888 he not only verified Maxwell's electromagnetic theory, he also started the scientific preoccupation with short wavelengths; after all scientists were specifically interested in the theory of light. However, radio pioneers such as Marconi turned to longer wavelengths and exploited them for communications and for broadcasting. But by the 1930s, coincident with the advances in semiconductor physics, radio engineers too were looking at the shortwave end of the spectrum. In addition, military development of radar speeded the exploitation of the shorter wavelengths and made demands for improved microwave equipment (including crystal rectifiers).

The development of electronics was largely the development of radio up to the late 1930s. The only active components available, apart from the copper oxide and selenium rectifiers, were the vacuum valves or tubes. These devices proved to be of little use at the shorter wavelengths then being investigated and interest returned to the now obsolete crystal detectors. Silicon detectors were known to be the most reliable of the old devices; germanium point-contact rectifiers had been manufactured from about 1925. One researcher started his new work on silicon by visiting a New York radio market to obtain some old silicon detectors.[3]

Efforts to produce better detectors for the shorter wavelengths created demands for purer germanium and silicon (as well as for improved theories) and brought together teams of physicists, chemists, and metallurgists. It is to these and subsequent groups that solid-state electronics owes a big debt. Without such teams of varied specialists uniting theory and practice, the work would never have been completed successfully. Semiconductor physics was not an area in which an individual inventor could hope to tinker his way to success.

As World War II approached, security considerations blocked the free passage of information aimed at producing better devices. Parallel developments took place in secret in various countries and the same thing could be discovered several times in different places. Americans and Germans made great strides in improving the purity of silicon and discovered how to make, almost to order, silicon of the excess and defect type, or n-type and p-type as they became known at Bell Laboratories. One of the melts of silicon made by J. H. Scaff and H. C. Theuerer at Bell, the pioneers in this area, turned out to be p-type at one end and n-type at the other. R. S. Ohl discovered in 1941 that the junction so formed—the first p-n junction to be investigated—was an excellent rectifier and gave a strong photovoltage, just as Davidov had suggested in 1938. A further development by the same group identified the impurities that caused the silicon to be p-type or n-type and showed the 'acceptor' impurities to be from column 3 of the chemical periodic table and the 'donor' impurities from column 5. Silicon itself lies in column 4. After the war semiconductor laboratories emphasized the more fundamental scientific studies of the two simplest semiconductors, silicon and germanium, and, at least at Bell, were soon able to produce samples for study with specified impurity concentrations.

Another group of scientists, under Karl Lark-Horovitz at Purdue University, began work in 1942 as part of a coordinated American effort; their task was to learn more about germanium so that it might be used more effectively as a radar detector.[5] Three areas were tackled: an investigation of the fundamental electrical properties of germanium, the preparation of purer materials, and the fabrication of detectors. It was the great success of their work that created the early lead enjoyed by germanium over silicon. One may speculate that it might have well been Purdue, not Bell, where the transistor was discovered if Purdue had not reverted to fundamental research after the war. Even so Ralph Bray, at Purdue, came very close in 1948.

The Transistor

The analogy between the semiconductor and vacuum diodes was an obvious one and ultimately led investigators to interpose (by analogy with the vacuum triode) a 'grid' in the semiconductor diode to produce a semiconductor or crystal triode. The technical problems were immense because of the size limitations. Nevertheless ideas were put forward and practical attempts made.

In the late 1920s J. E. Lilienfeld filed for American patents for what would have become field-effect and bipolar transistors; in 1935 Oskar Heil of Berlin obtained a British patent for what would now be described as an insulated-gate field-effect transistor. Though these ideas were sound, the technology did not then exist to support them. In 1938 R. Hilsch and R. W. Pohl published a description of an actual device using potassium bromide crystals which was, in principle, a working solid-state triode. It was not very practical—its cutoff frequency was of the order of 1 Hz or less.[3] O. V. Losev is said to have built crystal amplifiers in Russia in the 1920s and published work describing a 'detector-amplifier.'[4]

The forerunner of today's devices was the point-contact transistor made by John Bardeen and Walter Brattain at Bell Laboratories in December 1947, hailed by some as the most important invention of the 20th century. The story of its invention is a story of theory and experiment feeding on each other from the improvement of point-contact diodes for microwave use, through the improvement in purity of silicon and germanium and the understanding of the importance of minority carriers, to the Nobel Prizewinning culmination.

Social inventions also played a role.[7] International fellowships enabled promising young physicists to travel and work with eminent scientists of the time and were particularly useful in the late 1920s in the development of quantum mechanics. During World War II research became much more highly organized than before and modes of organizing research efforts were improved. One effort that was to play a part in the invention of the transistor was the organization of the American research to study semiconductors for radar applications. Under the leadership of the MIT Radiation Laboratory various industrial and academic institutions, including Bell and Purdue, were brought together to focus on one set of problems. Such wartime developments set the scene for the arrival of the transistor. Not only did they improve communications between industry and the universities, they provided a mixing of scientists, ideas, and missions and left many scientists with a desire to seek practical applications for their scientific work.

The wartime work led to a postwar situation in which theory and materials had reached a stage where further effort would have a good chance of paying off with practical devices. In July 1945 a solid-state research program was set up at the Bell Laboratories to seek "new knowledge that can be used in the development of completely new and improved components."[7] It was a subgroup in this program, the semiconductor group, that discovered the transistor effect three years later and produced a device that was to show the way towards solving the growing problems involved in telephone switching, problems to which vacuum tubes and relays had limited answers. William Shockley and Brattain had already made one early attempt to answer the problem in 1939–40 when they attempted to make a solid-state amplifier based on copper oxide, a semiconducting compound. That attempt failed. The next attempt, by John Bardeen and W. H. Brattain, was successful and produced the point-contact transistor.

The semiconductor group was interdisciplinary and included experimental and theoretical physicists, a physical chemist, and a circuit expert. It also enjoyed close collaboration with the metallurgical group. The group decided in January 1946 to concentrate on silicon and germanium, the two simplest semiconductors, and later to include a study of surface properties as well as bulk properties. Several problems with semiconductor surface effects had arisen, including a growing inadequacy of the theory of rectification.

Shockley had predicted that it ought to be possible to modulate the conductivity of a thin layer of semiconductor by the application of an external electric field (the 'field effect') and so produce amplification. But experiments failed to produce the predicted effects. Bardeen found he could explain the puzzle by assuming the presence of energy states on the semiconductor surface, such as I. Y. Tamm and Shockley had previously suggested for the free surface of a solid. That was the breakthrough. Many experiments were performed to test the surface-state theory. The transistor effect was observed during the course of this work on 23 December 1947 by Bardeen and Brattain. The first transistor (a term coined from 'transfer resistor' by J. R. Pierce), which was a point-contact version and not a junction transistor, was used as an amplifier the same day and as an oscillator the next day. Its manufacture was described by Brattain:

"After discussions with John Bardeen we decided that the thing to do was to get two point contacts on the surface sufficiently close together, and after some little calculation of his part, this had to be closer than 2 mils. The smallest wires that we were using for point contacts were 5 mils in diameter. How you get two points 5 mils in diameter sharpened symmetrically closer together than 2 mils without touching the points, was a mental block.

"I accomplished it by getting my technical aide to cut me a polystyrene triangle which had a small narrow, flat edge and I cemented a piece of gold foil on it. After I got the gold on the triangle, very firmly, and dried, and we made contact to both ends of the gold, I took a razor and very carefully cut the gold in two at the apex of the triangle. I could tell when I had separated the gold. That's all I did. I cut carefully with the razor until the circuit opened and put it on a spring and put it down on the same piece of germanium that had been anodized but standing around the room now for pretty near a week probably. I found that if I wiggled it just right so that I had contact with both ends of the gold that I could make one contact an emitter and the other a collector, and that I had an amplifier with the order of magnitude of 100 amplification, clear up to the audio range."[7] (Fig. 10.2.)

The transistor was fortunately not classified for security purposes, as it might have been, and the news became public six months later, on 1 July 1948 with a short announcement in the *New York Times*. The magazine *Electronics* acquainted its readers with 'The Transistor: A Crystal Triode' in a four page

(a)

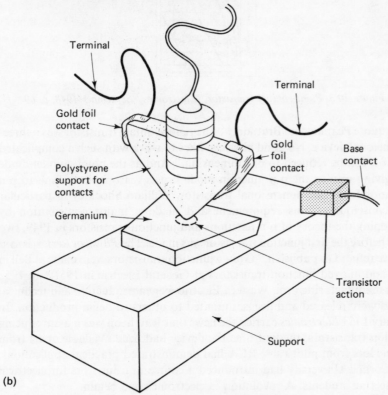

Terminal

Terminal

Gold foil
contact

Gold
foil
contact

Base
contact

Polystyrene
support for
contacts

Germanium

Transistor
action

Support

(b)

*Figure 10.2 (a) First point-contact transistor, December 1947; (b) sketch; gold foil
contacts were later called emitter and collector*

article in its September 1948 issue, promising that "It will replace vacuum tubes in many applications and open new fields for electronics." That promise has certainly been kept. By then the transistor had changed its appearance and was looking much more business-like (Fig. 10.3).

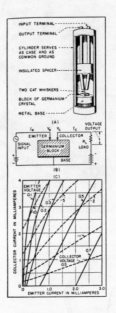

Figure 10.3 Point-contact transistor, 'Electronics,' September 1948, p. 69

To quote Pearson and Brattain,[3] "The point-contact transistor was a three-dimensional device. No good physicist likes to work with such a complicated case if it can be reduced to one dimension." And as the point-contact diode was giving way to the p-n junction diode it was reasonable to look to p-n junctions for a one-dimensional transistor. William Shockley in particular was conscious of this requirement and made a major contribution by developing the theory of p-n junctions and junction transistors in 1949, two years before the first junction transistor was made. The *Physical Review* is said to have refused to publish it![5] Grown junction transistors were made at Bell in 1950 and alloyed junction transistors at General Electric in 1951.[8] In 1952 Bell's manufacturing arm, Western Electric, began production and technical details were released and licenses granted to others to begin production. In January 1953 *Electronics* carried the news that Raytheon was massproducing junction transistors, the first time the device had been available apart from sample lots from pilot runs; RCA had demonstrated practical applications; and Cornell University had introduced a course in transistors for electrical engineering students. A revolution in electronics had begun.

The three inventors, Bardeen, Brattain, and Shockley, were awarded the Nobel Prize for physics in 1956, only eight years after performing their work—

quite a contrast to the reaction to Lee de Forest's invention of the triode in 1907 which, in lawsuits just a few years later, was referred to as being worthless.

The 1950s saw a phenomenal change in electronics as the transistor gradually ousted the electron tube from many old applications, found new applications for itself, and produced a new electronics technology. Within ten years the transistor itself underwent massive changes and emerged in new applications in integrated circuits (ICs). The path from bipolar transistor to integrated circuit began with an improvement in materials, specifically the application by G. K. Teal and J. B. Little in 1950 of the old Czochralski crystal-pulling technique to the preparation of single-crystal germanium, the semiconductor that started transistor electronics, and the purification of germanium by zone refining by W. G. Pfann in 1952. As p-n junction devices reached the market to join the point-contact diode (which had sales in the millions) and the point-contact transistor, the real revolution had begun because it was the much improved p-n junction that made the first integrated circuits possible. An integrated circuit based on noisy, fickle point-contact devices would be a nonstarter.

Bell's grown-junction transistor took its name from the fact that the junctions were formed by variations in the impurity content introduced as the germanium crystal was grown. It was quickly joined by General Electric's alloyed-junction transistor, in which junctions were formed by a process of alloying the impurity to opposite sides of a thin germanium wafer.[8] When both pnp and npn versions became available, they offered circuit designers the novel possibilities of complementary use, something not available to users of electron tubes. Junction transistors also gave improvements in gain and noise levels over point-contact transistors, although the cutoff frequency was not as high as for electron tubes at first.

Silicon proved to be less easily handled than germanium because of its higher melting point and its higher reactivity, and for a while germanium devices dominated the market. But in 1954 Texas Instruments (TI) surprised the industry by announcing the arrival of the silicon grown-junction transistor. At the same time a novel process called floating-zone refining, a variation of zone refining in which the liquid zone is supported by surface tension rather than by a crucible, paved the way for the application of the alloying technique to make silicon transistors. For a few years TI was the only company to manufacture silicon transistors. Meanwhile the frequency response of germanium devices was improved by novel techniques aimed at producing the extremely narrow base regions needed if performance above a few hundred kilohertz was to be achieved. In 1953 Philco developed a jet electrolytic etching technique by which both sides of a thin semiconductor wafer could be etched away so as to yield a very narrow base. This process produced the surface-barrier transistor, which gave the best frequency response so far.

In the mid-1950s the all-important process of diffusing impurities from the vapour phase into the semiconductor to produce the desired p and n regions was developed by Bell and by General Electric. This process, together with oxide masking, became and remained the standard process in the semiconductor industry for a quarter century or more. It proved to be the best method of controlling the doping of a semiconductor with impurities so as to give properly designed electrical characteristics to the resulting devices. Diffusion technology was quickly applied to the improvement of existing techniques and to the production of new transistor structures. The first was the mesa transistor, which pushed the performance of germanium transistors beyond that of silicon. Even more important, the diffusion technique made it possible to massproduce transistors by the process of fabricating them by the hundred on a single slice of germanium or silicon and with a better performance than previously achieved. The slice could then be cut up to obtain the individual transistors, and it could only be a question of time before someone would think of interconnecting the transistors while they were still on the slice so as to provide certain circuit functions. It was also in the mid-1950s, as the important new techniques arrived on the scene, that transistors really took off in the market place. According to one American report, in 1957 sales of semiconductors were doubling and even tripling annually and rose from "nothing in 1952 to $37 million last year."[9] Though $37 (£15) million looked like peanuts compared to the $855 (£350) million sales of the tube industry, sales of semiconductor devices were rising fast; Figure 10.4 illustrates the phenomenal growth of semiconductor electronics.

Oxide masking was another major improvement in transistor technology. Oxides on silicon were shown to impede the diffusion of the impurities and, coupled with photographic masking techniques, provided a very valuable tool in controlling the location of the impurity zones. Other methods were forthcoming about this time in improving the structure of the silicon crystals by almost eliminating dislocation faults and so maximizing the use of the silicon wafer. However, the key to the whole future of semiconductor work was the planar process developed by the new Fairchild Semiconductor Company from 1958 to 1960. A patent was awarded in 1962. The planar process uses the diffusion and oxide masking techniques to manufacture devices in a plane parallel to the surface of the semiconductor. The oxide is grown as the first step in the fabrication process and serves also as a protecting screen to the devices formed under it. The epitaxial deposition of a thin layer of controlled-purity silicon onto a silicon crystal substrate was a Bell innovation made in 1960. This process, together with the planar technique, permitted the fabrication of improved devices (with speeds of operation ten times greater than previous ones) in the epitaxial layer, and the substrate gave mechanical support.[10] To complete the product, plastic encapsulation was introduced by General Electric in 1963 to replace the metal cans previously used.[11] By 1961, then, the techniques essential to the production of the future integrated circuits were being applied to the manufacture of discrete transistors: the epitaxial, planar-diffusion, oxide-masking process. The mid-

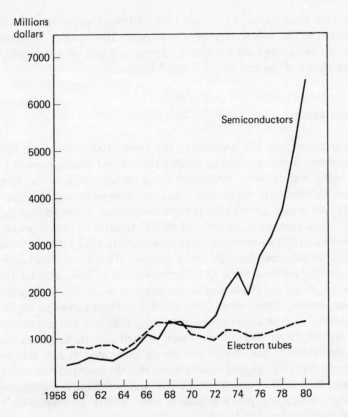

Figure 10.4 Estimated U.S. spending on electron tubes and semiconductor devices, 1959–1980 (based on 'Electronics' Annual Markets Forecast; only retrospective figures used)

1960s were to see a growth in sales of integrated circuits just as phenomenal as that seen for transistors a decade earlier.

Transistor improvements came from each successful innovation rather than from experience gained in manufacturing, and they brought their own rewards. The introduction of the mesa transistor, for example, pushed the sales of Fairchild Semiconductor from $500 000 in 1958 to $7 million in 1959.[11] New techniques also spread quickly, partly helped by the fluid movement of talented engineers from one company to another. This movement of young people not hidebound by concern for security and pension has been a particular characteristic of the American semiconductor industry and has probably contributed to its world success. It certainly produced vigorous new companies. When William Shockley left Bell to form Shockley Semiconductor Laboratories, and later Shockley Transistor, he attracted many talented men to join him. In 1957 eight of them left to form their own company, Fairchild Semiconductor. Since then over forty companies have been started by former Fairchild employees. Fairchild may be exceptional in this respect, but only in degree.[5] This fluidity of personnel is recognized as having been a major advantage enjoyed by American firms in keeping ahead of the rest of the world.[5,12]

Miniaturization

Miniaturization was not invented in the 1960s. Extra electrodes had been crammed into valves to produce double triodes and double pentodes. Some large bulky valves, with associated large bulky components, had been replaced by miniature valves and miniature components in the mid 1940s; these devices in turn were challenged by subminiature valves and components, and of course transistors, in the mid-1950s. As early as 1946 a pocket radio was produced from subminiature components.[13] In 1942 a decision was made to explore two-dimensional silk screen printing of conductive inks to produce electronic components for a U.S. Army proximity fuze, and by 1945 the Centralab Division of Globe Union, the original contractor, was massproducing these devices. After the war, work continued and turned to applying the technique to civilian applications in hearing aids and radio and television parts. By 1962 over 140 million screen printed circuits had been made by Centralab alone.[14] But it was the silicon integrated circuit that enabled miniaturization to be pushed to such limits that the essential components of a computer that would outperform ENIAC (one of the first digital electronic computers, which occupied a volume of some $3000\,\text{ft}^3$ or over $100\,\text{m}^3$) could be held easily in one hand.

The motivation to miniaturize seems always to have been associated with factors such as the ever-increasing complexity of electronic systems, the desire for improved reliability, reduction of power consumption, weight, and of course, cost. All these factors are inextricably interlinked. Automation crept in with printed wiring, dip soldering, etched copper tuning coils, and the like, and brought some improvement in reliability and size as some hand wiring and chassis building were eliminated.

The American military establishment played an important role in bringing about the miniaturization of electronics, but exactly how important is still debated. Electronics had proved its worth in World War II by providing radio communications, radar, navigation aids, and so on. Each B-29 bomber, for example, carried nearly 1000 vacuum tubes and associated electronics.[15] But World War II standards could not provide real-time radar capable of giving the American continent an early-warning system able to detect single fast moving aircraft, and after the Soviet Union detonated its first atomic bomb in 1948 that is what the U.S. Department of Defense wanted. A computer-radar system in which the computer alone used over 14 000 tubes not only spurred the development of computers but showed the need for even greater reliability.

As the 1950s got under way, and as the Cold War and its demand for missile systems developed, new ideas were suggested for miniaturization: double-sided printed-circuit boards, plated-through holes, multilayer etched circuits, and three-dimensional stacking to produce high density modules and micromodules. Tinkertoy (1951) was an early American example of modular electronics. Ceramic wafers with printed wiring supported solid or printed components. Four to six wafers could be stacked, and machine-soldered wires provided interconnections between wafers. Tubes could be mounted on top. Later, transistorized examples were encapsulated with epoxy resin to improve mechanical strength and give environmental protection. Such modules were early attempts at standardizing shapes and sizes for easy replacement.

Other techniques were also studied. Thin-film circuits, in which two-dimensional components were evaporated or sputtered onto a ceramic substrate, eventually achieved a sizable market with the aid of thin-film transistors. However, another technique called 'molecular electronics,' a bold concept from the U.S. Air Force, never saw the light of day. Blocks of semiconductor were envisaged performing circuit functions without any part-for-part equivalence with the equivalent discrete circuit. Despite the millions of dollars spent on such projects, in the end none of them could compete in terms of miniaturization with the integrated circuit.

By the late 1950s scientists and engineers were discussing the possibility that complete circuit functions, such as logic gates, could be formed within a single block of semiconductor, and as early as 1952 G. W. A. Dummer of the British Royal Radar Establishment took a "peep into the future" to see "electronic equipment in a solid block with no connecting wires." He speculated that the block "may consist of layers of insulating, conducting, rectifying and amplifying materials, the electrical functions being connected directly by cutting out areas of the various layers."[15] Though it was a remarkable vision neither Dummer, nor anyone else then, had much idea of how this goal could be achieved on a large scale.

Even so, the first patent for an integrated circuit was filed only a year after Dummer's suggestion, on 21 May 1953, by Harwick Johnson of RCA. Johnson's patent was for a "semiconductor phase shift oscillator and device" and a stated object was "to provide a novel phase-shift oscillator in a unitary semiconductor body."[16] The circuit consisted of a transistor and its circuit (an RC phase-shift network) all made in the same piece of semiconductor. Several p-n junctions, formed for example by the alloying technique, provided the capacitance; the resistors were filamentary portions of the semiconductor formed by abrasion or by etching (Fig. 10.5). The capacitance could be controlled by a bias voltage. Variations on the design provided for an integrated load resistor and a single elongated p-n junction for the RC network. However, Johnson's integrated circuit was ahead of its time. At that time discrete transistors were only just going into mass production.

Dummer, meanwhile, had not given up on his own idea. In April 1957 a contract was placed with the Plessey Company research laboratory in England to develop a "semiconductor integrated circuit."[17] Plessey appears

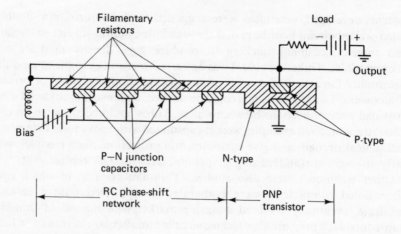

Figure 10.5 Johnson's integrated circuit phase-shift oscillator, 1953 (U.S. patent 2 816 228) (redrawn)

to have regarded the idea as a laboratory curiosity, or at most as an exploratory feasibility study. Consequently the "first real research project on silicon integrated circuits" was not placed with Plessey until May 1959. By that time Texas Instruments had just announced the arrival of the integrated circuit.

Despite Johnson's patent and Dummer's ideas it was J. S. Kilby of Texas Instruments who actually made the first integrated circuit. During July 1958, while TI was closed down for annual vacations, Kilby had his idea of the monolithic concept in which transistors, resistors, and capacitors are formed within one slice of semiconductor. Kilby virtually had the place to himself as he did not take a vacation, since he had just come to TI in May from Centralab where he had worked on the screen printing of components and on transistors.[15] Since other ideas for the miniaturization sought by the U.S. Forces involved only repackaging, and Kilby's idea involved much more than that, he was given the go-ahead to try to build a circuit entirely from semiconductors and, if that was successful, to build the first integrated circuit. Three handmade phase-shift oscillators were made in germanium on 12 September 1958. Kilby has been quoted as saying, "It looked crude and it was crude."[17] The first one oscillated at 1.3 MHz. An equally crude multivibrator followed on 19 September 1958, and although the concept was not the molecular electronics the U.S. Air Force had in mind, a contract was issued for investigation. Anything that might miniaturize the electronics for computers, missile guidance systems, and the growing space race was worth research funds.

About the same time J. W. Lathrop, one of the pioneers of photo-lithography as a semiconductor technique, also joined TI. Kilby developed his concept further and in October the design of a new germanium flip-flop was started in which mesa transistors, junction capacitors, and bulk resistors were employed, made with the help of the photoetching and diffusion techniques (Fig. 10.6).[15] The first working models were completed early in 1959 and were used in the public announcement of what was rightly claimed as the most significant development since the silicon transistor, also a TI first.

Meanwhile at the then small firm of Fairchild Semiconductor, essentially the same idea as Kilby's had occurred to Robert Noyce, in January 1959. Fairchild had pioneered the planar process and was using it to produce hundreds of transistors on one wafer of silicon. "But then people cut these beautifully arranged things into little pieces and had girls hunt for them with tweezers in order to put leads on and wire them all back together again; then we would sell them to our customers, who would plug all these separate packages into a printed circuit board."[17] Batch processing, and the planar technique, were the keys to integrated-circuit production.

The idea of integrated circuits was very much in the air. Kilby's TI patent[18] was filed on 6 February 1959 for Miniaturized Electronic Circuits, Noyce's Fairchild patent for a Semiconductor Device-and-Lead Structure was filed on 30 July 1959, and in between, Kurt Lehovec of Sprague Electric filed for a patent for Multiple Semiconductor Assembly on 22 April 1959, this last for the use of reverse-biased p-n junctions to isolate devices electrically on the semiconductor slice. The conception of the integrated circuit in 1959 was an idea, in Noyce's words, "whose time had come," and one "where the technology had developed to the point where it was viable." The development of ICs from transistors in the industry at that time was as natural as a plant coming into bud. In a sense nothing new was produced, yet what had been produced was something that had not existed before, and which would blossom in the very near future. By the end of 1961 both TI and Fairchild were producing integrated circuits in commercial quantities.

The integrated circuit was the answer to the question of how to miniaturize electronics and, though it still was not the molecular engineering envisaged by the U.S. Air Force, it quickly burrowed its way to the very heart of the U.S. military. Not that they were its only customers; the computer industry also played an important part, although the military impact was felt there also. In 1960 it was reported that IBM was probably the biggest customer of every American semiconductor company.[19] It has also been estimated that in the period from 1958 to 1974 the U.S. government spent more than $930 (over £400) million on research and development in the semiconductor industry. Private industry in America probably spent even more, around $1200 (£550) million in the same period, and the total technical effort in the 20 years from 1955 to 1975, including applications engineering, marketing, process control, and production areas, is said to have cost at least $3000 (£1400) million.[19]

The U.S. military proved to be a major market for electronics as well as a major source of finance as the nation went through the Cold War, the space race, and the Vietnam war. *Fortune* magazine as early as 1957 stated, "'Peace' if it came suddenly would hit the industry very hard."[20] A glance at Figure

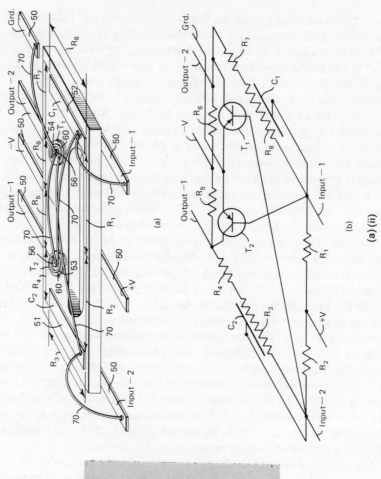

(a)

(a) (ii)

(b)

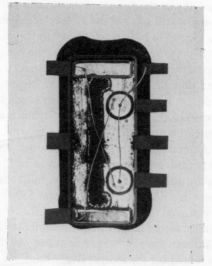

(a) (i)

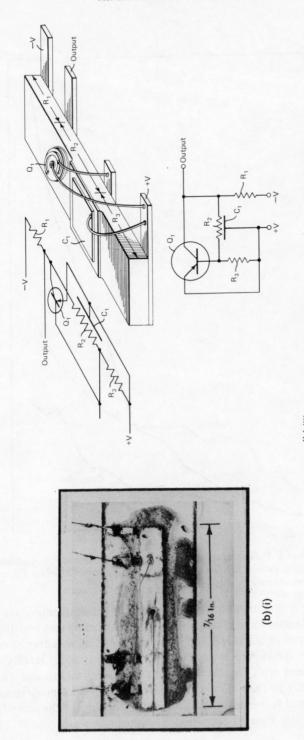

(b)(ii)

(b)(i)

Figure 10.6 The first integrated circuits: (a) (i) Kilby's IC multivibrator, 1959; (a)(ii) layout and circuit diagram; (b)(i) Kilby's IC phase-shift oscillator, 1959; (b)(ii) layout and circuit diagram (U.S. patent 3 138 743)

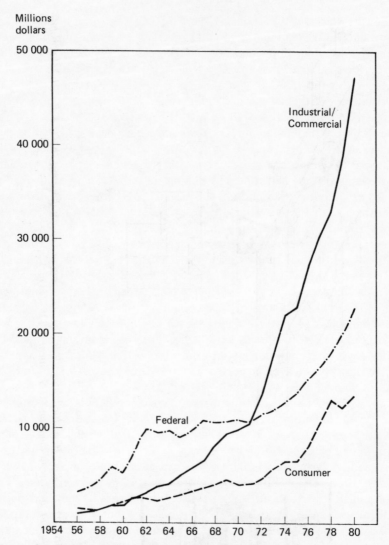

Figure 10.7 Estimated U.S. Federal, consumer, and industrial/commercial spending on electronics, 1956–1980 (based on 'Electronics' Annual Markets Forecast; only retrospective figures used)

10.7 shows why: the dominance of Federal spending on electronics over the industrial and consumer sectors, particularly during the early and mid 1960s. The relative importance of Federal and industrial spending was dramatically reversed though in the 1970s. One author commented in 1962, "Hardly any organization dealing with the electronics field remains untouched either directly or indirectly by the nation's guided missile and space programs."[21] That year, the second year of integrated-circuit production and the year of the Cuban missile crisis, Federal spending in electronics was about $10 000 million, of which $9200 million was spent by the Department of Defense and $500 million by the National Aeronautics and Space Administration (NASA). Industrial spending of $3200 million was small by comparison.[22] In one survey it was estimated that about 80 % of engineers and scientists working on electronics in the USA at that time were doing work supported by the government.[23] In the same article it was noted that for a military capability in space a mean-time-to-failure rate of 50 000 hr was needed. The state of the art gave about 1000 hr. Reliability was obviously a problem but new techniques were expected to help and to reduce size and weight into the bargain. Molecular electronics was still discussed but the definition given, "the build-up of circuits directly on surfaces by etching and depositing techniques," was not a very accurate description of the original concept or of the achievable planar technique. Integrated circuits were recognized as being very important but their full potential was only dimly grasped.

Several manufacturers were making integrated circuits under one name or another. Motorola actually called them integrated circuits, whereas at Westinghouse they were called 'molecular electronics' or 'molectronics;' at Fairchild, 'micrologic circuits;' and at TI, 'solid-circuit microelectronics.' The year 1962 saw the real start of mass production. Of these four firms, three received financial support from the U.S. Air Force; Fairchild's policy was to go it alone. The initial prices were astronomical by later standards, $450 (£185) for a simple flip-flop circuit from TI in March 1960, but such prices tumbled quickly. An average price of $50 in 1962 was down to $1.03 in 1972[5] and less than 80 cents by 1975.[15]

The advantages of microminiaturization were particularly felt in military systems and computers. By the late 1950s electronics was coming face to face with what seemed to be its inherent limitations. Systems were becoming more and more complex and more parts meant more expense and more to go wrong, yet reliability, particularly in military systems, needed to be increased and, if possible, costs reduced. Field maintenance was an increasing burden. For space and missile applications, any reduction of size, weight, and power requirements was welcome. Micromodules helped but the increased complexity of some systems offset such minor gains and it appeared that some tasks simply would not get done. Integrated circuits offered reductions in size, weight, and power consumption *and* increased reliability by a reduction in the number of soldered joints in a given system. They also offered a way out of the threatened impasse to further increases in complexity. Massproduction would reduce production costs; even more important, increased reliability and modular replacement should reduce maintenance costs. Little wonder the American military services were interested. They were encouraged in April 1963 by a memorandum from the Director of Defense Research and Engineering in which he stated, "This gain in reliability, coupled with reduction in size, weight and power requirements, and probable cost savings, makes it imperative that we encourage the earliest practicable application of microelectronics to military electronic equipment and systems."[24]

By 1964 the U.S. Department of Defense had spent around $30 (£14) million on research and development for IC technology. Sales had just taken off and the circuits were in use in military systems; the Minuteman II missile guidance system for example prompted a contract to TI for a family of twenty-two special circuits.[15]

Still, not everyone was overly enthusiastic in the early days. Some mistakenly compared the arrival of integration to that of transistorization and anticipated long delays in achieving high standards of reliability, similar to the delays involved in the change from tubes to transistors. However, integration was not a fundamental change and transistor technology, as already noted, had provided the necessary requirements. Yet there might have been more serious objections. Undeniably integrated circuits did not make optimum use of materials. Neither silicon nor germanium make the best capacitors or resistors. The change in circuit design that integrated circuits brought about was not foreseen by such objectors. In integrated circuitry, because active components are cheaper and take up less precious space than passive components, the maximum number of transistors and minimum number of resistors and capacitors are used—the opposite of what happens in discrete circuitry (Fig. 10.8). A more serious objection cast doubts on the ability of manufacturers to produce the actual circuits. Yields of functioning circuits would be low (and they were). If the probability of producing a good component were 90 %, then the chance of producing a good circuit consisting of twenty components would be only 12 %, more than a serious liability. It has been said that if all military components had received the same cossetting given to those in Minuteman the cost would have exceeded the Gross National Product.[5] Other arguments pointed to the difficulty of changing a design once made and some claimed that circuit designers would soon be out of a job.

The argument concerning the yield would appear to be the most serious, and the need to improve the yield, or the number of successful circuits per slice of semiconductor, was one of the driving forces for improvements in the industry's technology. Ultimately, as production lines came to churn out successful circuits in large numbers and at an economic price, such objections became, in Kilby's words, "simply irrelevant."[15]

Though it was in America that most of the advances in ICs took place, progress was also made elsewhere. In all the major West European countries integrated circuit manufacture began quickly after the American initiative, sometimes via offshoots of American firms and sometimes as a local effort (Table 10.2). In Britain the early work by the Royal Radar Establishment and Plessey has already been noted. A large research and development effort was also made on thin-film circuits. Work on ICs also began in other companies early in the sixties, including Ferranti, the British branch of TI, Standard Telephones and Cables, and Mullard. Annual sales of integrated circuits had reached nearly £12 million by 1969 and nearly £50 million by 1970. Other European countries including France, West Germany, the Netherlands, and

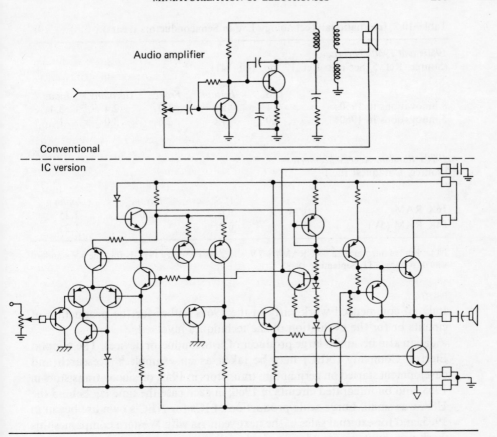

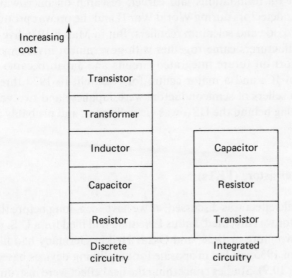

Figure 10.8 Design rule changes in the 1960s

Table 10.2 International Technology Lag in Semiconductors (years)

1950s and 1960s (average)
(Source: Ref. 5. See also Ref. 12, pp. 140–141)

		U.S.	Britain	France	Germany	Japan
8 innovations in 1950s		0.1	2.6	3.0	2.4	3.4
5 innovations in 1960s		0.0	1.6	2.6	3.0	1.2

1970s Random Access Memories
(Source: Dataquest Inc.)

		U.S.	Europe	Japan
16K RAM	1976	0.0	2.0	1.25
64K RAM (5V)	1978	0.25	—	1.2

Note: Fujitsu introduced a 64K RAM (+ 7 V, − 2 V) in February 1978, before the 5 V standard was introduced (TI, September 1978).

Italy all commenced work early in the 1960s, either for the production of circuits or for the evaluation of the techniques involved.[27]

Japan also became a large producer of semiconductor devices. The Nippon Electric Company (NEC) may be taken as an example.[28] Research and development started on germanium transistors in 1949, on silicon transistors in 1958, and on integrated circuits in 1960; in each case the time lag behind the USA was small. Large-scale production of ICs for NEC's own use began in 1965, and for external sales in the next year. As with Western companies this work was built on foundations laid earlier; research on microwave mixer diodes for radar detectors during World War II and the prewar production of copper-copper oxide and selenium rectifiers. But in March 1976 five or more Japanese manufacturers came together with government involvement for a cooperative effort on future integrated circuits and Japan became a major world source of ICs and a major centre for research. In 1977 three of the world's top ten sellers of semiconductors were Japanese and two years later the technology lag behind the USA was virtually zero, and probably ahead in some areas.[28]

Field-effect Transistors (FETs)

The so called field effect was discussed, as we have seen, long before the point-contact transistor was invented. Julius Lilienfeld had filed for a U.S. patent in 1926 and for two more in 1928, and Oskar Heil in Germany had filed for a British patent in 1935, all for proposals for amplifying devices based on the field effect (Fig. 10.9). Studies concerning the field effect were instrumental in leading up to the invention of the transistor and Shockley and G. L. Pearson

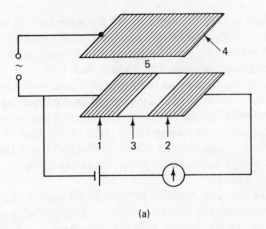

(a)

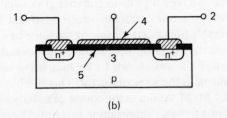

(b)

Figure 10.9 Similarities between (a) Heil's device patented in 1935 (British patent 439 457), and (b) a modern MOS transistor: 1, 2 ohmic contacts; 3, semiconductor; 4, thin metallic control electrode; 5, insulator

demonstrated the field effect in 1948. However, a practical field-effect transistor, in which the current between two terminals is controlled by an electric field resulting from the application of a voltage to a third terminal, was not available commercially until the early 1960s. Such slow development would have been normal in the 19th century but was somewhat unusual in the mid-20th century, at least in semiconductor devices. Why was there such a long delay? One obvious reason was the long wait for semiconductors of good quality; another was the geometrical restrictions that had to be met for such a device to work, namely that the ratio of the control electrode's surface area to the conducting channel's volume must be large.[24] This requirement was met by thin-film FETs in 1961.

After Shockley and Pearson had shown that the basic idea was valid, i.e., control of a current through a semiconductor by use of a transverse electric field to modulate the conductivity, Shockley proposed a device now called a junction-gate FET, in which he achieved this modulation by using the control electrode voltage to vary the width of the depletion layer in a p-n junction. Later, in the mid-1950s, another proposal was made to use an effect that had been observed in the early work on junction transistors. It had been noticed that the conductance between the emitter and base of a transistor was sometimes much larger than that to be expected from junction theory, and it was thought that a conducting channel might exist connecting the two. The channel was believed to arise from ions absorbed on the surface of the base which, in effect, were converting the surface of the base to the opposite polarity than that to which it was doped. In this way a pnp transistor might have a narrow p-type channel in its n-type base connecting the p-type emitter to the p-type collector.[24]

In 1955 I. M. Ross in America proposed that such a channel could be induced electrostatically by an electrode deliberately placed in the base region of the transistor. In the system now used this goal is achieved by a gate electrode separated from the semiconductor by a thin insulator; such a device is known as an insulated-gate FET. To achieve a satisfactory working system took four years, largely because of the difficulty in finding a suitable insulator. The insulator had to have a high dielectric strength so that it could be used in a very thin layer of just a few dozen micrometres, so that small voltages could produce a reasonably high electric field. At the same time the dielectric losses had to be small at the transistor operating frequencies. A suitable system for an insulated gate FET was shown to be that using silicon dioxide as the insulator and silicon as the semiconductor. This combination was discovered at Bell in 1959 by M. M. Atalla in the course of a study of silicon dioxide, the insulator that came to be so important in integrated circuits.[24] Meanwhile in France in 1958 a Polish scientist, Stanislas Teszner, had produced the Tecnetron, the first commercial FET in which germanium and the alloying technique were used. But the first metal-oxide-silicon (MOS) transistors using the $Si-SiO_2$ system were announced by Dawon Kahng and M. M. Atalla in 1960. Even so commercial production had to await the achievement of really clean oxide films and MOS transistors only became available about 1963. In 1963 only six companies were producing FETs of one form or another; three in Europe (Ferranti, Philips, Sesco) and three in the US (Amelco, Crystalonics, TI). The next year the list had risen to fourteen.[26] Field-effect transistors offered designers some real advantages including low noise levels, a high input impedance and, important to those concerned with circuit compatibility, electron-tube-like performance.

By 1970 the MOS IC had established itself. It was slower than bipolar devices but required less power, it was smaller and so gave a greater packing density, and it was also simpler to make and therefore cheaper. Since two types could be made, p and n channel, it was natural to seek ways of using the two together. This goal was achieved in 1963 by F. M. Wanlass and C. T. Sah, who filed for a U.S. patent. The result, known as complementary-MOS or CMOS, had the advantage of very low power levels.

Impact of Integrated Circuits

Integrated circuits have had an enormous impact on electronics and, indeed, through electronics on society. As yet they have not completed the permeation of society that may be expected from them. The progress of the micro-miniaturization of integrated circuits can be followed in several ways; through the increasing density of components per unit area and the consequent reduction of the cost of an 'electronic function,' through the rise in world sales of ICs, and through the ever widening applications being found for integrated circuits. Doubtless other methods of charting their progress can be suggested, for example the applications to computing, which are dealt with in the following chapter.

In the first twenty years of integrated circuits the number of components per circuit in the most advanced integrated circuits has virtually doubled every year, an achievement regarded by most people as astonishing. The trend has even been dignified with the title Moore's Law, named after G. E. Moore of Fairchild Semiconductor, who first noted it in 1964. By 1978 the number of components in the most complex IC was comparable with the number in the most complex electronic equipment of 1950 vintage and direct comparisons could be drawn between one of the earliest electronic digital computers, ENIAC, and a microcomputer consisting of a few ICs on one printed circuit board.[29] In twenty years the industry's most complex devices have changed from chips containing one component in 1959, to about 10 in 1964, about 1000 in 1969, about 32 000 by the mid seventies, and around 250 000 by the late seventies. Costs have also fallen dramatically, particularly when measured in terms of the cost of an electronic function, since the number of functions per chip has risen, as we have seen. The definition of an electronic function is a bit vague but can be regarded as one switching gate or bit in a digital IC, or one amplifying stage in a linear IC. If the random-access memory is taken as an example of the best that has been achieved then according to Noyce the cost per bit fell by an average of 35 % per year from 1970 to 1977.[29] In general, IC costs fell by around 28 % with each doubling of the industry's experience, a fairly standard figure compared to many industries if inflation is ignored.[29] The real drama lay in the fact that the semiconductor industry was doubling its experience almost every year through a stunning increase in complexity and sales of circuit functions, and therefore on a time scale the costs were tumbling. Circuits have gone from small-scale integration to medium-scale, from medium to large, and from large to very large, though the exact meanings of the terms are not precise. The industry may run out of adjectives if the trend continues, though some see signs that economic limits are being approached while others go on to research wafer-scale integration.

The increasing component density has of course been the product of improved technology as circuit designs, transistor design, and manufacturing techniques have changed. The diameter of the raw material, the silicon wafer, has increased from 1 to 4 in. (2.5 to 10 cm) and more as improvements have been made in growing bigger and more fault-free crystals. Diffusion furnaces have become larger and more accurate in the control of temperature and gas flow and have cost more, about $3000 per tube in 1966 to about $12 000 (£6000) in 1977.[19] Other equipment costs have also risen and will probably continue to rise as photolithographic techniques using light give way to electron beam systems, for example.

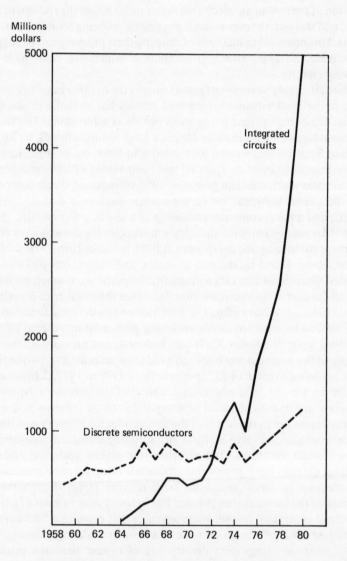

Figure 10.10 Estimated U.S. spending on discrete semiconductors and integrated circuits, 1959–1980 (based on 'Electronics' Annual Markets Forecast; only retrospective figures used)

Sales of ICs on first being marketed took off if anything even faster than those of the first transistors (Fig. 10.10). The computer manufacturers and the military were the first major users. By 1971 total world sales were around $1000 ($£500$) million,[15] and this figure continued to increase dramatically during the seventies except for temporary setbacks. Semiconductors had become big business. In the mid-1970s some 70 % were sold by U.S.-based companies, suitable reward for their pioneering work, though by the late 1970s the Japanese were steadily chipping away at this American lead. An interesting feature of semiconductor sales in America is the almost total absence from the list of the top companies of those who were the suppliers to the old electron-tube market. Of the ten leading American producers of vacuum tubes in 1955 only two (RCA and GE) were among the top ten semiconductor producers twenty years later.[29] In other countries, where the movement of personnel to form new companies was not as rife, the old manufacturers moved into semiconductors successfully.

While integrated circuits were still relatively primitive attention began to centre on the way electronics was increasingly pervading society. In the home this encroachment of electronics into new areas of use has meant that radio and television receivers have been joined by other electronic goods, and old items, such as cookers and washing machine controls, have been improved. Here again only the first ripples have been felt in the home. In industry, commerce, banking, hospitals, and so on, machine control and information gathering and processing have been dramatically improved by integrated electronics. And in some military fields electronics has become more important than the weapons. Integrated electronics was a vital step along the road taken by what has been called the electronics revolution, and it was a step which began with the invention and improvement of the bipolar transistor. In the 1970s the trend was for MOS integrated circuits to capture an increasing share of the market at the expense of bipolar circuits, probably because of their greater packing density and generally lower costs. By 1976 about six times as many electronic functions (i.e. uses of devices) were performed by MOS devices as by bipolar devices.[30] Such a trend could relegate bipolar transistors to a minor role in electronics.

Technical progress in ICs continues and fresh technologies, such as doping by ion implantation and electron-beam lithography instead of photo-lithography, might prove to be as important as the arrival of the planar technique in 1960. It is perhaps significant that some of the newer components, particularly those designed for the mass storage of data, have abandoned the transistor completely except for support circuitry. This step is seen in charge-coupled devices, which use long lines of MOS capacitors, and in magnetic bubble devices, which use cylindrical magnetic domains in ferromagnetic materials, and serves to remind us that one day the transistor era may be as dated as the valve era is today. Such a time seems to be a long way off. However, once the electronics industry again finds itself approaching a stalemate concerning complexity, such as was occurring in the late 1940s and late 1950s, the search will be on for improvement or replacement. Meanwhile design criteria are again changing to overcome complexity problems, this time to what may be called software electronics. The microprocessor points the way to standard yet versatile hardware where the job in hand is solved by programming rather than by circuit design. But even the software can be standardized and offered in hardware form.

The importance of microelectronics to electronics and to society has increasingly been recognized by all hands: the military, the business community, governments, and the public. Bad reporting to the public of 'silicone' chips that will change their lives and jobs can misinform and alarm instead of hold promise for the future. As is common with many great technological changes, the most important and most difficult task is not the technical one of how to do it, but the social one of how to use it well for the real benefit of humanity.

References

1. J. M. Early, *Proc. IRE* 50: 1006–1010, 1962.
2. E. W. Herold, *J. Franklin Inst.* 259 (No. 2): 87–106, February 1955.
3. G. L. Pearson and W. H. Brattain, *Proc. IRE* 43: 1794–1806, 1955.
4. E. E. Loebner, *IEEE Trans.* ED-23: 675–699, 1976.
5. E. Braun and S. MacDonald, *Revolution in Miniature*, Cambridge University Press, London, 1978.
6. E. E. Loebner, *Physics in Technology* 5: 147–148, 1974.
7. C. Weiner, *IEEE Spectrum* 10 (No. 1): 24–33, January 1973.
8. R. L. Petritz, *Proc. IRE* 50: 1025–1038, 1962.
9. W. B. Harris, *Fortune*, 135–138, 286–290, May 1957.
10. P. C. Mabon, *Mission Communications: The Story of Bell Laboratories*, Bell Laboratories, New Jersey, 1975.
11. Special Report, 'The transistor: Two decades of progress,' *Electronics* 41: 19 February 1968.
12. C. Freeman, *The Economics of Industrial Innovation*, Penguin, London, 1974.
13. W. O. Swinyard, *Proc. IRE* 50: 793–798, 1962.
14. S. F. Danko, *Proc. IRE* 50: 937–945, 1962.
15. J. S. Kilby, *IEEE Trans.* ED-23: 648–654, 1976.
16. H. Johnson, U.S. Patent 2 816 228, 10 December 1957.
17. M. F. Wolff, *IEEE Spectrum* 13 (No. 8): 45–53, August 1976.
18. J. S. Kilby, U.S. Patent 3 138 743, 23 June 1964.
19. J. G. Linvill and C. L. Hogan, *Science* 195: 1107–1113, 1977.
20. W. B. Harris, *Fortune*, 136–143, 216–226, April 1957.
21. S. Ramo, *Proc. IRE* 50: 1237–1241, 1962.
22. *Electronics* 36: 67–72, 4 January 1963.
23. *Electronics* 35: 52, 5 January 1962.

24. D. Kahng, *IEEE Trans.* ED-23: 655–657, 1976.
25. J. M. Cohen, *Electronics* 41: 120–123, 19 February 1968.
26. J. Eimbinder, *Electronics* 37: 46–49, 30 November 1964.
27. G. W. Dummer, *Proc. IEEE* 52: 1412–1425, 1964.
28. Private communication, Nippon Electric Co. Ltd., 1979.
29. R. N. Noyce, *Scientific American* 237: 63–69, September 1977.
30. G. E. Moore, Inst. Physics, Conf. Series No. 40, *Solid State Devices*, 1–6, 1977.

11 COMPUTERS

The definition of a computer given by Webster's dictionary was changed in 1955 from 'one who performs a computation,' to 'one or that which performs a computation.' The addition of three words reflected a change in mankind's methods of manipulating numbers that was to have many widespread yet unforeseen effects. The human computer, and his mechanical calculator, had been joined by the electronic computer.

Partly a product of the military demands of World War II and partly a result of the continuing advance of electronics, electronic computers have given us undreamt of powers for 'number crunching' and logical manipulation. Starting with the physically huge machines of the postwar era, a few of which were expected by some to fulfil the world's needs, computers have shrunk in size, grown in power, and proliferated so much that the household dog may soon lose its status as man's best friend.

From ancient times methods have been sought to mechanize the processes of arithmetic. One of the oldest answers, the abacus, is still with us, as any visitor to the Far East will quickly discover. Like the wheel, the abacus is one of the outstanding success stories of all time; and it is cheap. It has been in general use for 3000 years or more. In the West arithmetic was made easier by the introduction of the Arabic number system (1, 2, 3, etc.) some time around 1000 A.D. It slowly replaced the clumsy Roman one, although Roman numerals remained in general use for many centuries. Arabic numerals originally came from India and it was the Indians too who invented the notation for zero, "one of the greatest cultural achievements of all time."[1] The decimal point is said to date from 1492.

Even though arithmetic was becoming easier thanks to a better notation it was still far from easy, as Samuel Pepys found when working as a senior civil servant in England. In 1662, at the age of 30, he recorded in his famous diary his efforts to learn mathematics, "my first attempt being to learn the multiplication-table." However, help was at hand. The Scottish mathematician John Napier had published his tables of logarithms in 1614 and seven years later an Englishman, William Oughtred, invented the slide rule, only recently displaced by the electronic calculator. Calculating machines from Pascal and Leibniz followed later (Table 11.1) and were the crucial steps that eventually yielded the wide range of mechanical calculators used before the birth of the electronic variety. Precision mechanical engineering had progressed sufficiently for volume production of mechanical calculators to begin early in the 19th century, and by the end of that century Herman Hollerith in America had developed the punched-card system for use in data processing.

Table 11.1 Mechanical Calculators (sources: Refs. 1, 3, 4, 5)

	Pebbles as counters—Ancient civilizations.
c. 1000 BC	Abacus
1617	John Napier, Napier's bones (multiplication aid)
1621	William Oughtred, slide rule
1623	William Schickard, mechanical calculator ($+$, $-$), destroyed by fire and project abandoned
1642	Blaise Pascal, first practical calculator ($+$, $-$), over 50 built. Wheels and gears
1671	Gottfried Leibniz, reliable calculator ($+$, $-$, $\times$, $\div$)
c. 1820	Charles Thomas de Colmar (Alsace), commercial production, 1500 machines built over 60 years
1875	F. J. Baldwin, variant of Leibniz wheel. Manufacture begun by W. T. Odhner. Odhner-type machines were made in large numbers
1885	William S. Burroughs, printing calculator, key-set for numbers, handle for operation
1886	Dorr E. Felt, comptometer, keyboard machine
1890	Herman Hollerith, punched card machines, 45 columns, round holes (80 columns, rectangular holes, adopted in 1928)

Note: calculus is Latin for pebble.

Besides performing arithmetic operations computers must store information. The use of punched cards for the storage of information for control purposes dates back to 1725, when they were first used to achieve the automatic weaving of patterns in the French silk industry, just four years before Stephen Gray discovered electrical conduction. One improved version, the Jacquard loom of 1808, was especially successful and more than 10 000 were operating by the end of its first decade.[2] J. M. Jacquard's punched cards may have contributed to Hollerith's ideas, though he is also said to have been inspired by a conductor punching a railway ticket. Hollerith contracted to supply a tabulating machine, with data stored on punched cards, to help with the U.S. census of 1890 (Fig. 11.1). It was a great success. Later an adding facility was designed and automatic card handling machines developed. A company was formed in 1896 and merged with two others in 1911 to form the Computing-Tabulating-Recording Company, registered in New York State. A change of name in 1924 produced the now more familiar International Business Machines Corporation (IBM), a company that eventually came to dominate the electronic computer industry.

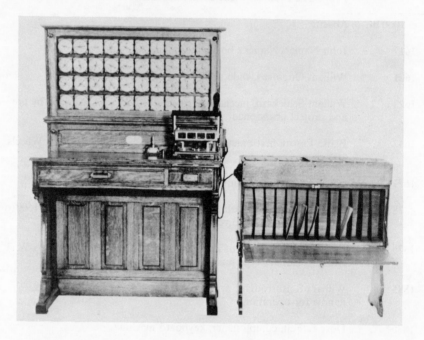

Figure 11.1 Hollerith tabulating machine, 1890

Mechanical calculators and tabulators were extremely useful machines, but they were not computers. At the very least it must be possible to 'program' a computer to carry out various series of arithmetical operations without human intervention, apart from the programming. It should also be able to store its program and data and manipulate both.

The first person to come up with a design for an automatic calculating machine was Charles Babbage in England. In 1822 he began by demonstrating a small 'difference engine' that could calculate difference tables for quadratic functions (Table 11.2). Government support was secured for the design and construction of a larger machine to handle sixth-order polynomials. Though such a machine was never built the published details inspired the Swedish

Table 11.2 A Table of Finite Differences

$$y = 2x^2 + x + 4$$

x	y	D_1	D_2
0	4		
		3	
1	7		4
		7	
2	14		4
		11	
3	25		4
		15	
4	40		4
		19	
5	59		4
		23	
6	82		

Note: $D_1 = y_{(x = n)} - y_{(x = n - 1)}$, i.e., the difference between successive values of y.
D_2 is the difference between successive values of D_1. For a second-order
polynomial, D_2 is a constant. Knowledge of this constant enables values of y
to be calculated by use of addition only and errors can quickly be detected.

engineer George Scheutz to experiment. With his son, he built a successful
machine in 1843 that could handle fourth-order polynomials. By that time
Babbage was at work designing an 'analytical engine,' the first design for a
genuine computer. It was to be a full-scale, general-purpose mechanical
computer with a memory, arithmetic unit, Jacquard-type punched cards for
input and output, and card-controlled programs that allowed iteration and
conditional branching. The memory was intended to store 1000 decimal
numbers, each with up to 50 digits. This machine was never built either, partly
because of Babbage's hunt for perfection and partly because of the limitations
of the mechanical engineering of the day; but the design incorporated the
major features of today's digital computers except that they were to be
achieved mechanically instead of by electronics. The design, and the incredible
vision behind it, still stand as a remarkable tribute to a man who was about a
century ahead of his time. Lady Lovelace, daughter of Lord Byron, worked
closely with Babbage and today is honoured with the title of the world's first
computer programmer.

Although Babbage died a disappointed man, the world has now recognized
the intellectual achievement of the man who designed the first program-
controlled general-purpose digital computer. Randell has said of this
achievement, made a century before the first electronic computers were built:
"The earliest program-controlled computers, namely those of Zuse and Aiken
in the early 1940s, were conceptually hardly a match for Babbage's engine."[4]

After Babbage at least three more attempts were made to design a mechanical computer.[4] The last attempt was probably that by Louis Couffignal of France in the 1930s. He planned to use the binary number system, based on radix 2, as an alternative to decimal but, like the others, his machine was never completed. With that the dream of a Babbage-like mechanical computer was almost at an end.

The Development Period

By the time of World War II the idea of a machine to perform calculations automatically was an idea whose time had come. Vast amounts of repetitive calculation were needed to compile ballistic tables for shells and bombs. Previously a handful of people had dreamed of machines to produce or check mathematical tables; now many more wanted such machines for military as well as civilian use. Previously the technology of the day had posed serious problems; now electrical engineering and electronics were sufficiently advanced to be used. Previously, one or two people had attempted the near impossible; now teams of people attempted the possible. In Germany, Britain, and the USA various machines were built, some to perform special types of calculation or logical processing, others to perform more general tasks. Some were largely mechanical; several contained electromagnetic relays. Then came the first electronic machines, which depended on thermionic tubes. At first they were automatic calculators rather than computers; that is, they could not store and manipulate their own programs. All these machines, and the computers that followed shortly after World War II, belong to what is called the Development Period of modern computers. From that period issued the embryo computer industry, which was to progress through successive 'generations' of computers. Details of some of the more important machines of the development period are given in Table 11.3.

During the development period the fundamentals of digital electronic computers, regarded for a time as scientific or engineering oddities, became established. General and special-purpose machines became operational. Thermionic valves proved to be reliable enough to be used in large quantities and operated about 1000 times faster than the relays they displaced. Memories, or stores, were developed along various lines and the stored-program concept became established. By 1950 general-purpose stored-program digital computers were a fact of life and commercial exploitation could begin.

For our purposes we shall consider only digital computers because of their far greater importance and impact compared to those of analogue or hybrid (analogue-digital) machines. Analogue machines, which set up some electrical analogy of the problem to be solved, continue the tradition laid down by instruments such as planimeters and slide rules; whereas digital computers are the successors to the abacus and Babbage's Analytical Engine. Developments that took place in Germany, the USA, and Britain will be considered.

Table 11.3 Some Development Period Computers (figures given may not be directly comparable and are given as an indication only)

Computer	Date Start–Operational	GP or SP	Radix	Relays	Valves	Stored Program	Memory Type	Memory Capacity	Word Length	Speed +	Speed −	Speed ×	Speed ÷	Remarks
Zuse, Z1	1934–1938	G	2	Mechanical		No	Mech.	16 words	24	1.0 s	1.0 s	5 s	5 s	Working for tests only, floating point
Zuse, Z2	1938–1939	G	2	Yes	No	No	Mech.	16 words	16	0.2 s	0.2 s	3 s	3 s	Electromechanical fixed point, working for tests only
Zuse, Z3	1939–1941	G	2	2600	No	No	Relay	64 words	22	1.0 s	1.0 s	4 s	4 s	Floating point keyboard/lamps I/O
Bell/Stibitz Model I Complex Number Calculator	1937–1940	S	2	450	No	No	Crossbar switches	10 registers						Remote access
Bell Model II Relay Interpolator	1940/41–1943	S	Bi-quin	440	No	No		6 registers						Self-checking arithmetic
Bell Model III Ballistic Computer	1942–1944	S	Bi-quin	1335	No	No		10 registers						100% self-checking

Table 11.3 (contd.)

Computer	Date Start–Operational	GP or SP	Radix	Relays	Valves	Stored Program	Memory Type	Memory Capacity	Word Length	Speed +	Speed −	Speed ×	Speed ÷	Remarks
Heath Robinson	1942–1942	S		Yes	30 to 80	No								Gifford line printer O/P
Colossus Mk 1	1943–1943	S	Bi-quin	Yes	1500	No								Specialized towards Boolean calculations
Colossus Mk 2	1944–1944	S		Yes	2500	No								Conditional branching
Harvard Mk I ASCC	1939–1944	G	10	2000–3000 wheels	No	No	Relay, tape, switches	72 accumulators 60 constants	24	0.3 s	0.3 s	6 s	11.4 s	Largely mechanical
IBM, Pluggable Sequence Relay Calculator	–1944			Yes	No	No								
ENIAC	1943–1946	G	10	1500	19000	No	Selector switch "PROM" Vacuum tube	3600 digits 200 digits	10	0.2 ms	0.2 ms	2.8 ms	26 ms	First GP electronic computer built
IBM, SSEC	1945–1948	G	2	21400	12500	No	Electromag. paper tape, electronic	150 words 20000 words 8 words	20	<1 ms	<1 ms	20 ms		

274

Name	Dates				Memory			Capacity	Word length					Remarks
Manchester University Mk I	1946–1948	G	2	Yes	Williams tube	500	No	32 words	32				1.2 ms	First GP electronic computer based on stored-program concept
Manchester University Enhanced Mk I	1948–1949	G	2	Yes	Williams tube, drum	1300	No	128 words, 1024 words	40	1.8 ms	1.8 ms	10 ms		First use of index registers
EDSAC	1946–1949	G	2	Yes	Delay line	3000	No	512 words	"35"	1.5 ms	1.5 ms	6 ms		First stored-program computer to offer a user service. Division by subroutine
EDVAC	1945–1951	G	2	Yes	Delay line	3600 to 5900	150	1024 words	44	0.05 ms	0.05 ms	2.1 ms	2.1 ms	
ACE Pilot	1945–1950	G	2	Yes	Delay line, drum (1954)	1081	No	361 words, 4096 words	32	0.54 ms	0.54 ms	2 ms	N.A.	
UNIVAC I	1947(?)–1951	G	2	Yes	Delay line, mag. tape	5400		1000 words	84	0.52 ms	0.52 ms	2.2 ms		
IAS	1946–1952	G	2	Yes	Williams tube	2300		1024	40	62 µs	62 µs	720+ µs	1100 µs	
Whirlwind	1947–1951	G	2	Yes	Storage tube, electrostatic	5000		1024 words, 4096 bits	16	22 µs	22 µs	37.5 µs	71 µs	About 11 000 crystal diodes, mag. core in 1953

Germany can claim the first general-purpose program-controlled computer (Z3), America the first general-purpose electronic computer (ENIAC), and Britain the first special-purpose electronic computer (Colossus) and the first practical stored-program computer (EDSAC).

Germany

The pioneering work on program-controlled computers in Germany was carried out largely by one man, Konrad Zuse, the son of a Berlin civil servant. At first he worked more or less alone and at home, more in the manner of a 19th century inventor than a 20th century engineer. Unlike the Babbage-type mechanical computers Zuse's design was based on the use of mechanical relays. He later explained that his object was to "develop a mechanical analogue for the electrical relay," and that mechanical switching element techniques resulted. Switching algebra could then be applied to the design of a calculating machine.[4] The first result was the Z1, a mechanical floating-point binary computer controlled by a program tape. Zuse designed it when he was twenty-six years old. Two years later, in 1938, it was built and working, though it proved to be unreliable in operation.

Work then began on a second machine, the Z2, an electromechanical model in which second-hand telephone relays and the Z1's mechanical store were used. The following year simple formulas were calculated and program control was demonstrated. More important was the work carried out on the Z3, an electromechanical relay computer that was the world's first general-purpose computer, predating the more famous American Harvard Mark I, which for many years was thought to have been the first. The Z3 was financed by the German Aeronautical Research Institute and constructed between 1939 and 1941.

The Z3 was program-controlled by a standard punched tape with eight bits used for each command. Its relay memory could store 64 words with a word length of 22 bits (sign, 7 bit exponent, and 14 bit mantissa). The floating-point binary arithmetic unit had built-in operators for the four basic mathematical functions, and for square-roots and multiplication by five fixed constants. Input was via a keyboard and allowed four decimal places; output was via a lamp display panel. About 2600 relays were used in its construction. The dataflow of the Z3 is illustrated in Figure 11.2. After the Z3 came the Z4, an improved machine with a 32-bit wordlength and the only Zuse machine to survive the war. The Z4 was completed after the war and rented to the Technical University in Zurich, where it served from 1950 to 1955 before being transferred to St. Louis.

Zuse was helped in the construction of the Z1 by Helmut Schreyer. In 1939 while Zuse was working on his relay computer and had been called up for military service (from which he was eventually released to begin work on the Z3), Schreyer set about the design of electronic computer circuits containing

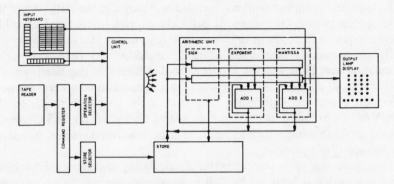

Figure 11.2 Counting circuits of Z3 (1941) (after Ref. 4)

thermionic valves and neon tubes that were meant to simulate the action of a relay, only faster. According to Zuse, writing in 1962, "the valves had the function of the coil of an electro-magnetic relay, and the neon tubes the function of contacts."[4] The neon tubes with their on-off facility could also be used for storage purposes. Schreyer constructed a 100-valve machine which, like the Z machines, was destroyed in the war. A proposal for a 1500-valve machine was abandoned after the idea was rejected by the German government.

Besides the general-purpose computers Zuse also built two special-purpose relay machines to aid the design of aircraft wings, and a small logic computer "for operations in propositional calculus."

The work carried out in Germany by Zuse and Schreyer has several parallels with that performed in America a little later. First, a machine largely or entirely mechanical in design was built, closely followed by machines that contained electromagnetic relays to increase the operating speed. A hesitant start was also made on electronic machines. Others in Germany pursued work on electronic accounting machines and magnetic storage systems, but their work was not combined with that of Zuse. When work resumed in Germany in 1949 Zuse again used relays since "electronics were [sic] still prone to uncertainties."[4] The Z4 was followed by a faster Z5, which represented the end of the development series that had began with the Z3. Production runs instead of isolated machines came later. At least thirty of the Z11 Relay Computer were put into use.

United States

In many respects the USA is the home of the electronic digital computer. It was there that the very important stored-program concept originated in its most detailed form, that electronic vacuum tubes were first used on a vast scale, and that the magnetic ferrite core memory was developed.

At least four groups of workers operated in America in the early part of the development period. One group, at Bell Laboratories, investigated Zuse-like relay computers from about 1937 onwards. Another, at Harvard University, built the previously mentioned Mark I, or ASCC. At IBM more relay calculators were built, and the Moore School of Electrical Engineering at the University of Pennsylvania constructed the first operational general-purpose electronic computer. During the later part of the development period the UNIVAC I was hailed as the first commercial computer, the IAS computer helped establish basic principles, and the MIT Whirlwind set new standards for the speed of operation.

During the war years several analogue calculating or computing machines were designed and built at the Bell Laboratories, mainly for military fire control purposes; but work also began on what became a series of digital relay machines that progressed from calculator to computer.[4,7] All the digital machines were designed to aid the work on anti-aircraft fire-control equipment, though postwar work produced more general-purpose machines. The Bell engineers had extensive experience in telephone switching networks on which to base their work.

Bell's involvement with digital calculators began in 1937 when G. R. Stibitz sketched out a design for a machine to perform complex number arithmetic to help in the design of filter networks. This Model I, or Complex Number Calculator, was built under the direction of S. B. Williams as a feasibility study and became the first in a series of six. From January 1940 it was in daily use for nearly ten years and has a special claim to fame as probably the first machine to offer users remote access via teletypewriters.

Meanwhile Stibitz studied ideas for automatic sequencing and error-detection codes. Other special-purpose wartime machines followed and used paper tape for program storage. The number system used was bi-quinary, an unusual system in which one decimal digit is represented by two digits, one for 00 or 5 (bi) and one for the numbers 0 to 4 (quinary); somewhat like the abacus. Model III was the big advance as far as power and size were concerned; it came much closer to being an automatic computer than the previous two. It was especially remarkable as it featured 100 % self-checking of all operations. Paper tape could be moved backwards as well as forwards to hunt for instructions, and as this action was independent of the calculation, faster operation was achieved. These wartime models proved to be very reliable and gave several years of service during and after the war, even after faster electronic computers had become available.

The postwar machines (Model V, and a simplified version, Model VI) were full program-controlled general-purpose relay computers. Two Model V machines were built; each weighed about ten tons. One of them continued in use at least until 1964, probably a record for a computer built immediately after World War II. Each machine used 9000 relays and was designed to handle up to an astonishing fifty-five pieces of teletype equipment, though in practice far fewer were used.

Much more famous than the Bell machines is the general-purpose computer known as the Harvard Mark 1, or the Automatic Sequence Controlled Calculator (ASCC). One of the many machines of the development period designed by university staff, the ASCC was supported by funds from the IBM advertising budget and built at IBM's Endicott Laboratory. IBM personnel are acknowledged as co-inventors and standard IBM components, as well as special mechanisms, were used in its construction. For a long time it was thought to be the first general-purpose computer to have been built, hence its fame. Its principal designer, Howard Aiken, was well aware of Babbage's work; the ASCC machine, which was largely mechanical with electrically driven shafts, gears, chains, and wheels, was pronounced by others to be Babbage's dream come true. Like some others of the period it was a massive machine, over 50 feet long and 8 feet high. It was considerably bigger than the Bell machines and the Z3 and was first demonstrated in January 1943, about two years after the German Z3. It remained in active use at Harvard until 1959.

Aiken first visualized the machine in 1937 as "a switchboard on which were to be mounted various pieces of calculating machine apparatus," each panel of which was dedicated to a specific operation.[4] The completed machine came close to the original idea. It usually operated on 23 significant decimal digits, plus one sign digit, but could be doubled up to handle 46. The calculator had 60 constant registers, each consisting of 24 ten pole hand-set switches; and 72 accumulator registers able to perform addition and subtraction and made up from 24 ten-segment electromechanical counter wheels. There was also a central multiplication and division unit, and sets of electromechanical mathematical tables. The program was presented on 24-hole paper tape and proceeded in sequence. Output was via card punches and two electric typewriters. Much of the machine was driven and synchronized by a chain-and-gear-connected mechanical system driven by a shaft that ran nearly its full length, powered by a 5hp motor. It all sounds a far cry from the later electronic computers. ASCC was first used to calculate ballistic and other military tables. After the war it spent much of its time calculating mathematical tables, paralleling Babbage's hopes for his analytical engine.

After ASCC, the Harvard Mark I, Aiken and IBM went their separate ways. Aiken turned to relay calculators and built the Mark II, said to have used some 13 000 relays. This device was completed in 1947 and was followed by other machines.

Meanwhile at IBM attention was turned to a pluggable sequence relay calculator. More important, though, was the IBM Selective Sequence Electronic Calculator (SSEC), whose design began in 1945 and which was unveiled to admiring gazes in January 1948 (Fig. 11.3). Based on a combination of electromagnetic relays and vacuum tubes, the SSEC was able to "compute detailed instructions as it goes along from general outlines presented to it."[4] That meant that the sequence being followed could be changed by the machine (a feature reflected in its name, Selective Sequence,

Figure 11.3 Selective Sequence Electronic Calculator (SSEC), 1948, marked IBM's move from electromechanical to electronic machines

which led to IBM holding some important patents). This was the stored-program concept in a primitive form, though the SSEC is not regarded as a stored-program computer in the nature of the later EDVAC and EDSAC.

The SSEC helped move IBM away from the adding wheels of the ASCC to the electronic vacuum tubes of the future. Yet it was not the first general-purpose electronic computer. That honour goes to the Electronic Numerical Integrator and Calculator, more affectionately known as ENIAC, which was built by John Mauchly and J. P. Eckert at the Moore School of Electrical Engineering of the University of Pennsylvania.

In his book on the origins of digital computers, Randell points out that the first known attempt to build an electronic digital calculator was made in the mid-1930s at Iowa State College by J. V. Atanasoff.[4] Small capacitors were used for binary storage and a prototype computing element was demonstrated late in 1939. Work on a full machine remained uncompleted after the leading lights left Iowa for jobs elsewhere. A third of a century later, in October 1973, in a $200 million suit between Sperry-Rand and Honeywell, the Eckert-Mauchly patent was ruled to be invalid. By the end of the 135-day trial Atanasoff had been legally credited with the invention of the electronic digital computer.[8]

The first digital electronic circuit was almost certainly the 'one-stroke relay' or flip-flop invented by W. H. Eccles and F. W. Jordan and patented in 1918 by the British Admiralty (Fig. 11.4). It is still an essential computer element. In the 1930s counting circuits were designed, usually with an eye to counting pulses from Geiger-Müller tubes for use by the growing band of nuclear physicists. Once counting circuits had been made, adding circuits could not be far behind, and if a circuit can add then it has the potential for subtraction, multiplication, and division by complementary arithmetic and repeated addition. Also in the mid-1930s E. W. Phillips in Britain advocated the use of binary arithmetic with octal (base 8) notation, a system now long familiar to

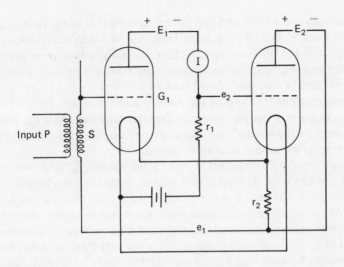

Figure 11.4 Eccles–Jordan flip-flop (1918) ('Radio Review', Vol. 1, October 1919, p. 145)

any computer engineer. Phillips wanted the "whole civilised world" to "discontinue counting in tens and to count in eights instead,"[4] but limited his immediate evangelistic efforts to the scientific and business communities where machine calculators were more common. He was quick to point out historical uses of binary and 'octonary' systems. In Germany, Zuse was already using base two. While mentioning such basics we may note that a Bell Laboratories book has claimed credit for the basic electronic logic gates for A. W. Horton Jr. (OR gate, 1939) and W. H. T. Holden (AND gate, 1941).[7]

Despite the efforts of Atanasoff in the USA and Schreyer in Germany to construct electronic computers, ENIAC remains as the first operational general-purpose electronic computer, though some special-purpose electronic code-breaking machines were built in Britain shortly before ENIAC.

ENIAC was another monster machine. It was constructed between 1943 and 1946; contained some 1500 relays, 19 000 vacuum tubes, and 70 000 resistors; and consumed around 150 to 200 kilowatts. Some stories claim that the lights of Pennsylvania dimmed when it was switched on.

The parentage (or perhaps we should say the grand-parentage) of ENIAC can be traced back to the electromechanical analogue calculators or differential analyzers of the early 1930s (Vannevar Bush, MIT, 1930) and to the ideas of Atanasoff. By 1942 the Moore School engineers were deeply involved in the calculation of ballistic tables for the U.S. Army's Ballistic Research Laboratory and were using a differential analyzer for this work. Mauchly was one of the people who had worked with the analogue analyzer but he was also aware of the possibilities for digital machines, since he had considered computing from the point of view of weather prediction, had tinkered with electronics, and had visited Atanasoff at Iowa. He also knew of Stibitz's work at Bell, though apparently not of Aiken's at Harvard nor of Babbage's in the 19th century. Eckert meanwhile had been involved in replacing some of the analogue machine's mechanical amplifiers with partially electronic ones. From this general background was born the idea of a digital electronic machine to compute the Army's ballistic tables. Indeed ENIAC's original name was the Electronic Difference Analyzer. As time passed the design outgrew the original concepts until the completed machine had become the first general-purpose electronic computer.

Though it was not completed in time for war service, ENIAC went to work solving problems in ballistics and atomic physics. Development of the machine continued until it was retired in 1955, after ten years of service.

ENIAC was about one thousand times faster than the relay machines of the same period and finally proved that thermionic valves were reliable enough to be used in really large-scale projects. With vacuum-tube reliability as it then was, "no one believed that a machine with hundreds, let alone thousands, of vacuum tubes would ever be reliable enough."[9] It was thought the designers would be lucky if it worked for ten minutes without a failure. By careful circuit design, for which Eckert took the major credit, and by operating well below the rated values of components, the designers proved this to be untrue. Tube life of 25 000 to 50 000 hours was aimed for, and "pretty much" achieved.[9] After ENIAC, in the long term, the relay machines were doomed to extinction.

ENIAC's major drawback was its method of programming. To set up a new program meant throwing a vast number of switches and plugging and unplugging a large number of wired connections, which could take days. The time involved, and the detailed knowledge of the machine required to do all that successfully, made programming (or coding) a major stumbling block. The ENIAC team recognized the problem and came up with the answer—the concept of the stored program—even before ENIAC was completed, and design work began on its successor, the EDVAC.

The stored-program concept is usually credited to John von Neumann and to Mauchly and Eckert, and in general terms to the ENIAC group.[4, 10] Von Neumann was associated with the Institute of Advanced Study (IAS) at Princeton University and was one of the leading mathematicians of his day. He became involved with the ENIAC group in 1944, by which time both sides had already come up with the idea of storing the program, just as data would be stored. The first document to discuss the idea was von Neumann's draft report on EDVAC, dated 30 June 1945.[4]

The EDVAC design, therefore, was probably the first to incorporate a stored program and since the design received wide publicity, this fundamental idea spread. But the EDVAC team broke up at the end of the war and work was delayed. Eckert and Mauchly left to start their own computer company and went on to design and build the UNIVAC I as a commercial venture. Remington-Rand acquired this company, which became the Univac division and led the American market for several years. By the time EDVAC was completed a British computer, the Cambridge University EDSAC, had claimed the status of being the first practical stored-program computer. At the same time John von Neumann and his colleagues were designing the IAS computer which, though it was completed after both EDSAC and EDVAC, proved to be very influential on future designs. The IAS progress reports have been described as "textbooks on logical design and programming,"[4] and as "the most important tutorial documents in the early development of electronic computers."[10] The IAS design was quickly repeated in other computers, one of which is said to have been used in producing and testing the first hydrogen bomb. This machine had possibly the most appropriate acronym in the history of computers: MANIAC.[11]

Another important American computer of this period was the Whirlwind 1, an appropriate name for a fast computer that was probably the first to aim at real-time applications. Work began in 1947 at MIT and, as with EDVAC and the IAS computers, documents and diagrams achieved wide circulation and served as an education medium. At first an electrostatic store was used but in 1953 Whirlwind got one of the first magnetic-core stores.

Britain

British work on computers began during World War II, when some special-purpose electronic machines were constructed to aid intelligence operations at Bletchley aimed at breaking German codes and ciphers. Even now details of these machines are sadly lacking, although they have been claimed as the first electronic computers.[12]

The first design was for an electromechanical machine containing relays and 30–60 thermionic valves. The Royal Navy's women operators (WRNS, or Wrens) nicknamed it the Heath Robinson (after the famous cartoonist's weird mechanical 'inventions') because of its whirling tapes and pulleys, which carried the punched-paper tape input.[12,13] It was followed by a design for a Super Robinson, but far more important was the resultant electronic machine; the Colossus Mark I. The specifications for Colossus, probably the first special-purpose electronic computer (Fig. 11.5), were drawn up by M. H. A. Newman, a Cambridge mathematician who had moved to Bletchley and who had also been responsible for starting work on the Heath Robinson. The engineering design was the responsibility of T. H. Flowers of the Post Office Research Station in London, where the machine was built between February and December 1943. Colossus Mark I used at least 1500 thyratron triode valves, far more than any other piece of electronics built up to that time, and just about the same number that Schreyer had suggested in the proposal that had been rejected by the German government. Flowers overcame objections that valves were unreliable by pointing to Post Office and BBC experience that they were very reliable provided they were never switched off.[12] A Mark II version of Colossus used 2500 valves. Ten were ordered and the first became

Figure 11.5 Colossus, the first special-purpose electronic computer; see also Fig. 1.4

operational in July 1944. One anecdote relates the reply from a Ministry of Supply official to yet one more secrecy-shrouded request from the Post Office team for another couple of thousand valves, "What the bloody hell are you doing with these things," he asked, "shooting them at the Jerries?"[33]

No two Colossi seem to have been exactly the same, though they all used photoelectric paper tape readers which operated at an incredible 5000 characters per second. Output was somewhat slower, 15 letters per second on an electric line typewriter for the Mark II version. Their arithmetic power is thought to have been somewhat limited, although their logic capabilities were extensive.[4]

The ultimate fate of the Colossi seems to be shrouded in even more mystery than the work they performed. Their importance for cryptoanalysis in the later part of the war is thought to have been enormous and they certainly influenced at least some of the British postwar work on computers. Like the ENIAC (though on a smaller scale) they showed that use of a large array of electronics was practical, but they were also important in providing a breeding ground for computer designers. Special mention should be made of the British mathematician A. M. Turing, who published his theoretical ideas for automatic computers in 1936. Turing had been involved in the design of the Heath Robinson and influenced many at Bletchley who went on to postwar computer work.

As in the USA subsequent work took place in several centres in Britain, including Manchester and Cambridge Universities, and the National Physical Laboratory (NPL).

Manchester University was fortunate in acquiring the services of F. C. Williams as professor of electrical engineering and M. H. A. Newman as professor of pure mathematics. Williams had been working at the Telecommunications Research Establishment on a technique for using a cathode-ray tube (CRT) as a storage device for radar signals, and Newman had at one time led the Colossus team. Both men took others to Manchester with them. Turing also joined Manchester later, after he left NPL.

Williams, assisted by Tom Kilburn, developed the CRT electrostatic storage system into the first practical electrostatic random-access store for computers. The technique was to store data as dots and dashes on the screen and to read it out via capacitive coupling to a metal plate on the glass face of the tube. A single electron beam could be used to refresh the data and to assist in reading and writing. A small prototype computer, yet another Mark 1, was built primarily to test the Williams Tube memory. As the program was stored the Manchester Mark 1 is claimed as the first (albeit experimental) stored-program computer, although EDSAC was the first really practical example. The Mark 1 was soon subject to extensive development and in 1949 two Enhanced Mark 1's were commissioned. During the first three years of the design period the team had averaged around four people, plus two technicians, and taken out some thirty-four patents.[14] There is a story that when Williams visited IBM in 1949 he was fascinated by the way desks and doors were adorned with the word "Think." When asked how "two men and a dog" had been able to achieve such success at Manchester, he replied tongue-in-cheek that it was because they had not stopped to think too much.[14]

From November 1948 close co-operation began with the Ferranti company in Manchester that eventually led to the commercially available Ferranti Mark 1 and Mark 1 Star computers as production versions of the university machine. Kilburn went on to design a Mark 2 computer from which Ferranti produced another commercial machine known as Mercury. Other, faster machines came along later and continued the cooperation between Manchester University and Ferranti. Probably the best known example was ATLAS.

At Cambridge University, mercury acoustic delay lines were used for storage and were the only real alternative to the Williams Tube. Unlike the Williams Tube they gave serial access. EDSAC, the Cambridge machine, was built by a team led by M. V. Wilkes, who had attended lectures at Pennsylvania on the EDVAC machine. The first program was run in May 1949. EDSAC was relatively small when compared with some of the American monsters. Besides being the first practical stored-program computer on the EDVAC style, it was also notable for having a wired set of initial orders, a sort of basic assembler and loader. A year later the National Physical Laboratory's own experimental computer, ACE, was completed. Other experimental computers were also built in Britain shortly after the war by various government research establishments and educational institutes. Birkbeck College in London, and Manchester University deserve mention for their early use of magnetic-drum memories.

By 1950 the development period was ending. It began with work on mechanical, electromechanical, and relay machines and ended with the basic principles of general-purpose stored-program electronic computers well established. The two leading countries, USA and Britain, were about to begin commercial exploitation, more or less from the same starting point. The most basic concepts of computer hardware had evolved and vacuum electronics had proved itself. However, memory techniques had not stabilized. Though delay lines and Williams-type storage tubes were established, the magnetic-core store was yet to make its debut.

The Computer Industry

When the computer industry began the thermionic valve was the basic active circuit component available, backed up with resistors, capacitors, soldered joints and so on, and, perhaps most important, the first germanium diodes. A typical computer might have 1000 valves and 50 000 diodes.[10] Input and output devices included teletypewriters, punched cards, paper tape, lamp displays, and (with UNIVAC I) magnetic tape. Large computers were built by many electrical companies and by companies that had been involved in the older mechanical tabulating industry. Inevitably some got their fingers burned. Parallel to the demand for production-line computers, demand grew for a small number of supercomputers, machines that would be orders of magnitude more powerful or faster than those commercially available. On the commercial side manufacturers came to categorize their machines into generations, First-generation computers were built with vacuum tubes, second-generation computers with transistors. The third generation was not quite so clearly defined. Some companies defined it as machines based on integrated-circuits; others said the performance of the machine provided the dividing line. In general the two criteria often amounted to the same thing and any machine on the market after 1965 can be regarded as third generation.[10] The criterion of judgment is even more fluid with fourth-generation definitions, again split along design and performance considerations versus the scale of integration. Some claim microprocessor-based machines as the fourth generation; others see the generation concept as being completely obsolete.

Alternative methods of classifying computers are by application or by size. Especially in the early years computers were designed either for business or for scientific use. Business computers had more input and output facilities, whereas scientific ones had more processing power. When classified by size computers came large, medium, or small, rather like soap powder. Additional classes have been added at each end: supercomputers at one extreme, minis and micros at the other. In less than thirty years the industry had progressed from offering only a few machines to having a vast range, a range that stretched from machines only a government could afford to those children played with at home.

First Generation

The first computer built by the Eckert-Mauchly Computer Corporation was the BINAC. It was not a success but it was followed by the Universal Automatic Computer, or UNIVAC, which was an outstanding achievement and remained as probably the best large-scale computer for data processing for nearly five years. Its development proved to be much more expensive than originally expected which, together with the death of Eckert-Mauchly's chief financial backer, led to financial problems that were not resolved until the company became a division of Remington Rand.

The UNIVAC I computer, the first of which was delivered to the U.S. Bureau of Census in June 1951, introduced magnetic tape recording as a computer input and storage device. Tape recording had progressed slowly from Poulsen's wire recorders in the early 20th century and steel-tape recording was used in broadcasting before World War II. The big development came with the use of magnetic iron oxides on a flexible tape, pioneered in Germany during the war, which appeared as a revelation to the Allies in 1945.[15] UNIVAC I also used a mercury delay line memory. Other memories of the period used electrostatic storage as in the Williams Tube, or magnetic drums. By about 1953 the magnetic core random-access memory threatened to make delay lines and the Williams Tube obsolete.

Commercial machines began to appear in Britain also in 1951 and were based on the experimental university computers. Ferranti produced the Mark 1 modelled on the Manchester University machine; the London bakery firm, Lyons, produced LEO, the first machine designed solely for business work, which was based on the Cambridge EDSAC. The Lyons firm moved into the computer field after deciding it needed a computer but found none was available commercially. English Electric stepped into the field with a computer based on the NPL's ACE Pilot model, and the British Tabulating Machine Company with one derived from the Birkbeck College computer. With a bakery, office machine manufacturers, and electrical engineering companies beginning manufacture of computers, the British computer industry was set to grow. Other companies followed: EMI, Elliott Automation, GEC, Plessey, and Powers Samas Accounting Machines Ltd. By the end of the 1950s, with annual sales around £10 million, there were too many companies chasing too specialized a market. Mergers and takeovers followed and some ungainly names, such as English Electric-Leo-Marconi Computers, resulted. By the late 1960s the field had narrowed to two major concerns, English Electric Computers Ltd. and International Computers and Tabulators Ltd., plus Plessey. With help from the Ministry of Technology the three combined to form International Computers Ltd. (ICL) in 1968. The rationalization of the British computer industry had produced a company which it was hoped would combat the success of IBM as the world leader (Fig. 11.6).

In America, IBM entered the electronic calculating arena in 1946 with the model 603 and in 1948 with the more powerful 604, a calculator that combined

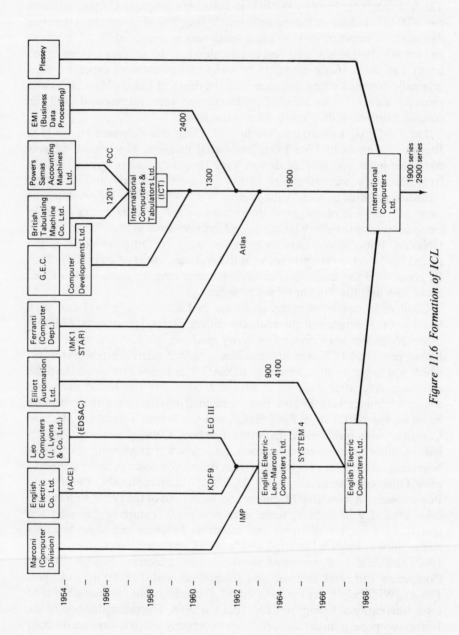

Figure 11.6 Formation of ICL

electronics with the company's previous expertise on card-punch machines. Over 5000 of the 604s were delivered in the next ten years.[16] However, these and similar subsequent machines were not computers. With the start of the Korean War in 1950 work began on the Defense Calculator, later known as the IBM 701. This was the first IBM computer; the first model was delivered in 1953. It was a large-scale machine designed for scientific use. A 36-bit, 2048-word Williams Tube memory was used and supported by magnetic tape and magnetic drum storage. The machine was faster than the UNIVAC. Another computer, the IBM 702, was announced for the business sector but was generally regarded as inferior to the UNIVAC I and was fairly quickly withdrawn from the market. According to Rosen[10] this episode presented IBM with a crisis to which, in characteristic fashion, they responded quickly and with vigour. The result was the faster IBM 705 in which a core memory replaced the electrostatic storage of the 702. A replacement was also announced for the scientific 701. The 705, a large-scale vacuum-tube computer, was the computer that established IBM as the leader in the field. Univac responded with the UNIVAC II, also with a core memory, but deliveries were delayed for various reasons and IBM gained a two-year lead from which it has never looked back. An intermediate-size computer, the IBM 650, was announced in 1953, a year before the 705. Over 1500 were eventually installed.[16] By the end of the first-generation period, about 1959, IBM had achieved a dominant position throughout the computer industry.

IBM and Univac were not the only American manufacturers. Raytheon Corporation was active in the early days and in 1954 set up a joint company, Datamatic, with Honeywell, a manufacturer of control equipment that was also seeking a way into the computer field. Eventually Honeywell acquired Datamatic. RCA was also in from the beginning, particularly in the design of memories, and may have been the first manufacturer to produce a coincident-current magnetic core memory. Some of the smaller companies, which had the medium scale computer market much to themselves until they were hit by the IBM 650, were absorbed by mergers. One merged with the National Cash Register Corporation, for example, and another was absorbed by Burroughs.[10] Burroughs had entered computer research in 1948. Librascope and Bendix corporations were also successful with small computers. The pattern of merger and absorbtion was probably inevitable in an industry that expanded so quickly while its product was still under rapid development. One anecdote from Britain, dating a little later than this period, told of an unfortunate engineer who quit his job because of personality clashes with his superiors. A few weeks later he was back. His new company had been taken over by the old and the two divisions had been 'rationalized' into one. Although the story may be apocryphal, it is an interesting comment on how some engineers viewed the mergers.

By 1959, the year Texas Instruments announced the first integrated circuit and thereby sowed the seed of the future third generation of computers, the first generation was at an end and the second had begun.

The first generation showed what could be achieved with the vacuum tubes of the fifties. Specialized computers had been made and sold for both scientific and business use and large, medium, and small machines were available. IBM had emerged as a world leader for large and medium computers and Britain had fallen behind America. The large magnetic-core memory, backed up by magnetic tape, drums, and discs, had rendered CRT memories and mercury delay lines obsolete. Program interrupts had been introduced (by Univac) and internal buffering exploited to permit simultaneous input, output, and computation. In the mid-1950s there were probably fewer than 1000 computers in the USA. By the end of the first generation the figure was around 5000 (mid-1960s), and had rocketed to some 220 000 by the end of 1976, by which time the third generation was well established.[10,17]

Second and Later Generations

Transistors became commercially available in small quantities from about 1953 onwards and Manchester University grabbed another first with their experimental stored-program computer based on point-contact germanium transistors. In the USA, MIT's Lincoln Laboratories set new standards with their transistorized TX series of computers.

Transistors promised faster switching with much smaller size, lower power requirements, and very little heat dissipation, which in turn offered the possibility of computers with 10 to 100 times more active components than the biggest vacuum-tube machines. At first the promise was not kept. Switching times were poor and transistors lacked uniform characteristics. Technical advances in transistor design, such as Philco's surface barrier transistor, soon remedied the situation. By the end of 1959 the computer industry was into its second generation and, in addition to transistors, ferrite core memories had become common.

The transistor generation saw the introduction of further new concepts, like the thin-film memory from Univac in 1960,[17] but the most important was the beginning of the concept of families of compatible computers. As software became relatively more expensive, standardization of the hardware and design concepts within a given company grew more important. Architectural features such as instruction sets and addressing schemes would henceforth be compatible within a family of computers. And with a degree of standardization came the threat of new competition as rivals moved in with compatible items of equipment.

The proliferation of computer manufacturers and computer models make it impractical to survey the whole field of activity. Instead two IBM families only will be mentioned. Other manufacturers ought not to be forgotten, though. The Elliott 803 was the first commercial British transistorized computer.[18] In America RCA, NCR, Burroughs, and Univac were still there, and Honeywell bounced back despite rumours that it was to pull out of computers. Philco entered the fray, and in 1957 a group of Univac employees broke away to form a new company that was to become very important, Control Data Corporation (CDC).[10] In Germany Siemens entered the computer arena for the first time with the transistorized 2002, and in Japan Fuji was first to move from valves to transistors.

IBM strengthened its commercial position with the 1400 series of small to medium business computers, the 1620 scientific computer (of which over 1000 were installed), and the 7000 series of large scientific computers.

The 1400 series began with the 1401 in 1960 and it was followed by others with fairly compatible hardware. The 1401 brought stored-program computers to the smaller end of the business world. More than 10 000 units were eventually delivered; it became the most widely used computer of its time.[16] Several versions of the 1400 family were introduced through the 1960s. At the other end of the scale the IBM 7000 series began with a military request in 1958 for very large and fast computers for the missile early warning system. IBM won the contract with the 7090, the development of which, it seems, stretched them to the limit. It had eight multiplexed data channels, a 32K-word core memory with a $2.18\mu s$ cycle time, and could perform 229 000 additions per second. For a time its reliability was in doubt but once the bugs were out the 7090 proved to be both reliable and successful.[10] Hundreds were eventually used. Many were later converted into a slightly faster version, the 7094. Cheaper scaled-down versions, the 7040 and 7044, were introduced in the early 1960s.

The next generation of computers, the third, again depended on the next generation of components, but also brought about much stronger families of computers with a greater degree of standardization. Many also regard time sharing as a third generation feature.

The IBM System/360 is the outstanding example. It was hailed by the IBM chairman as the "most important product announcement in company history." System/360 took the family concept to clan status. It was a family of compatible small, medium, and large processors with common architecture, together with peripheral equipment. Five basic System/360 computer models were introduced in 1964 in nearly 100 countries around the world. The largest processor was about 100 times more powerful than the smallest and calculating speeds ranged from about 33 000 to more than 2.5 million fixed-point additions per second.[16] Nineteen new memories were introduced and main storage ranged from 8000 8-bit bytes to more than 8 million. A standard interface was used for the attachment of peripherals. *Electronics* magazine commented, "With a single new system, IBM has made every one of its commercial computers obsolete."[17] IBM's transistorized second-generation computers had consisted of about seventeen processors arranged into seven families aimed at the different needs of users. While the processors were compatible within a family the families themselves were largely incompatible. The third generation attempted to replace all that with one big happy family.

The electronics at first used hybrid integrated circuits, before the monolithic variety became widely available, and IBM decided to make their own. A new $100 million components plant was opened. All in all something like $5000 (£1800) million was spent on developing the System/360. After six years, and over 30 000 computers, the family was superseded by the compatible System/370, one of which contained the first commercial integrated-circuit main memory.

Other early third-generation systems include the ICT 1900 series, the RCA Spectra 70 series with Fairchild chips, GE's 600 series, and several more. Computers now lived in families and the monolithic integrated circuit ruled supreme. The third generation also saw CDC confirm its grasp on the very large scale computer market, while at the other end Digital Equipment Corporation (DEC), a fairly new company, introduced the minicomputer with its PDP-8.

New companies still entered the field, and some old ones left. Two gaints, GE and RCA, pulled out in 1970 and 1972, respectively. Hewlett-Packard entered in 1966. Data General was founded in 1968, by three ex-DEC designers, and offered the Nova computer at a mere $8000 (£3400). In 1970 their Supernova SC was the first minicomputer with semiconductor memory. Outside the USA potential rivals formed. In the Netherlands, Philips formed a computer division in 1962 and produced the first machines in 1968, and in Japan the government, as in microelectronics, encouraged the growth of a domestic industry. As already noted, ICL became the British giant in 1968. Three mutually incompatible ranges were inherited by ICL: the ICT 1900 series, English Electric's System 4, and the 4100 series. The lives of the first two could be extended and the 1900 series in particular was very successful in providing a range covering a considerable spectrum of power and facilities. In fact the 1900 series, in its various guises, has proved to be a long lasting and very important product range for the company, even into the 1980s. However, a new family was planned right from the start, the 2900 series, and was launched in 1974. Though the 2900 series took ideas from several previous British machines, Manchester University's MU5 is acknowledged as having been the single most important external influence.[19]

Computers had come a long way since the days of the Z3 and ENIAC, yet the industry was still a young and dynamic one. By 1970 the computer revolution was in full swing with plenty still to come. The microprocessor and microcomputers were just around the corner. At the other extreme were the so-called supercomputers, machines that were orders of magnitude faster or more powerful than the majority of commercially available computers. The transistor generation saw the first supercomputers; Univac's LARC and IBM's Stretch were early examples. Both were built for U.S. government atomic research centres and were first delivered in 1960 and 1961.[10] Stretch (Fig. 11.7), which contained 150 000 transistors,[16] was renamed the 7030 but neither it nor LARC succeeded in the market place. The success they did enjoy lay in stimulating the industry. It has been suggested, for example, that the very successful IBM 7090 would have been delayed by two years if it had not been

Figure 11.7 *IBM Stretch (7030), 1961; containing 150 000 transistors it was one of the first supercomputers, but a failure in the market place. Only seven were installed*

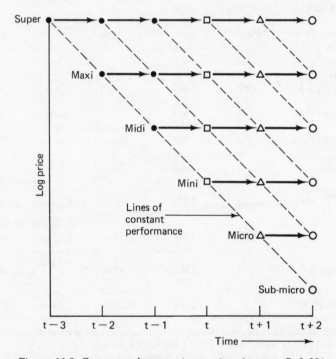

Figure 11.8 *Computer classes, price vs. time (source: Ref. 20)*

for Stretch.[10] The most famous British supercomputer was the Manchester University-Ferranti Atlas of 1962. Even after twenty years mention of it seems to produce a warm glow in the hearts of some of those who were associated with it. Contributing to its fame are the invention of 'firmware' (software frozen in hardware) and paging, or virtual memory, a technique by which separate fast and slow memories are made to appear as a single fast memory. ICL, the inheritors of Atlas, introduced the virtual-machine concept about 1972.

The power capabilities of the supercomputers, the racing cars of the computer industry, have a way of percolating down the line to the mass market machines. The power of yesterday's supercomputer is found in today's large-scale machine and in tomorrow's midi or mini, a feature illustrated by Figure 11.8, and of course the supercomputers themselves undergo continual

Table 11.4 Supercomputers

Computer	Started	First Delivery	Memory Cycle Time	Processor Cycle Time	Millions of Instructions/ Operations per sec.
IBM, NORC	1951	1955	$8\,\mu s$	$1\,\mu s$	0.015
Univac, LARC	1956(?)	1960	$4\,\mu s$		
IBM, STRETCH[3]	1956/7(?)	1961	$2\,\mu s$		0.67
Manchester University/ Ferranti, ATLAS	1956	1962	$2\,\mu s$	N.A.	0.5
CDC, 6600[2]	—	1964	$1\,\mu s$	100 ns	3
IBM 360/91[3]		1967	780 ns	60 ns	16.6
CDC 7600[2]	—	1968	220 ns	27.5 ns	20–25
Burroughs[1] ILLIAC IV	1966	1972	250 ns	62.5 ns	200
Cray, 1	1972	1976	50 ns	12.5 ns	80
CDC, Cyber 200 (203)[2]	—	1980	80 ns	$40 + 20$ ns	800
CDC, Cyber 205[2]	—	1981	80 ns	20 ns	800

Note that the figures given may not be directly comparable and are given as an indication only.
Notes: 1. ILLIAC IV memory used 256×1 Bipolar RAM IC.
  2. CDC, start dates are impossible to pinpoint as one design evolves from another.
  3. Millions of additions per second.

improvements. Table 11.4 gives a few examples. Some of the newer companies were especially successful, such as CDC and Cray Research; the latter was founded in 1971 by Seymour Cray who left CDC to do just that.

Minis and Micros

The term 'minicomputer' first appeared (in hyphenated form) in 1968 and its definition, often vague, has had several shades of meaning. It began as an indication of small size, short word length, limited software support, and a low price. As a large, even vast, market was established, and as integrated-circuit technology improved, things began to change. Physical size remained small, word length increased to range from 8 to 32 bits, software support increased, and the price dropped; as much as 20 to 25 % per annum through the 1960s. By 1974 word length had increased so much that it ceased to be a distinguishing feature. The mini became established not merely as a computer, but as a general system component to be treated as a building block. Instead of being the eighth wonder of the world the computer, in its mini form, became simply another box of electronics which, although it was the hub of a system such as a large automatic test system, could be placed out of the way at the top of a rack. With its small size and cost the mini took computing power to numerous areas never reached before, a trend that the next stage, the microcomputer, was to continue.

The Digital Equipment Corporation (DEC) claim their PDP-8 as the first true minicomputer.[20] DEC was founded in 1957 by two brothers, Kenneth and Stanley Olsen, and for a couple of years the company existed by designing and manufacturing logic modules. In 1959 they produced their first computer, the Programmed Data Processor 1 or PDP-1, a small transistorized 18-bit machine which sold, in its standard system, for a then cheap $120 000 (£43 000). It was one step along the road from the large and expensive giants to the tiny, relatively inexpensive minis.

The road to the mini, at least to DEC's mini, began back in 1951 with the MIT Whirlwind which had a small memory, a limited instruction set, and a short 16-bit word length. Other small experimental computers followed from MIT's Lincoln Laboratory, notably the TX-0 in 1956, containing 3600 transistors, and the TX-2 in 1957 containing 22 000 transistors. DEC's commercial PDP-1 was a direct descendant of these MIT machines.[20] A total of 50 PDP-1s were made and half the sales went to ITT, a success which helped to put DEC onto its digital feet. Burroughs, Librascope, and Bendix also moved into the small-computer area and marketed machines for less than $50 000 (£18 000) in the mid-1950s.

The early 1960s saw further moves towards smaller and cheaper processors and the eventual emergence of the true minicomputer. The PDP-1 was followed in 1963 by the 12-bit PDP-5, which sold as a small general-purpose computer for around $30 000 (£11 000). Better transistors, and mass produced transistor logic and core stores, brought down both size and cost. In 1965 manufacturing techniques made a redesign of the PDP-5 possible. The result was the PDP-8, a computer that could be built into other equipment. The cycle time had been halved and the cost cut to around \$18 000 (£6500). The PDP-8 minicomputer turned out to be a huge success. A whole family of PDP-8 instruction set computers, the 8-family, was developed from it as a result of the ever-changing technology, including an Intersil LSI processor-on-a-chip in 1976. By 1978 the PDP-8 had been "reimplemented 10 times with new technology over a period of 15 years."[20] Costs had continued to fall. For the basic system, consisting of a processor and 4K-words of memory, costs fell about 22 % per annum up to the late 1960s and about 15 % per annum in the

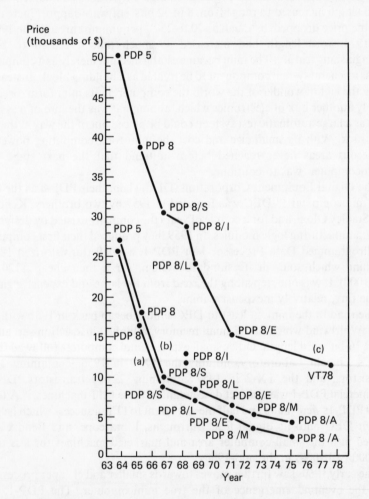

Figure 11.9 Microcomputer costs vs. time: DEC PDP-8. (a) CPU + 4K-words memory; (b) a + hard copy and paper tape; (c) user system—8K-words, 2 tapes, hard copy (source: Ref. 20)

early 1970s (Fig. 11.9). By 1978, 40 000 computers of the PDP-8 family had been manufactured.

Other manufacturers also decided to manufacture minicomputers and the area grew to become an important subset of the computer industry. New companies included Data General and Interdata but the more established companies also brought out minis: Hewlett-Packard, Texas Instruments, Honeywell, and many more.

After the minicomputer with its assembly of functional modules came the microcomputer with its assembly of LSI chips. As Figure 11.10 shows, the holding structure or physical support for a computer has dramatically decreased over the years as a result of the miniaturization of electronics. In the vacuum tube days an entire room was needed; then a large cabinet or several large cabinets containing modules whose capacity increased from about one bit per module to a whole register per module. As the degree of integration advanced a register could be placed within a single integrated circuit and the computer reduced in size so that it fitted inside a single rack or bench-mounted box. This was the minicomputer. With further advances in microelectronics and with greater packing densities becoming possible, a whole processor (the central arithmetic and logic unit of a computer) could be placed on one chip of silicon. The choice of name for the new chip was obvious, the microprocessor. With the addition of a controller, memory, and input-output (I/O) ports, a microcomputer could be assembled from a handful of integrated circuits.

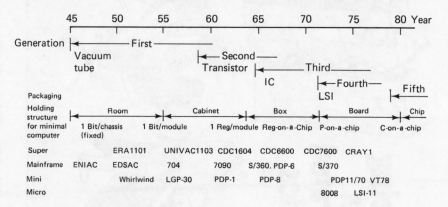

Figure 11.10 Evolution of packaging as seen by Digital Equipment Corporation (source: Ref. 20)

The first such microprocessor was announced by Intel Corporation in America in 1971 with advertisements claiming 'a new era of integrated electronics.' A patent was issued on 23 June 1974. Using P-channel silicon-gate MOS technology, Intel made its first microprocessor on a piece of silicon that measured $1/8 \times 1/6$ in. (about 3×4 mm). It contained 2250 transistors, had a 4-bit word length, and was called the 4004. With three other chips, a read-only memory control unit (4001), a random access memory for data

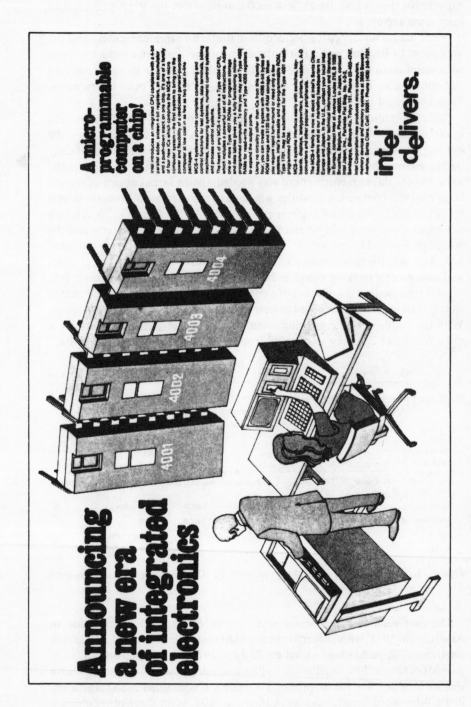

Figure 11.11 First advertisement for a microprocessor appeared in 'Electronic News', 15 November 1974

storage (4002), and a special shift register for I/O expansion (4003), it was offered as a microcomputer system kit, the MCS-4 (Fig. 11.11).

Intel Corporation had been formed in 1968 by R. N. Noyce and G. E. Moore, one-time founding members of Fairchild Semiconductor (and ex-Shockley Transistor). The name stands for Integrated Electronics. The first successes came with semiconductor random access memories (RAM) and with the introduction (in 1970) of the first 1K-bit RAM. They can also claim, incidentally, to have been first out with the 4K-bit RAM in 1973 and the 16K-bit RAM in 1976. (The standard 64K-bit RAM (5V supply) was first produced by TI and then by Motorola in 1978.[21])

In 1969 a Japanese calculator manufacturer approached Intel asking them to design a set of eleven chips for a family of electronic calculators. Each calculator was to be specified by read only memory (ROM) programs. The Intel designer who was given the job, M. E. Hoff, said later, "Instead of making it look like a calculator with some programming capabilities, I wanted to make it look like a general purpose computer programmed to be a calculator."[17] He succeeded. A smaller number of larger chips with more complex circuitry were designed, one of which contained the entire central processing unit (CPU). Babbage's calculating mill, the CPU of a computer, had become an electronic component far smaller than a vacuum tube. The first microprocessor had been made but was wrongly hailed as a computer on a chip, something still a few years away. Nevertheless, the increasing degree of miniaturization had meant that the semiconductor component industry and the computer industry had now met on a small chip of silicon. In 1979 Hoff was awarded a Franklin Institute medal to mark his development of the microprocessor.

The 4004 microprocessor had a 10.8 μs instruction cycle time and the MCS-4 system could have up to 4K of 8-bit ROM words, 1280 4-bit RAM characters, and 128 I/O lines without the aid of interface logic. A year later the 8008 was announced, the first 8-bit microprocessor, and it could address 16K bytes of memory. In 1974 Intel turned to the faster NMOS technology and introduced a new 8-bit microprocessor, the 8080, employing 5000 transistors (Fig. 11.12). With an instruction cycle time as low as 2 μs and the ability to address 65Kbytes of memory the 8080 became, in the words of *Electronics* magazine, one of the most sought-after LSI devices in the history of the business.[17]

Other manufacturers moved into the market and sales rose as the microprocessor left its pocket calculator image behind. Fairchild, National Semiconductor, and Rockwell had products out in 1973. Masatoshi Shima, who designed the 8080, left Intel to join Zilog, where he designed the Z80. The pattern of labour movement, new companies, and new products that had characterized the American semiconductor industry was also evident in the microprocessor field. In 1974 Texas Instruments produced a 4-bit microcontroller and Motorola unveiled its 6800 microprocessor. After the 8-bit chip came the 16-bit chip, the first of which came from National Semiconductor.

Further developments within semiconductor technology led to more

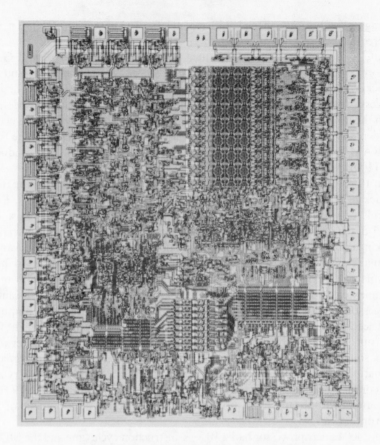

Figure 11.12 Intel 8080 microprocessor, 1974

circuitry being placed onto one chip of silicon until, with the Intel 8048, the single-chip microcomputer was born. The CPU, RAM, ROM, and I/O were all in the same integrated circuit.

The microcomputer, whether built with one chip or several, and whether using 4, 8, 12, 16 bits or more, made the biggest impact on electronics since the invention of the integrated circuit. "Today's microcomputer," wrote Noyce in 1977, "at a cost of perhaps $300, has more computing capacity than the first large electronic computer, ENIAC. It is 20 times faster, has a larger memory, is thousands of times more reliable, consumes the power of a light bulb rather than that of a locomotive, occupies 1/30,000 the volume and costs 1/10,000 as much. It is available by mail order or at your local hobby shop."[22] Such has been the pace of change in 30 years.

Microprocessors themselves have developed in three directions. At one end of the spectrum they became more advanced as computers to produce single boards that could be grouped together as multiple microcomputers communicating with each other through a common bus. At the other extreme there are simplified versions for use in applications where great versatility is not required, as in cash registers or electronic scales. The third development is known as the bit/slice approach. In this technique the chip is not a complete system but merely a part of a central processor. The word length is just two bits, but the strength lies in using bit/slices in parallel so as to achieve any word length desired, even as great as the large mainframe computers. High performance is the goal.

The second half of the 1970s saw microprocessor sales take off, somewhat after the manner of transistor sales in the 1950s and IC sales in the 1960s. Dataquest, the American consultant firm, listed about ten models to choose from at the end of 1975 with total sales of around 1.1 million units. By the end of the decade they were listing approximately 37 varieties with total annual sales of 75 million units. Sales were increasing at a rate of around 20 % per quarter. By mid-1980 one single family, the 8048, was enjoying annual sales of around \$7 million.

Microprocessors have been busily engaged in changing many aspects of electronics. These tiny computer elements, with their associated circuitry, moved into old established areas of electronics, such as instrumentation, and at the same time created new areas of electronics: personal computers, electronic games, multifunction electronic watches, and a thousand-and-one industrial uses. Universities and colleges offered crash courses to train practising engineers in the use of the mighty micro. Many engineers would have a sense of déjà-vu, having already undergone one or more fundamental retraining exercises during their careers: from thermionic valves to transistors, transistors to ICs, linear electronics to digital. To them the micro was one more step along the path of electronics progress.

Digital Logic Circuitry[23,25]

The basic building blocks of computers and other digital electronic products are the electronic circuits now called gates; the ANDs and ORs, NANDs and NORs, and so on. In the early days the AND circuit was often called a gate and the OR circuit was called a buffer.[23] Although gates can be treated as functional black boxes, the actual circuitry involved in building them has changed dramatically over the years. In discrete component circuitry the emphasis was on using cheap and reliable resistors as much as possible in preference to bipolar transistors. Diodes, when they became cheap enough, replaced some of the resistors so as to reduce the power consumption and raise the operating speed. The first integrated-circuit gates naturally followed some of the circuitry of the discrete component approach, until changes occurred that made better use of the advantages of integration; more transistors and fewer resistors, for example. MOS technology provided an alternative approach to bipolar, and eventually circuit techniques were used that would be impossible with discrete components.

At almost any given time in the evolution of logic circuitry the user has been offered a choice of techniques, sometimes a bewildering choice, each with its own peculiar advantages and disadvantages. Families of logic circuits were developed by use of different circuit designs to achieve a given gate or other circuit function. The designer would then choose the family whose characteristics best suited his requirements, such as logic flexibility, speed, availability of complex functions, noise immunity, acceptable temperature range, power dissipation, noise generated, and cost.[24] Of these families some proved to be more popular and durable than others and some, such as TTL, spawned their own offspring.

One of the early families was known as transistor-resistor logic (TRL). Built from discrete components the design maximized the use of cheap resistors and minimized the expensive transistors. It was introduced in 1956–57 and may have been the first to use the term NOR. A variation on TRL was to replace some of the resistors by diodes. The result was DTL, diode-transistor logic. It gave a lower power consumption and a higher speed than TRL, but still economized by depending on only one switching transistor. DTL became a very popular discrete-component technique and was one of the first to be adapted for integrated-circuit use. Signetics used it about 1961, right at the start of the IC era.

Philco made use of its fast surface-barrier transistor by introducing direct-coupled transistor logic (DCTL) in 1955 in a bid to reduce power consumption and increase reliability. Very simple circuits were used but an expensive transistor was employed for each input. DCTL suffered from a problem called 'current hogging' but the effects could be reduced by the addition of a resistor to the base of each transistor, which resulted in yet another family; resistor-transistor logic or RTL. Fairchild's integrated Micrologic family of 1961, one of the first IC logic families, used RTL and the technique was still around late in the 1970s.[25]

In the summer of 1961 most people in the Device Development Section of Fairchild's R&D laboratories were busy characterizing the new Micrologic and had little time to break in new recruits. One such new recruit was Heinz Rüegg. Rüegg was given the task of further developing an invention made by R. H. Beeson of Fairchild in which transistors were used instead of diodes as the input of inverter transistors so as to form a sort of modified diode-transistor logic. The result was transistor-transistor logic or TTL, which has become the most popular of all the logic families. Beeson was the original inventor. TTL was announced at a conference in 1962 and a year later Sylvania was one of the first companies to market it.[26] TTL was the first logic circuitry to use a technique that could not be provided with discrete components. In this case it was the multiple-emitter transistor, in which two or more emitters share a common base and collector. TTL was to become the workhorse of digital electronics and spawned many variations: low-power TTL, high-speed TTL, Schottky TTL, and several more. Schottky TTL was one of two answers to limitations on the speed of operation. In it a Schottky metal-semiconductor diode prevented the transistor from saturating. The other answer involved a circuit with a differential amplifier and was called emitter-coupled logic (ECL) or common-mode logic (CML). ECL was introduced by General Electric in 1961 and by Motorola in 1962. It turned out to be extremely fast indeed, with an average gate delay of 1 ns in one of Motorola's later families.

The first experimental integrated circuit did not appear until 1958, which means that some of the important bipolar logic families were used quite early in commercial ICs. The most significant technique not dating from the early sixties is integrated-injection logic (IIL or I^2L), probably first reported at a conference in 1972 by workers from both IBM and Philips. Yet "it is said that the newest of techniques are sometimes the oldest of techniques, and this is true in integrated-injection logic."[25] I^2L can be seen as a development of DCTL, one of the oldest of logic families, but a development that could only be accomplished in integrated-circuit form. Resistors are almost completely eliminated and circuit densities are eight to twelve times greater than in either TTL or CMOS. Multiple collectors and bases are used and transistors are built both horizontally and vertically. The popularity of I^2L grew through the late 1970s following early use of it by TI in a LED watch announced in 1975.

MOS digital logic, the alternative to bipolar, first became easily available in the early 1970s. NMOS offered greater speed than the earlier PMOS; CMOS gave very low power consumption. CMOS was invented at RCA Laboratories in 1963 under a U.S. Air Force contract and was known as COS/MOS. The first commercial chips were announced by RCA in 1968, although CMOS did not really catch on until plastic packaging brought the price down after about 1971. Since then it has become a popular technology for uses in which low power requirements are important, as in battery-powered applications.

With so many logic families available it may appear to be a difficult problem choosing the best one for a given application, but it is generally true that specific families have tended to dominate specific applications. DEC, for example, uses TTL for their mid- and high-sized computers, ECL for the larger-scale machines, MOS for memories and microprocessors, and CMOS especially for battery-driven microcomputers.[20] This specialization is typical of the industry.

Programming Languages

No computer is complete without its programs and no chapter on the development of computers would be complete without a few words on programming languages. There have been, and still are, many of them. In 1974 one writer, under the title "A Load of Boole," light-heartedly suggested that the proliferation of computer languages "must already be on a par with the number of human ones, all with local dialects depending on manufacturer, machine and compiler."[27]

Programming a computer involves breaking down the problem to be solved into small steps and then instructing the machine how to perform them. Around the time of the French Revolution a famous French engineer called M. R. de Prony, who is also remembered for his dynamometer, was given the task of calculating a vast set of mathematical tables that were to be bigger and better than any previously made or even conceived. He solved this immense problem by using a handful of mathematicians who broke the work down into relatively simple tasks of addition and subtraction, which were then performed by a small army of mathematical slaves.[2] In a sense Prony had programmed a computer of around 90 people. Babbage later used similar mathematical techniques in his mechanical designs and Lady Lovelace began to write programs.

In the early days of electrical computers, programming was achieved by use of the computer's own machine language in which instructions were given as sets of 0's and 1's. The difficulty of writing a program in this way was soon alleviated by use of mnemonic codes and decimal rather than binary numerals. Eventually, 'higher-level' languages were developed in a bid to make programming easier by making the language, at the human end of the chain, as mathematical, or as English-like, as possible. A special program, the compiler, is then used to get the computer to do its own machine code.

The first literature on modern programming was contained in reports on the early experimental computers, sometimes called the zeroth generation. The Harvard Mark I Manual (1946) and the EDVAC report are examples. Another is a classic book on programming the Cambridge EDSAC.[28] With these reports, the concept of a subroutine library became the basis of programming. When computers eventually became a commercial venture, programming groups were set up by manufacturers. At the Eckert-Mauchly Computer Corp., Grace Hopper was the leader of what was probably the first such group. She held a firm belief that programming languages should be oriented towards people and problems rather than an individual computer, and a whole series of languages were developed for UNIVAC I. At least one of them, the Algebraic Translator AT3, was finished after its machine had become obsolete, an early example of a problem that was to become familiar throughout the industry. Another, FLOWMATIC, was the ancestor of COBOL. For the IBM 701, languages such as SPEEDCODE and PACT were developed; the latter shared the same obsolescence problem as AT3. Some of these early languages were to contribute ideas to the later more widely used languages such as FORTRAN and COBOL, and some should be remembered for their names if for nothing else: SOAP, IT, and BACAIC for example. The latter, which seems to suggest a need for medical attention, was the acronym for the Boeing Airplane Company Algebraic Interpretive Coding System for the IBM 701. A few later languages had equally interesting acronyms. MAD was the Michigan Algorithm Decoder, and LOLITA was not V. V. Nabokov's nymphet but a Language for the On-Line Investigation and Transformation of Abstractions.

FORTRAN

FORTRAN was the first of the major higher level languages and has been suggested as the most important milestone in the development of programming languages.[29] It has become one of the most widely used and accepted programming languages and is almost as much taken for granted as silicon is as a semiconductor or the telephone as a communications aid. Informed comment on its early days are therefore especially interesting and are worth repeating. "Like most of the early hardware and software systems FORTRAN was late in delivery, and didn't really work when it was delivered. At first people thought it would never be done. Then when it was in field test, with many bugs, and with some of the most important parts unfinished, many thought it would never work. It gradually got to the point where a program in FORTRAN had a reasonable expectancy of compiling all the way through and maybe even of running."[30] FORTRAN of course did eventually work and the considerable number of high level language sceptics were proved wrong.

FORTRAN started life as a FORmula TRANslating system for the IBM 704. It was initially developed by a team made up mostly of IBM employees and led by J. W. Backus, who deserves most of the credit. Backus was to become famous from his contributions to programming languages. Work began in 1954 and the first version came out in 1957 after something like 25 man-years of effort. It was not easily accepted, however. A new version, FORTRAN II, came out the next year with significant language additions and FORTRAN III was developed for internal IBM use. Other manufacturers began to use the language about 1960 or 1961, and by 1963 many were committed to it. As its use spread, along with some confusion, there were demands for further additions and changes. The end result was FORTRAN IV whose preliminary bulletin was issued in 1962. The same year saw the formation of a committee to produce an American standard, two of which eventually resulted; FORTRAN and Basic Fortran. These are roughly equivalent to FORTRAN IV and FORTRAN II.

ALGOL

If FORTRAN was the most important practical milestone, ALGOL 60 has been described as the important conceptual milestone.[29] It grew out of a bid by a European organization concerned with applied mathematics and mechanics, called GAMM, to set up a common algebraic language that could be used on various computers. The Association for Computing Machinery (ACM) in America was invited to contribute; the first committee meetings took place in 1958. The language was to be known as the International Algebraic Language, but in fact it became known as the ALGOrithmic Language or ALGOL. Not only was ALGOL the first computer language designed by an international committee (but luckily did not result in a camel, supposed to be a horse designed by a committee), it was the first to use a block structure and, in its 1960 form, to have a precisely defined syntax. This was also largely the work of IBM's John Backus and set an important precedent.

The first ALGOL report came out in 1958 from a meeting held in Zurich; that version of the language is known as ALGOL 58. Many variants followed such as the Burroughs version (Balgol), the Bendix Algo, and other "dialects." Meanwhile the ACM-GAMM work continued and in 1960 a report was published defining ALGOL 60, a substantial improvement on ALGOL 58. Arguments raged over ambiguities and obscurities and a revised report was issued. The debate was such that participants became known as ALGOL lawyers and ALGOL theologians.[29]

COBOL

In May 1959 a meeting was held in the Pentagon under the auspices of the U.S. Department of Defense. About 40 representatives from computer users, government installations, manufacturers, and others agreed on the need for a common business language for data processing and three committees were set up to examine the short, intermediate, and long-range possibilities. The short-range committee, consisting of representatives from two military agencies and six manufacturers, went well beyond their brief and actually wrote a new language: the COmmon Business Oriented Language, or COBOL. Then occurred what has been called the battle of the committees.[30] The Intermediates suggested that a new Honeywell language, FACT, was better than COBOL but the Shorts fought back and won the day. COBOL was published in 1960; revisions followed in 1961 and 1962.

Some manufacturers began to use COBOL while others avoided the still developing language, preferring their own developments. The U.S. government then decided to back COBOL; as a major user of data processing it wished to avoid giving advantage to any one manufacturer. Maybe more than anyone else the government needed a standard language. In 1960 it was announced that the government would not buy or lease a computer if a COBOL compiler was not available, unless the manufacturer could prove that his equipment was better without it. After that COBOL was here to stay and became one of the most widely used computer languages.

PL/1 and Others

FORTRAN, ALGOL 60, and COBOL may be thought of as the first generation of high level languages. Many others have followed. In 1963 IBM and SHARE, an IBM users' organization, got together to develop an advanced language that would avoid the problems of existing languages and be available to a far wider range of users. The first official manual of the New Programming Language

(NPL) was issued in 1965. The name was subsequently changed to PL/1 to avoid confusion with the British Government's National Physical Laboratory, which is also known as NPL. PL/1 can be regarded as a synthesis of FORTRAN, ALGOL 60, and COBOL.

About the same time as PL/1 was being developed J. G. Kemeny and T. E. Kurtz at Dartmouth College (USA) developed a simple language to help in the teaching of programming to large numbers of college students. The result was the Beginner's All-Purpose Symbolic Instruction Code, better known as BASIC (1964–65), a language that was later to see widespread use in the microcomputer and hobby-computer fields. Towards the end of the 1960s two languages were designed as successors to ALGOL 60. They were ALGOL 68 and PASCAL. ALGOL 68 was a more generalized version of ALGOL 60 and perhaps suffered by being too general. PASCAL on the other hand, like BASIC, aimed at simplicity and has proved to be popular. It was designed by Niklaus Wirth, who had also worked on ALGOL 68.

Many other languages besides those mentioned here in this brief bird's eye account have been devised, often for specific applications. Other histories, on which this account has been based, give more thorough reviews.[29,30,31] Some of the classics, such as ALGOL 60, have spawned offspring, yet two of the oldest, COBOL and FORTRAN, despite defects, have remained two of the most popular. Their durability is not only a tribute to their original writers, it is a mark of the vast amounts of money that has been sunk into their application.

Conclusion

Since the heady days of the late 1940s, when some of those in the know felt that a few electronic computers would satisfy the world's needs, computers in huge numbers have become part of our way of life, so much so that they could soon be the world's third largest industry. They have been frequently hailed as harbingers of a second industrial revolution. Men have long used tools to help in performing physical and mental tasks. The first industrial revolution brought about a fundamental change in the physical tools; the second is bringing a change of similar magnitude to our mental tools. The computing age is the information age, the data-processing age, the number-crunching age. The present products of that age may seem almost trite before too long. Yet, like any tool, computers can be used for good or evil. The existence and security of remote-accessed databanks containing confidential personal information has been one area of concern.

Computers, together with other modern-day marvels, are changing our society. To many this change is a promise, to some a threat. Some of those who feel overawed by electronic computers may find comfort in words attributed to Wernher von Braun: "Man is still the fastest computer that can be produced with unskilled labour."[32]

References

1. S. H. Hollingdale and G. C. Toothill, *Electronic Computers*, Penguin, London, 1965.
2. M. F. Wolters, *The Key to the Computer*, MTP-Oxford, 1972.
3. R. E. Lynch and J. R. Rice, *Computers, Their Impact and Use*, Holt, Rinehart and Winston, New York, 1977.
4. B. Randell, Ed., *The Origins of Digital Computers*, Springer-Verlag, New York, 1975.
5. T. I. Williams, Ed., *A History of Technology*, Oxford University Press, Oxford, 1978, vol. 7, pt. 2.
6. B. V. Bowden, Ed., *Faster Than Thought*, Pitman, London, 1953.
7. M. D. Fagen, Ed., *A History of Engineering and Science in the Bell System, 1925–1975*, Bell Laboratories, New Jersey, 1978, vol. 2.
8. G. D. Kraft and W. N. Toy, *Mini/Microcomputer Hardware Design*, Prentice-Hall, Englewood Cliffs, New Jersey, 1979.
9. J. W. Mauchly, *IEEE Spectrum* 12 (No. 4): 70–76, April 1975.
10. S. Rosen, *ACM Computing Survey*, vol. 1, 7–36, March 1969.
11. J. M. Adams and D. H. Haden, *Computers*, Wiley, New York, 1973.
12. B. Johnson, *The Secret War*, BBC, London, 1978.
13. A. C. Lynch, 6th IEE Weekend Meeting on the History of Electrical Engineering, Nottingham, p. 46, 7–9 July 1978.
14. S. H. Lavington, *A History of Manchester Computers*, National Computing Centre, Manchester, 1975.
15. B. Lane, *Wireless World* 81: 102–105, 161–164, 222–225, 283–286, 341–342, 1975.
16. J. Belzer, A. G. Holzman, and A. Kent, Eds., *Encyclopaedia of Computer Science and Technology*, vol. 15 (Supplement), Marcel Dekker, New York, 1980.
17. *Electronics*, Special Issue, 53 (No. 9), 17 April 1980.
18. P. Stoneman, *Technological Diffusion and the Computer Revolution*, Cambridge University Press, London, 1976.
19. J. K. Buckle, *ICL Technical J.* 1: 5–22, 1978.
20. C. G. Bell, J. C. Mudge, and J. E. McNamara, *Computer Engineering: A DEC View of Hardware Systems Design*, Digital Press/Digital Equipment Corporation, Bedford, Mass., 1978.
21. Dataquest Inc., Cupertino, Calif.
22. R. N. Noyce, Chapter 10, Ref. 29.
23. G. G. Langdon Jr., *Logic Design: A Review of the Theory and Practice*, ACM monograph, Academic Press, New York, 1974.
24. L. S. Garrett, *IEEE Spectrum* 7 (No. 10): 46–56; (No. 11): 63–72; (No. 12): 41, October, November, December 1970.
25. C. A. Harper, Ed., *Handbook of Components for Electronics*, McGraw-Hill, New York, 1977.

26. R. H. Beeson and H. W. Rüegg, IEEE Int. Solid State Circuits Conf., Philadelphia, Digest, 10–11, 104, 1962.

27. C. Brown, *Electronics and Power* 20: 407, 1974.

28. M. V. Wilkes, D. J. Wheeler, and S. Gill, *The Preparation of Programs for an Electronic Digital Computer*, Addison-Wesley, Cambridge, Mass., 1951.

29. P. Wegner, *IEEE Trans.* C-25: 1207–1225, 1976.

30. S. Rosen, Ed., *Programming Systems and Languages*, McGraw-Hill, New York, 1967.

31. J. E. Sammet, *Programming Languages: History and Fundamentals*, Prentice-Hall, Englewood Cliffs, New Jersey, 1969.

32. H. Young, B. Silcock, and P. Dunn, *Journey to Tranquility*, Jonathan Cape, London, 1969, p. 21.

33. B. Randell, 'The Colossus,' in N. Metropolis *et al.*, Eds., *A History of Computing in the Twentieth Century*, Academic Press, New York, 1980.

For a more detailed account of early mechanical calculating machines see Refs. 1, 4, 5, and 6.

12 A TECHNOLOGICAL SOCIETY

If we accept that technology concerns mankind's methods of doing or making things, then from the time that men first used stones to kill in order to eat they have lived in societies in which technology has played a role.[1] The importance of that role has increased slowly through the centuries. The arm that threw the stone was eventually aided by a sling; the stone gave way to a spear. The bow and arrow became dominant. Guns fired bullets, howitzers hurled shells, and now intercontinental ballistic missiles carry nuclear bombs. At whatever time we consider in the history of the human race men have been using technology and the technology has been changing. However, for most of our history the rate of change has been very slow, perhaps barely perceptible. The effects of the changes were probably felt as isolated events separated by many years in which little or nothing altered.

The rate of change has been increasing. It accelerated towards the end of the last century and has increased immensely during the present one. In the 20th century the developed nations of the world have experienced the eruption of a vast technological society, which is spreading in all parts of the globe. For an increasing number of people a shift has taken place, at first slowly but later almost at a rush, from what by comparison was a nontechnological society to a way of life dominated by technology. According to one's viewpoint this change, to which we have become so accustomed, can be said to have begun at almost any time after the start of the Industrial Revolution. The introduction of the spinning jenny, the advent of the steam engine, the coming of the railways, the demise of sail, the generation of electrical power, all these and many other events mark a fresh impetus to that change.

If we were to ask an octogenarian to tell us of the changes in society that he has witnessed the list would be impressive. In his childhood he lived in a world that now seems to have had only its surface scratched by technology. A few people might have seen an electric light at an exhibition or in a large city, and they might have read in a newspaper of Mr. Marconi's exploits with the wireless. They may have sent a telegram, but most had never used a telephone. Yet they would have been familiar with mechanical contrivances such as the steam-driven machinery used in factories and on the railways. If the octogenarian could recall a similar conversation that took place when he was a boy, the list then formed would have been shorter and less impressive. Go back another couple of generations and the list might hardly be worth making at all. A technological society, as we understand the phrase, did not exist. Yet the society of those days was indeed a technological society. The essential difference is that it was one whose technology was largely based on skills and crafts. Our society is one whose technology is largely based on science.

It has become common to talk of technology and society as if the two are quite separate entities, somewhat like strawberries and cream, that are brought together one to enhance the other; or for those who believe the cream has turned sour, one to ruin the other. Such a view is too simplistic. Technology and society are not two separate entities coexisting, harmoniously or otherwise. We live in a technological society, not in a society with technology. Technology is part of the very fabric of our society. As in strawberry-flavoured ice cream, the flavour runs right through the very substance itself. Whether it be the bus, train, or car we take to work, the television that entertains us, the cutlery with which we eat, or the telephone on our desk, technology is a part of our life, not an adjunct to it. And so it has been for centuries. The craft-based industries of previous centuries, such as weaving, paper making, or even steel making, and the craft- or science-based ones of today, such as electronics, ensure that our society is one permeated by technology.

A few moments' thought should produce our octogenarian's list, a catalogue of industries that have contributed to our present style of technological society. Most of them are either products of the 20th century or products of the 19th century that have been substantially developed in the 20th. Whether we think of our modes of transport, or the technologies that have moulded our home or work life, or the things we fear, we find that electrical or electronics engineering are common threads running through or touching upon them all. The petrol or diesel engine, aircraft and their control, our buildings and architecture, the chemical and petroleum industries, plastics, pharmaceuticals, rocketry and nuclear power; all employ some or many forms of the electrical engineer's art in their production or use. Many and varied are the technologies that have moulded our society and it would be foolish to attempt to order them into some degree of priority. We may simply assert that electrical engineering in its many forms has contributed as much as most, and more than many.

Drucker has pointed out that early in the 20th century the beginnings of science-based technology were restricted to a few people in Europe, the USA and, to a lesser extent, Japan.[1] Since then those technologies have grown and have become worldwide both in terms of where their products are consumed and where they are produced. Essentially the same large cities can be found around the globe where mankind lives in what is virtually a man-made environment. There, and in the towns and villages of the more developed nations, the material society produces and consumes the world's goods.

Western man can live as western man in any large city of the world, comforted by his Japanese television, his Hong Kong radio, and his European or American domestic appliances. The components he uses, even integrated circuits, those hallmarks of high technology, are used and produced in countries as far apart as the Philippines and El Salvador.

To a lesser degree such a situation touches upon most of the human race. Remote villages in many undeveloped regions have locally generated electricity to power lighting and maybe a Super-8 movie projector. Some have a communal television set, which brings images of technologies not to be felt in the village perhaps for many years. Men, women and children around the world have cheap electronic calculators which offer mathematical functions few of them understand, and even fewer actually need. Twentieth century technology is hawked around the world. So too were previous technologies; British and German railways and Edison's light bulbs in the last century for example. But it is the 20th century, with its science-based technologies, that has witnessed mass production for a world market as a great leveller.

Many of the old craft-based industries are still with us but it is the new science-based ones that are the more dynamic. The electrical industry was one of the first of these science-based industries and its development reflects the general trend. This trend may be summed up as an increasing dependence on scientific knowledge, which in turn demands more organization and a better-educated workforce.

The first important industrial application of electricity was electroplating. Although it was based on the infant science of electrolysis, it demanded no scientific understanding to put into practice. At most it needed an introduction to the principles, a few rules, and a little practice. After the discovery of electromagnetic induction, small hand-operated electricity generators were offered by instrument makers either for scientific work or for fun. Again, once the trick was known, and provided that the rules were understood and obeyed, little in the way of scientific knowledge was needed in order to make a machine that worked. In the case of commercial telegraphy, some scientific knowledge was needed. The first electromagnetic telegraph was built for scientific purposes by two eminent scientists, Gauss and Weber. Neither Cooke nor Morse, the first commercial telegraphers, were scientists and they needed scientific help to solve their problems. Thereafter, with the initial problems beaten, understanding the early telegraphs was not difficult and a number of inventors attempted to leap onto the bandwagon. Later, when the electric lighting and electrical power industries began, they were again founded on scientific understanding and engineering ingenuity, even though there was still plenty of room left for inventors and tinkerers, especially in the early years. Despite his research laboratory, Edison is not remembered as a scientist but as a genius of an inventor who could buy the science he needed.

It is obvious that the first benefits of the move from a new science to a new technology are likely to be those that are the easiest to obtain. With his building blocks a child erects the simplest structures first. More sophisticated ones may follow later when he learns to build with experience as well as blocks. So too with the move from a science to a technology. As time passes, the greatest achievements, if anyone can quantify such things, become more dependent on science, and in turn on organized science, and less dependent on tinkering. Still, we must be careful not to oversimplify what has been a very complex process. The pursuit of science does not follow a straight path. There is still room for inspired tinkering in scientific work as elsewhere. Nor are there always clear-cut demarcations between what is science, engineering, or technology. The usual ranking of the three, from science down through engineering to technology, is not the only philosophical view of the trio, nor is it necessarily the most accurate. Each can dramatically influence the other two; think of the effects of new instrumentation on the pursuit of science, for example. One can merge into another, and there is a flow of new knowledge between all three. Nevertheless, it is generally true to say that the science-based industries, of which electrical and electronics engineering are our examples, depend increasingly on an organized approach based on scientific knowledge and a highly educated workforce. Inspired tinkerers, where they exist, are now likely to have a university degree and a research team behind them and government or industrial funds available to them.

The increasing dependence of electrical engineering on science rather than on tinkering can be seen in several instances. A scientific push, and personal ambition, got the first landline telegraphs operational, and for a while they were able to develop almost without further scientific help. However, successful submarine telegraphy was achieved only after the loss of many expensive cables brought about a scientific enquiry. Lord Kelvin, one of the leading British scientists of the last century, will always be associated with submarine telegraphy. The invention of the telephone also had close links with the world of science. Bell had researched the science of sound, studied Helmholtz's work and experimental equipment associated with electric tuning forks, and had spoken with M. G. Farmer and Joseph Henry about electrical matters.[2] His nearest rival, Elisha Gray, had visited the Royal Institution in London, where he enjoyed the co-operation of some leading British scientists. In the design of generators of electricity the scientific understanding of the magnetic circuit, by men such as Gisbert Kapp and Oliver Heaviside, lifted the design of generators out of the era of trial and error into an era of design based on a good comprehension of the scientific principles involved. Heaviside was also responsible for the heavy mathematical attack on the effects of inductance and its critical application, which enabled long-distance telephony to become a reality. Well before the end of the last century electrical engineering, which has always been rooted in electrical science, was thus increasingly using science not only in its roots but in its daily blossom.

This fact has been even more firmly established since the turn of the century. The early radio industry offers another example of a technology which, after an early scientific push from the likes of Hertz and Lodge, was developed into a profitable industry by a nonscientist, Marconi. For further substantial development consultant engineer-scientists, such as Fleming, were needed. The outcome, the electronics industry, has been from its earliest days a science-intensive industry, to use Süsskind's phrase.[3] The vacuum triode was developed into an efficient electronic component in industrial laboratories, not in someone's backyard. The resulting family of vacuum valves required research in such disparate areas as vacuum technology, electron ballistics, and chemistry. The replacements for the triode (the transistor and the rest of the semiconductor family) depending as they do on a knowledge of atomic and quantum physics, are based on even deeper and more sophisticated scientific knowledge.

A comparison of some of the electrical inventions of the last century with some of those made this century brings out many features of the general trend apart from the increasing sophistication of the science available. Individual or small team efforts have been replaced by organized teams with large back-up facilities, such as technical support and libraries. Attics and backyards have largely given way to research laboratories with expensive equipment. Helpers with mechanical skills have made way for degree-carrying professional engineers and technicians. Compare, say, the invention of Morse's telegraph or Bell's telephone with that of RCA's colour television or the Allies' centimetric radar. The inventions of the incandescent and the low-voltage fluorescent lamps make another interesting comparison, as do the inventions of the triode and the transistor. Outside the field of electrical engineering there are many more examples; the first aircraft and the first rockets, the machine gun and a surface-to-air missile. Where the former was performed by what may be chauvinistically regarded as almost amateurs, the latter was carried out by highly trained, highly educated specialists. The trend has been towards increased professionalization, specialization, and institutionalization;[1] one might simply say, towards ever more organization.

This trend is felt in virtually all walks of life and it is nothing new to have it pointed out. However, because it is so widespread and accepted, the full force of what has been happening is easily missed. Drucker has stated that from 1900 to 1965 the percentage of Americans who earned their living by manual work fell from around 90 to around 25 %.[1] It is that sort of change that has shattered the pattern of tens of centuries, and such statements serve to heighten our awareness of what has been happening. One of the root causes has been the swing from craft-based to science-based industries and the corresponding need for a better educated workforce. As a means of earning a living, physical work is being replaced by mental work for an incredibly large percentage of the population. As always care has to be taken when interpreting statistics. In Drucker's example a check-out job in a supermarket would be classed as mental labour because of the mental tasks involved. The definition of mental labour is taken in its widest sense. With electronics taking over many mental tasks, including calculation of the change at the check-out counter, one wonders if that job, and many others besides, might have to be reclassified as physical labour. As more intelligent vending machines appear perhaps the task will cease to be either mental or physical work, and simply become machine work.

The widespread increase in organizational and educational levels can be seen in many examples. The old craftsman who could read and write, and who had a skill, was well up the educational tree of his day, somewhat equivalent to a university graduate of today. The individual inventor with a small workshop, and one or two colleagues, would now probably be a salaried employee working in a research laboratory. The bill that was settled by a cash transaction involving two people is now paid by cheque or credit card, involves several people and a computer, and possibly carries a surcharge. The slogan and photograph or sketch that sufficed for a newspaper advertisement is now an audio-visual experience on thirty seconds of videotape.

The rise in the educational level of the average person has not only contributed to more specialization, it has also brought about more generalization as well. A popular saying has it that the specialist now knows more and more about less and less. Whatever the merits of that statement, it is also true that the nonspecialist knows more and more about more and more. Sometimes this extra knowledge is incomplete, only partially understood, or of no practical value to the individual. Movies, radio, and television have educated millions about the geography of the world, its politics, climate, great cities, and so on. Before the 1970s nobody knew what the surface of the moon looked like from close range. Before the atom was split only a tiny percentage of mankind even knew it existed. Much scientific and engineering terminology has become public property even if not properly understood. Words, abbreviations, and acronyms such as amps, volts, kilowatts, FM, VHF, and radar have passed into the public domain. Nobody is likely to be afraid of a 9V battery, yet enough people take heed of a model railway exhibitor's warning of a 9000 mV live rail to make the trick worth while. Like the learned judge who failed to understand a patient lawyer's explanation, the public are better informed, even if none the wiser. Yet many are wiser, as well as better informed. Handymen can rewire their homes, housewives can fit a plug, electronic circuitry is a popular hobby, and microcomputers have become consumer goods.

Like all our inventions, and like we ourselves, modern technology can be both good and bad. Before 1900 electricity was used both to save human life (radiotelegraphy at sea) and to take it (the electric chair). Lasers are used as surgical scalpels but are also under investigation as potential weapons. Environmental groups have made us all aware of the damage modern technology can do to our world, much of it needless, and science and technology have long since lost their unwanted claim to being the keys to utopia. Their threats, especially when applied to warfare, remain. Even so, relatively few would wish to turn back the clock very far and return to a time when the problems of today, such as urban decay, chemical pollution, and the like, did not exist. If they could do so they would only rediscover the forgotten problems of a previous age, such as a visit to a pre-electrical age dentist. "The electric motor raised the tone of the (dental) profession," we are told.[4] "Instead of appearing as an organ grinder or mechanic using the pedal engine, the dentist took on the appearance of a medical profession." Whatever one's feelings towards the present-day dental profession, the threat of a mid-19th century "organ grinder" let loose inside one's mouth sounds far from reassuring.

All too often in the design of a project it is only the benefits that are considered and costed, and the problems, if foreseen, are ignored or their consideration is deferred. The nuclear industry has been accused of such conduct more often than most. What to do with disused nuclear reactors could be an even bigger headache than the disposal of nuclear waste, especially for those in ships and submarines. In an ideal world such questions would be examined at the design stage, when design, construction, use, decommissioning, dismantling, and disposal—the project as a whole—would be considered. In the real world such idealism is unrealistic, particularly if the project is expected to have a long operational life. The politics, economics, and ambition involved in the early stages of a project deny such considerations. Even NASA failed to plan ahead and engineer a controlled break-up of its giant space laboratory; instead half the world was left to wait and wonder where the pieces would fall. Although accidents will happen, some can be foreseen and averted. Science-based industry has brought enormous benefits to mankind, but there will always be risks. The pharmaceutical industry has a good record of minimizing the risks of a new drug but, almost inevitably it seems, a disaster such as thalidomide can occur. Despite criticism, especially after the Three Mile Island incident, the nuclear industry can also claim a good safety record. The human race lives with risks daily, yet we are easily fooled into believing the risks do not exist. When doctors are sued for failing to produce a miracle it is time to re-examine what we expect from society. Too often, although we know that perfection cannot be achieved, we demand it anyway and are aggrieved when it fails to materialize. Science and technology get the blame for human ineptitude. On a personal level, and on a national level, we have an ambivalent attitude towards them. While one ancient nation, China, has been actively pursuing modern technology and aiming for a technology-dependent society by the year 2000, whatever the ramifications, another, Iran, has been torn apart by a reaction against it.

Electrical engineering, electronics, the other science-based industries, and the trend towards more organization have changed our lifestyle. Hours of work are no longer dictated by the sun, strikes by organized labour can bring down the government of a stable country, electrical power failures bring chaos to industrial and domestic life. Organizations, social services, appeals to authority, and third-party intermediaries have replaced simpler one-to-one consultations. Rules, obedience to them, and circumvention of them, have become a way of life. The level of 'communications' has reached astronomical proportions. At one extreme it has stirred public opinion so as to bring changes of government policy with regard to war, be it in the Crimea or in Vietnam. At the other extreme half the world hears of some news trivia such as two giant pandas refusing to mate. The man who once stood on cold windswept terraces now sits at home in an armchair to watch a football match. Broadcasting brings the world, with all its beauty and violence, into our own living room. And when television begins to bore, the fully paid-up member of the technological society can always play with his latest electronic game.

As science, engineering, and technology have developed so too has the technological society. However, it is not driven by technical considerations alone. At the start of this chapter we referred to technology as man's way of doing things; in this modern technological society what man does, and how he chooses to do it, can be decided in quite old-fashioned ways. Personalities, money, and politics all have their role to play. Technology may offer new and sophisticated tools but they are used by the hand and brain of good old Man Mark 1, and he still has his oldest asset and problem—his own peculiar nature.

The personality of the individual inventor or pioneer can obviously have an enormous impact on the progress of his invention or discovery, though it need not necessarily do so. The success of a major discovery, such as Oersted's discovery of electromagnetism, does not depend on the discoverer's personality to the same extent as does, say, that of a major invention. Several men observed the effects of electromagnetic waves, for example, before Hertz performed his pivotal work, and the effect was well known before Marconi made its application his own. Perhaps radiotelegraphy could have been born some fifteen or more years before it was, if either David Hughes or Amos Dolbear had possessed that extra drive that might have carried them forward through adversity after they had made their initial discoveries. Similarly, the Edison effect had been a subject for learned discussion for 24 years before Fleming had his 'sudden very happy thought' that led to the invention of the vacuum diode. The personalities of men like Edison, Ferranti, Sprague, and Marconi provided the drive and ambition that was either absent, or present in smaller amounts, in most other inventors.

Hounshell has studied the effects of the quite different personalities of Bell and Gray and how this difference affected their style, politics, and etiquette as they pursued their respective telephone inventions.[2] He has shown how two men, who essentially invented the same thing at the same time, saw the invention in quite different ways. Bell clearly foresaw the commercial possibilities of the telephone and how it could lead to fame and fortune. He worked virtually in secret so as to protect his ideas and he skilfully used his scientific connections to support his priority. Gray saw only its scientific interest, a reward which he, as one of that band of late 19th century professional inventors, did not actively seek. For him the rewards lay in multiple telegraphy (which Bell abandoned to pursue the telephone) and which he felt had greater practical and commercial possibilities. Both men eventually received scientific acclaim, but only Bell achieved fame and fortune from the telephone.

Similarly, at a corporate level, the personality of a leader can have dramatic effects on the fortunes of a company. Edison again comes to mind for the way he founded a variety of companies to exploit his DC lighting system, thus further displaying his enormous drive, ambition, versatility, and enterprise. Later he displayed extraordinary stubbornness in refusing to accept the potential of AC systems. As long as AC was unable to offer an electric motor and an energy consumption meter, DC systems would be the preferred medium. In 1888 both of these missing links in the AC chain became available. A year earlier copper prices had doubled, another critical factor since copper costs were a major item in establishing any system but were dramatically less for AC than for DC.[5] Instead of accepting AC systems Edison launched his final attack before fading from the scene, switching his assault from technical to safety aspects. Four years later his companies merged with Thomson-Houston to form the General Electric Company (GE). In America, Thomson-Houston was then second only to Westinghouse in the production of AC equipment. Edison's former companies had adopted AC, but not Edison himself. The future now lay in the hands of Charles Coffin, the first president of GE.

Other companies have had their dynamic leaders. Jules Thorn founded a lamp manufacturing company that has become the giant Thorn-EMI. Bill Hewlett and David Packard set up shop in 1938 in a garage behind the Packard's rented home and now lead one of the world's foremost electronic instrumentation and computer companies. Anton and Gerard Philips took a family firm from being a small manufacturer of incandescent lamps to become one of the world's most important electronics companies with one of the best research laboratories. Paul Galvin took Motorola to a position of leadership. These individuals, and many more, have had the rare ability to carry a company forward to greater things and make it a lasting memorial to their own enterprise. None of them, however, nor Edison nor Siemens before them, could have done it without money.

Money is essential to the pursuit of both brilliant and poor research and invention. The skill of the successful financier lies in selecting the good from the bad, the more worthy from the less worthy. Almost immediately after the primary battery became established as a research tool, it is said that Napoleon was told that the English were making more scientific discoveries because they had more money.

How quickly, one wonders, would Marconi have progressed without wealthy family connections, or Edison without the backing of major commercial banks (it cost nearly $500 000 to bring his lighting system to the commercial stage). Morse needed government backing in order to build his first telegraph line from Washington to Baltimore, and came close to bankruptcy before he got it. A major contract with the Russian government helped put the young firm of Siemens & Halske on its financial feet. These men, and many more besides them, Mauchly with the ENIAC computer project, for example, spent their own money before spending that of others.[7]

In recent years we have become accustomed to hearing of vast sums of money being spent on the development of technical products. The $9 million spent by RCA in the 1930s to get television operational, or the $1 million that Edwin Armstrong took out of his own pocket to prove that FM radio was a reality, indicate that this is no new phenomenon. Major domestic products continue to consume vast amounts of capital in their research and development phase, as evidenced by the rivalries between formats for the video cassette recorder and the videodisc. Such sums however are paltry compared to those that major national enterprises such as the development of Concorde or a space effort can consume.

Often the money is well spent and can bring financial and other rewards. The success of the telephone brought financial riches to Bell as well as the deep satisfaction of having been personally responsible for making a major contribution to the modern way of life. The transistor has been worth every penny that Bell Laboratories spent on developing it. Yet sometimes money can be wasted or may have been better spent some other way, especially when a project has proved to be too ambitious and has become a financial drain. The first phase of submarine telegraphy was brought to a halt because of the large amounts of cable that had been lost at sea. Staite, a major early proponent of commercial arc lighting, was probably bankrupted because there was no suitable generator to supply the electricity. Although Ferranti's ambitious attempt to provide a 10kV supply to London broke new technological ground, it was also an investor's nightmare. More recently quadrophonic sound was stifled because too many companies had invested too much money to allow a rival to win the format war. Instead, they all lost. When vested interests are at work, even the voice of the experts can be ignored. "There is nothing a mere scientist can say that will stand against the flood of a hundred million dollars," was one expert's comment on the money spent on the development of the programming language PL/1.[6] "At first I hoped that such a technically unsound project would collapse, but I soon realized it was doomed to success," wrote C. A. R. Hoare.

Economic considerations, however, do not always only relate to finance. Politics, with all the shades of meaning of that word, can be equally, if not more, important.

Siemens & Halske got their Russian contract because a telegraph link from Moscow to the Crimea was needed to speed communications after the outbreak of the Crimean War. Edison's business sense ensured that his first central power station could provide electric lighting to the Wall Street financial area. The funding for the development of American microelectronics was strongly influenced by the politics of the Cold War and the space race.

War, or the threat of war, has long been the most extreme of political tools and a spur to technological invention and innovation. Whether the task has been to make a better shield to deflect a spear thrust, better armour plate to protect a tank, or a better guided missile, the threat of hostilities can produce funds and a workforce where none existed before. The electric telegraph served in Crimea and in the American Civil War. Radio communications and vacuum tubes were improved during World War I and paved the way for radio broadcasting to begin shortly after the Armistice. World War II witnessed the development of radar, radionavigation, and the first computers. "Peace," said *Fortune* magazine during the Cold War, "if it came suddenly would hit the electronics industry very hard." World War II especially helped to establish some of the major companies of today.[9] Hewlett-Packard for example, was founded in 1939. A year earlier college friends Bill Hewlett and Dave Packard had built an audio oscillator named the Model 200A, "because the number sounded big," (Fig. 12.1). An important order for a redesigned version, the 200B, followed from Walt Disney Studios; this model was used in the making of the famous film *Fantasia*. The war helped the new company to a steady, if undramatic growth, prompted a move into the microwave field, and left it in a healthy position from which it expanded in the 1950s. Another American engineer, Howard Vollum, was sent to England during the war where he trained on radar development. In January 1946, with associates, he founded another new company: Tektronix Inc., now known throughout the world as a major manufacturer of electronic instruments, especially oscilloscopes. The Keithley Instrument Co. is another major electronics instrumentation company that was founded by a returning serviceman, Joseph Keithley.

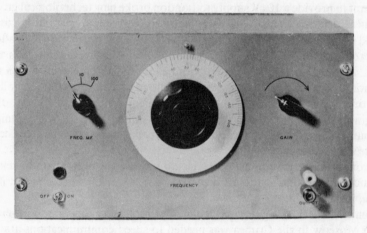

Figure 12.1 Hewlett-Packard's first product, model 200A audio oscillator (pre-production version)

Philips of Eindhoven benefited from the Dutch neutrality in World War I. Although it lost overseas supplies of materials because of the hostilities, the company was forced into greater self-reliance; for example, it started producing its own argon gas and glass bulbs for incandescent lamps. It also began to manufacture radio valves and even started its own shipping fleet.[8] During the Cold War large development funds were made available to the fledgling American semiconductor industry and helped establish American dominance in that field.

Lesser political acts than a declaration of war can have just as dramatic effects on the fortunes of a company. The Marconi company lost its commercial monopoly of radiotelegraphy as a result of an international agreement on safety at sea, a monopoly that had guaranteed the company success in its early years. Two giant companies, RCA in America and ICL in Britain, came into being as a result of political pressure from the Establishment. RCA was founded to ensure that American radio communications would not be dominated by a British company (Marconi); ICL was founded under somewhat similar fears for the British computer industry, which faced the might of IBM.

National politics can be critical in establishing an industry, but even when this is not the case politicians and some industralists would like to think it were, either to claim the credit or apportion the blame. The Electric Lighting Act of 1882, passed by a British Parliament anxious to protect the public from exploitation, is still blamed for stagnating the growth of the British electrical industry of the day and so opening the door to American and German imports. In a similar vein the British Post Office monopoly of internal communications, only recently ended, has been criticized by some industrialists as discouraging enterprise within the industry.

From the very earliest days of electrical technology politicians and civil servants have had their say. Early in the 19th century both Ronalds and Wedgwood had their offers of electrical telegraphs rejected by the British Admiralty. Morse almost gave up hope that Congress would grant him the cash to build his first telegraph line. In many countries today governments take an active role in encouraging the growth of an electronics industry. Japan provides the outstanding example of a national effort pursued with dedication and foresight. In 1970 few would have believed that Japan would in ten years have acquired a better reputation for reliability in LSI semiconductor devices than America, yet that was the position in 1980. What the Japanese have done with cars and semiconductor devices they have also done with items of consumer electronics such as hi-fi and video recorders, and may do with computer technology.

On a smaller scale other nations are treading similar paths. Taiwan and South Korea encourage investment in electronics. In Europe, the Irish government introduced such tax and other incentives to attract industry that many of the important international electronics companies from America, Britain, Europe, and Japan are now established with major plants in that country. At the same time Ireland has greatly expanded higher education in technical subjects so as to provide a locally trained work force.

International agreements within the industry also play a role. The international lamp cartel prior to World War II was one of the biggest, if not *the* biggest, cartel seen by the electrical industry. Virtually all the important manufacturers of incandescent lamps throughout the world were bound together for the common goal: fixing prices and markets and maintaining high profits. So successful was this arrangement that many old hands in the electrical industry maintain that profits from lamps provided the bread and butter for the whole electrical industry.

Some international agreements are forced by necessity and often reached only because one manufacturer or system achieves dominance—a position that every developer seeks. Radiotelegraphy could not have been successful without an internationally accepted version of the Morse code. ASCII (the American Standard Code for Information Interchange) has been accepted worldwide as a standard for data exchange; similarly, the IEEE 488 data bus (formerly the Hewlett-Packard Interface Bus, HP-IB) is gaining wide acceptance. However, a single system does not always achieve dominance and sometimes two, three, or even more can agreeably share the spoils. A radio receiver will work anywhere in the world but not a colour television receiver; three standards have come to be accepted. In the home videotape-recorder market a similar situation is developing. At a more mundane level the world accepts two standard frequencies for the domestic AC supply, 50 and 60 Hz. The voltage level is less clearly defined; 110, 200, 220, and 240 V systems are well established. Some agreements cover even more fundamental matters. Consider the confusion that can arise over the different colour codes used for the live, neutral, and ground wires supplied with imported domestic equipment, or the problems that would plague the electronics industry if different colour codes were used to mark resistor values. Perhaps the most fundamental agreements have been on the naming and definition of the scientific and engineering units we use. We are familiar with and have tried to cope with the problems caused by the clash between metres and feet, but far worse problems would have arisen if there had been multiple systems of units for current, voltage, resistance, etc. To achieve international standardization of our electrical and magnetic units was one of the great achievements of the last century, but it was a long struggle involving many international conferences, meetings, and agreements.

The establishment of an international system of units was one major step in the institutionalization of a field of interest which, by the turn of the century, had progressed from science to industry. Other steps included the establishment of professional institutions, the increasing number of conferences, the refinement of standards, the growth of a specialized jargon, the rise in the number of papers and books published, and the establishment of specialized education and training courses. As technical understanding spread the need to teach the basics moved down from the universities to the schools. The man in the street may think he knows what a volt is, but the schoolboy can recite a definition.

And so the technological society has spread its sphere of influence until it pervades our lives, and electrical engineering, in its many facets, has made a fair contribution. From the discovery that damp twine will conduct electric charge to an age of immediate international communications has taken approximately 250 years. What the next 250 years will bring we cannot guess. At the present rate of progress we cannot even predict what the next 25 years will bring, but whatever it is we can be sure that electrical engineering, based as ever on electrical science, will be making its contribution to society—and baffling as many people as ever. Somebody will still want to know why, if one wire brings it in and another takes it out, they actually have to pay for electricity. Somewhere there will be the modern-day equivalent of the reassuring message that electric lighting will not damage your health. Students will still find electromagnetic theory difficult to comprehend, and scientists will still be trying to figure out just what that thing is that we call the electron.

References

1. P. F. Drucker, Chapter 6, Ref. 13.
2. D. A. Hounshell, *Proc. IEEE* 64: 1305, 1976.
3. C. Süsskind, *Proc. IEEE* 64: 1300, 1976.
4. T. I. Williams, Chapter 11, Ref. 5.
5. T. S. Reynolds and T. Bernstein, *Proc. IEEE* 64: 1339, 1976.
6. C. A. R. Hoare, *Byte* 6: 414, 1981.
7. J. W. Mauchly, *IEEE Spectrum* 12 (No. 4): 70, April 1975.
8. P. J. Bouman, *Growth of an Enterprise: The Life of Anton Philips*, Macmillan, London, 1970.
9. *Electronics*, Special Commemorative Issue, 53 (No. 9), 17 April 1980.

NAME INDEX

SUBJECT INDEX